The Molecular Organography of Plants

The Molecular Organography of Plants

Quentin Cronk

OXFORD
UNIVERSITY PRESS

Great Clarendon Street, Oxford OX2 6DP

Oxford University Press is a department of the University of Oxford.
It furthers the University's objective of excellence in research, scholarship,
and education by publishing worldwide in

Oxford New York

Auckland Cape Town Dar es Salaam Hong Kong Karachi
Kuala Lumpur Madrid Melbourne Mexico City Nairobi
New Delhi Shanghai Taipei Toronto

With offices in

Argentina Austria Brazil Chile Czech Republic France Greece
Guatemala Hungary Italy Japan Poland Portugal Singapore
South Korea Switzerland Thailand Turkey Ukraine Vietnam

Oxford is a registered trade mark of Oxford University Press
in the UK and in certain other countries

Published in the United States
by Oxford University Press Inc., New York

First published 2009

British Library Cataloguing in Publication Data
Data available

Library of Congress Cataloging in Publication Data
Data available

Typeset by Newgen Imaging Systems (P) Ltd., Chennai, India
Printed in Great Britain
on acid-free paper by
CPI Antony Rowe, Chippenham, Wiltshire

ISBN 978–0–19–955035–7 (Hbk) ISBN 978–0–19–955036–4 (Pbk.)

10 9 8 7 6 5 4 3 2 1

Now it must be recorded that the botanist Pfitzer[1] had a favorite subject commensurate with his pedantic mind—the difference between thorns and spines. The popular saying, "There is no rose without thorns," is false because a rose has no thorns but spines (dermal products), while the hawthorn has thorns (transformed stems). The pedant never gave an examination without asking about this piece of morphology. Like every student I was prepared for this question, and, true, it was the first thing Pfitzer asked. But there must have been a gremlin in my cockpit.[2] I answered giving the definition of thorn for spine and vice versa. Pfitzer looked deeply insulted, bit his lips, and all but fainted. With subdued and reproachful voice he continued for the prescribed half hour, asking me questions, all of which I answered. Then he dismissed me with the grade C, which was the end of my hope for summa cum laude.

Goldschmidt, R. B.[3] (1956). *The Golden Age of Zoology*. University of Washington Press, Seattle (originally published as: *Portraits from Memory: Recollections of a Zoologist*).

[1] Ernst Hugo Heinrich Pfitzer was born on 26 March 1846 in Königsberg and died on 3 December 1906 in Heidelberg. In 1872 he succeeded Hofmeister as Professor and Director of the Botanical Garden in Heidelberg. He is the author of *Grundzüge einer vergleichenden morphologie der orchideen* (1882) and *Morphologische studien über die orchideenblüte* (1886).

[2] This gremlin, with whom we are all familiar, is remarkably tenacious. A rose has prickles (dermal products), whereas spines generally refer to transformed leaves.

[3] Richard Benedict Goldschmidt (April 12, 1878–April 24, 1958), zoologist.

Preface

J. Alfred Prufrock measured out his life in coffee spoons,[1] and so the life of the biosphere may be measured out in the organs of plants. They mark the seasons as they are iteratively produced, and as they are shed or die. Their decay drives the earth's geochemical cycles. Their evolution and increasing complexity marks time in the great geological intervals, and the nature of available plant organs has determined how the terrestrial biosphere has changed over time. It is plant organs that provide the diversity of plant life. Plant diversity is in the organized riot of green organs turned to the sun to form the thin photochemical envelope investing the earth, transducing the energy that drives the biosphere. Plant diversity is also in the tangle and soar of plant structures that provide the food and living space for millions of species, microbiological, invertebrate and vertebrate. The survival of these species depends on a successful outcome, day-to-day, of their interaction with particular plant organs. This is true for the eagle building a nest, the leaf-cutter ant on the forest floor or the rhizobial bacterium inducing root nodules far belowground. The study of the plant organ involves bringing together evolutionary history, plant form, developmental processes and genomic information. This book is not a textbook of plant morphology—there are many fine ones already written. Neither can it offer more than a snapshot of molecular-developmental genetics. Such a fast moving field is constantly changing. This book is, rather, an eclectic and personal tour through both subjects, with the intention of indicating how these two subjects may be brought together, and also pointing to what we do not know, but need to know in order to achieve that unity. Despite the title, much of plant organography is still without a molecular basis and thus large parts of the book lack accounts of genetic mechanisms. These parts are included to give some idea of the large amount we do not know, and I hope they will be a stimulus for research. If molecular mechanisms are not given, the unwritten subtext "we need to know the molecular basis for this" should be assumed. Another subtext that should be assumed is that covering exceptions. In a group as diverse as the land plants there are likely to be some exceptions to even the most general statement. "Flowering plants have leaves" (as a generalism) is true. "All flowering plants have leaves" (as a definition) is not true (certain angiosperms such as *Cuscuta* are essentially leafless). Statements of this kind should be assumed to be generalisms rather than definitions unless stated otherwise. The subject of this book is a large one and I have had to be selective. I apologize to all those whose work I have not included. I would be grateful for information about other work that should have been included. Even in a selective and short book such as this it is probably inevitable that there will be some errors, or at least points of contention. I would be grateful to have those drawn to my attention by email at: quentin.cronk@ubc.ca. I would like to thank all those, too numerous to mention, who, through conversation or their research work, have changed the way I think about plants. For stimulating discussion of the contents of this book, and/or helpful comments on parts of the manuscript, or other assistance I would like to thank especially Paula Rudall,

[1] Eliot, T. S. (1917). The love song of J. Alfred Prufrock. In *Prufrock, and Other Observations*. The Egoist Ltd., London.

Peter Endress, Philip Benfey, David Baum, Wilf Schofield, Guenter Theissen, Paul Kenrick, Julia Nowak, Barry Tomlinson, Richard Bateman, Stefan Gleissberg, Mike Whitlock, Pat Gensel and an anonymous reviewer. I thank Lindsay McGhee for preparing the figures and copyright holders and publishers for permission to redraw some originals, particularly the Company of Biologists (figs. 2.11, 2.13, 4.17, 4.30, 5.17, 5.23). The work was written in part during the tenure of a sabbatarian fellowship at the National Evolutionary Synthesis Centre (NESCENT) at Duke University and I would like to thank the National Science Foundation and NESCENT. I thank the National Tropical Botanic Garden (Kampong) for facilitating the photographs in the plates. Research in my laboratory is funded by the Natural Sciences and Engineering Research Council (Canada). I would also like to acknowledge gratefully the editorial promptings of Ian Sherman, Helen Eaton and Carol Bestley of Oxford University Press. Finally, I affectionately dedicate this book to Laura, Sophie, and William Cronk and to the memory of my parents.

Contents

CHAPTER 1

Introduction

1.1 The start

The study of plant evolution, at its grandest, can be seen as the study of how mutations in genes and genomes have affected the way in which the planet functions. For instance the evolution of oxygenic photosynthesis (involving two **photosystems**) from anoxygenic photosynthesis (using a single photosystem) changed the history of life. It involved gene changes in prokaryotic life forms (Allen & Martin, 2007) and was the first step toward plant evolution. This event, at around 2700 million years before the present (MyrBP), changed the atmospheric chemistry profoundly. The oxygen so produced permitted the evolution of complex life forms. It even changed the color of the earth, as oxidation tinted the land. Without the evolution of photosystems, geological formations with oxidized iron, such as the Devonian "new red sandstone" of Britain, would not be as they are.

Another example of a key innovation of planetary importance is the evolution (around 600 MyrBP) of **parenchymatous cell division** in plants from its precursor, **filamentous cell division**. This ultimately permitted the evolution of complex organization in plants. Filamentous cell division is still present in most algae including some which appear to have, at least superficially, quite complex organization, like the pseudoparenchymatous *Codium*. Parenchymatous cell division is found in certain green algae where it has evolved at least twice, in the ulvophytes and in the charophytes. Control of cell division in three dimensions, as in the parenchymatous system, implies greatly increased control over the cell division process.

1.2 Evolutionary innovations and the diversification of plant life on land

The colonization of the land by plants (c. 450 MyrBP) was a major event in earth's history, and therefore were the evolutionary innovations that permitted this: **cuticle**, **stomates**, **lignin**, **sporopollenin** (Table 1.1). However, the focus of this book is on those features that permit terrestrial ecosystems to diversify. The first of these (detailed in Chapter 2) was the **axis** (450–400 MyrBP), the radial, strengthened, tip-growing cylinder of tissue that allowed sporangia to be lifted out of the boundary layer (for spores to disperse over vast monoclonal mats), and the plant body to be lifted toward the sun. In this is the beginning of competition for light that would culminate in giant redwood forests rising over 100 m above the ground.

The next is the **leaf** (Chapter 3) at c. 400 MyrBP, for which there are at least two separate origins, microphyll and megaphyll. Lateral, dorsiventrally flattened organs increased the efficiency of carbon cycling, and greatly increased ecosystem productivity and the drawdown of carbon from the carbon-dioxide-rich atmosphere. The origin of the **root** (Chapter 4) at c. 400 MyrBP was also important in allowing the evolution of complex plants. Roots provided anchorage for taller organisms and a mechanism for water and nutrient absorption from soil directly and through mycorrhizal symbiosis. However, most importantly for planetary history, the origin of the root allowed the phenomenon of **soil** to come into existence for the first time. Soil formation is engendered by the

Table 1.1 An overview of levels of land plant evolution.

Level and date	Activity	Characters
Level −1 (prior to 450 MyrBP)	Prior to colonization of the land (increase in cellular complexity)	Parenchymatous growth, differentiated cell types
Level 0 (c. 450 MyrBP)	Colonization of the land	Cuticle, stomates, sporopollenin
Level +1 (c. 450–420 MyrBP)	Early evolution of land plants	Sporophyte building, axis building
Level +2 (c. 420–380 MyrBP)	Diversification of reproductive and vegetative organs	Roots, leaves, heterospory
Level +3 (c. 365 MyrBP)	Tree building, seed building	Massive construction and seeds
Level +4 (c. 150–80 MyrBP)	Early angiosperm diversification	The flower and associated features
Level +5 (c. 80–30 MyrBP)	Later angiosperm diversification	Evolutionary innovations of derived clades such as grasses, legumes, orchids, Asteraceae

injection of carbon and organic acids deep into weathered geological strata, irrevocably changing geochemical cycles, patterns of weathering and sedimentary processes.

Another innovation of major impact was heterospory and the **seed** (Chapter 5) at around 365 MyrBP. The seed, as an agent of dispersal that was both resistant and well-resourced, allows plant regeneration under conditions that would otherwise be hostile. The seed appears to have allowed plants of the late Devonian to colonize the uplands, so completing the terrestrial plant conquest of the land and enhancing plant-induced changes to global geochemical cycles. Finally, the origin of the **flower** (Chapter 6) at around 150 MyrBP commenced the rise to dominance of the angiosperms (flowering plants). The flower promotes efficient and specialised gene flow and that, in turn, promotes diversification. The flower and especially its component, the carpel, can therefore be seen as the essential evolutionary innovations that led to the immensely species-rich biomes of the present, such as the lowland tropical rainforest.

1.3 Plant versus animal development and evolution

It is striking that in animals complex multicellularity appears to have evolved only once, whereas in plants it has evolved several times. For instance it evolved in the clade that includes the unicell *Chlamydomonas* and the multicell *Ulva* (sea lettuce). It also evolved in the clade that contains the unicell *Mesostigma* and the multicell *Arabidopsis* (thale cress). In contrast, the colonization of the land by complex multicellular animals has happened several times, whereas in plants, complex multicellular algae colonized the land only once. This contrast tells us a lot about the different way of life of plants and animals and their different evolution.

The predisposition to multicellularity in plants points up a difference in cell division: in plants there is a sticky cell wall and division is by formation of a specialized microtubule structure (a **phragmoplast** in charophytes and land plants, a **phycoplast** in other green algae) on which the new cell walls form in intimate contact as a **cell plate**, so clumping of daughter cells is hard to avoid. Kinesin-like calmodulin-binding protein (KCBP) appears to be essential to this kind of cytokinesis in both land plants and green algae (Dymek et al., 2006), suggesting that this form of cytokinesis is a green plant innovation. In animals, cell division is by scission, which has the effect of separating the cells more cleanly, although the "midbody" of the animal cell can be viewed as having an analogous role to the phragmoplast (Otegui, Verbrugghe & Skop, 2005).

The substantial plant cell wall is a distinguishing feature of plants, a secreted lattice of carbohydrates and proteoglycans. It has numerous important functions in plant biology. First, it acts like a sponge in holding water by capillary action, creating a water and ion store called the apoplast. Its comparative stickiness and rigidity prevents cell movement in plant development. Animal cells can, and do, slide past each other to take up programmed developmental positions.

In plants, however, cell–cell positional signaling generally determines cell fate during development (Berger et al., 1998), rather than developmental history, that is, cell lineage effects, which are less common. One result is the lack of distinction in plants between germline and soma. Instead, cell position determines the development of germ cells (male pollen and megaspore mother cells) rather than cell lineage.

Land plants, unlike animals, have an **open morphogenesis** in that they grow continuously during their life cycle, adding **modules** in an iterative fashion (Tomlinson, 1984). This **iterative growth** is made more complex by interacting with the environment, causing a persistent environment-development signaling system. Furthermore, the many modules (e.g. leaves, roots and stems) produced by an individual plant must be physiologically integrated. By contrast, most animals have **closed morphogenesis**, with a single developmental trajectory. Animal development, when completed at maturity, generally produces a fixed number of organs which is largely independent of the environment. Humans raised on a high protein diet are often taller, but they do not have excess arms and legs. There is thus no continuous feedback between the major features of animal development and the environment. Instead, and in contrast with plants, animals have the power of changing their location and hence their environment by motility.

Animals are often separated from the environment by exoskeleton, scales or fur, or by the physiological separation of homoeostasis. Changing environments, even moving into the hostile environment of the land from the sea, is therefore relatively straightforward. For plants, which are much more in equilibrium with their environment, such transitions are rare. This may be the reason why complex multicellular plants colonized the land only once. The evolution of the cuticle, a semi-impermeable surface layer made up of a lipid polymer called cutin (Li et al., 2007), was the key innovation that gave plants sufficient separation from the hostile aerial environment to dominate the land.

An extraordinary feature of plants, at least when seen from an animal perspective, is the **alternation of generations** (Bell, 1989; Kenrick, 1994; Sheffield, 1994). Many plants (and all land plants) have a life cycle in which **haploid** and **diploid** generations alternate with each other. These generations most often have very different developmental trajectories, and so a given genome has within it the potential to create two phenotypically different organisms. The different developmental paths undertaken by haploid spores and diploid zygotes therefore indicate that radically different transcriptomes are possible depending on **epigenetic control** (such as **chromatin remodeling** and **DNA methylation**) triggered by ploidy level, or at least by ploidy level transition.

1.4 The plant genome

Ultimately, it is the changes in plant genomes that have created the ecosystems on which human life depends. Although these changes took place hundreds of millions of years ago, the genomic basis of revolutions in the phenome are discoverable by comparative and functional morphology and genomics. Although sequence divergence between the major clades of land plants is extensive, only some of these divergences are responsible for the key innovations that have built the biosphere. Molecular-developmental genetics must therefore sieve through millions of genomic changes to discover those that have been critical in plant evolution.

Plants have three genomes. The mitochondrial genome is generally about 10 times the size of the animal mitochondrion but is very variable in length (~200–2500 kb); it is conservative in sequence but highly labile in gene arrangement. The chloroplast genome is generally 120–160 kb, but in *Pelargonium* it is 217 kb (Chumley et al., 2006). The chloroplast is fairly conservative in gene sequence and arrangement, and, although with limited effect on overall plant morphology, this genome does encode much of the machinery for autotrophy.

It is the nuclear genome, however, that is responsible for most of the morphological diversity of plants. It is enormously variable in size (115 Mb in *Arabidopsis*, yet 250,000 Mb in *Psilotum*). It is also variable in gene sequence and arrangement. Nuclear genomes can increase in size by doubling through polyploidy (Adams & Wendel, 2005), or be hugely bloated by multiplication of transposable elements.

There are three major types of change relevant to the evolution of plant development. First, gene duplication by polyploidy or tandem duplication allows two gene copies, at first identical, to diverge in function. This is called **neofunctionalization** if one copy retains the old function and the other copy acquires a new function. The alternative is **subfunctionalization**, when each copy retains part of the ancestral function. Second, existing genes can change their expression patterns, usually by mutations in regulatory regions (***cis*-regulatory change**). This is an important possible mechanism of heterochronic and heterotopic change. Third, there may be a direct change in protein function. If this change occurs in a regulatory gene, such as a transcription factor, major changes can occur in the regulation of other genes (***trans*-regulatory changes**).

Whole genome sequencing is of great importance in the elucidation of developmental evolution, not least because it allows the diversification of developmentally important gene families to be assessed (Leebens-Mack et al., 2006). For evolutionary-developmental genetics it was somewhat unfortunate, although inevitable, that early genome sequencing focused so heavily on angiosperms, a limited phylogenetic spread. This was not the case for animals, for which early sequences of yeast, nematode, fruit-fly and human quickly provided a wide phylogenetic sampling. However, the plant community is catching up rapidly. In 2005, the moss *Physcomitrella* entered the Community Sequencing Program (CSP) of the Joint Genome Institute (JGI), and genome sequencing is now completed for *Physcomitrella* (Rensing et al., 2008).[1]

The liverwort *Marchantia* has been added to the CSP of the JGI for 2008 as a "small project" for a mixture of conventional Sanger sequencing and pyrosequencing. The sequencing of the clubmoss *Selaginella moellendorffii* (Wang et al., 2005) is now substantially completed and the sequence is currently in annotation, the reads having been released in 2007. Sequenced genomes of a fern (for instance *Ceratopteris*; Nakazato et al., 2006) and a hornwort (such as *Anthoceros* or *Notothylas*) would also be of great scientific value.

[1] http://genome.jgi-psf.org/Phypa1_1/Phypa1_1.home.html

1.5 Homology and serial homology

A necessary objective of comparative biology is to compare like structures with like structures, which in an evolutionary sense means that such structures derive from developmental programs (genes and modes of gene action) inherited from a common ancestor that also had that developmental program. Another way of putting this is "resemblance caused by a continuity of information" (Van Valen, 1982). Such resemblance is termed homology. It is quite different from similarity (with which it is sometimes confused) as similarity can be superficial with little biological significance. However, homologous structures may not look particularly similar. A palm leaf and a cactus spine are not particularly similar but they are homologous as phyllomes, and, at a deep level, share a developmental program that they have inherited from a leafy ancestor with that program.

There are a number of ways of assessing homology. First, there is the ascertainment that resemblances are not merely superficial, as they would be if the structures were analogous rather than homologous. A powerful way of assessing deep as opposed to superficial resemblance is to look at development. Development (or "embryology") was central to Darwin's definition of homology as

> that relation between parts which results from their development from corresponding embryonic parts, either in different animals, as in the case of the arm of man, the fore-leg of a quadruped, and the wing of a bird; or in the same individual, as in the case of the fore and hind legs in quadrupeds, and the segments or rings and their appendages of which the body of a worm, a centipede, etc., is composed. The latter is called serial homology. The parts which stand in such a relation to each other are said to be homologous, and one such part or organ is called the homologue of the other. In different plants the parts of the flower are

homologous, and in general these parts are regarded as homologous with leaves. (Darwin, 1872)

Second, and more recently, molecular data has added an important means of assessing homology. The involvement of homologous genes, acting in similar networks and with similar expression patterns, points to homology (Jaramillo & Kramer, 2007).

Finally, and also unavailable to Darwin, there is the phylogenetic criterion. The presence of structures that result from inherited information from a common ancestor should be a synapomorphy, or shared derived character, for the clade (monophyletic group) that derives from that ancestor. Homologous structures should therefore map perfectly to the structure of the phylogenetic tree. Thus the presence of microphylls maps to a clade within the lycophytes (and including all extant lycopsids). This is evidence that the microphyll arose only once, and that all microphylls are homologous.

These criteria can be used together in cross-species comparisons of organs. However, in assessing serial homology (for instance between petals and leaves within an organism) the phylogenetic criterion is unavailable. By contrast in assessing gene homology, phylogeny is the only usable criterion (Patterson, 1988). The phylogenetic criterion, when available, is the strongest, as it is the surest way to distinguish homology from analogy. Similarity alone (even detailed similarity) may always be, at least in theory, the result of parallel evolution of analogous structures. This is an enduring problem for serial homology for which the phylogenetic criterion is not available. However, the developmental and molecular criteria are particularly powerful in assessing serial homology as the development can be studied in the same organism, with the same genotype. Furthermore transitional organs often occur between serial homologues (such as transitions between stamen and petal in the water lily *Nymphaea*). Traditionally this has been taken to indicate that the developmental pathways of both are sufficiently similar to intergrade, and this in turn has been taken as evidence of homology. With molecular genetics, this hypothesis can now be tested directly. However, it should be borne in mind that the relationship between gene pathway and phenotype may be complex, and gene pathways are constantly evolving, giving rise to the potential for co-option and dissociation of genes to and from a particular structure (Jaramillo & Kramer, 2007).

As phylogenetically determined homologies are equivalent to synapomorphies, it is clear that they will be phylogenetically nested. It is therefore important to be clear about the level at which homology between particular organs is being discussed. It is not controversial that leaves, petals and stamens are homologous at the level of phyllomes. What is less clear is whether petals and stamens share homology at the level of microsporophylls, from which leaves and sepals are excluded. The staminopetal interpretation of petals of the eudicots suggests that petals evolved as sterilized microsporangia. This view is widely supported, although it has been challenged (De Craene, 2007).

1.6 Latent homology

In discussing homology it should be noted that features that are independently derived, and therefore not homologous, may result from homology of an underlying feature at a deep level. This has variously been called latent homology, phylogenetic prepatterning or synapomorphic tendency (Cronk, 2002). As an example, floral zygomorphy in asterids (such as *Antirrhinum*) and rosids (such as Leguminosae) may be mentioned. It is apparent from phylogenetic evidence that zygomorphy has an independent origin in the two clades. However, the same genes, homologues of *CYCLOIDEA*, are involved in both *Antirrhinum* (Corley et al., 2005; Luo et al., 1996) and legumes (Citerne et al., 2003; Feng et al., 2006). This is not coincidence but results from the presence of a key gene of similar basic function in both rosids and asterids. The homology (or latent homology) is "genetic competence to evolve zygomorphy."

Another example is nodulation by actinobacteria or rhizobial bacterial in the eurosid 1 clade (Table 1.2). Nodulation has arisen specifically in this clade, but it has arisen many times

Table 1.2 Distribution of nodulation in families of the eurosid 1 clade, largely representing independent evolutionary events but indicating a latent homology for the nodulation competence within this clade.

Type of nodulation	Family	Genera
Rhizobial	Leguminosae (Fabaceae)	Various
	Ulmaceae	*Parasponia*
Actinobacterial (*Frankia*)	Betulaceae	*Alnus*
	Casuarinaceae	*Allocasuarina, Casuarina, Ceuthostoma, Gymnostoma*
	Coriariaceae	*Coriaria*
	Datiscaceae	*Datisca*
	Eleagnaceae	*Eleagnus, Hippophae, Shepherdia*
	Myricaceae	*Comptonia, Myrica*
	Rhamnaceae	*Ceanothus, Colletia, Discaria, Kentrothamnus, Tetanilla, Talguenea, Trevoa*
	Rosaceae	*Cercocarpus, Chamaebatia, Cowania, Dryas, Purshia*

independently (Swensen, 1996). There are developmental differences between actinobacterial nodules, with root-like vasculature, and rhizobial nodules with stem-like vasculature (Sprent, 2001) although both have a first stage, which is the formation of a "prenodule." More importantly, the association of nodulation with a single clade indicates that there is an underlying latent homology for competence to evolve nodulation that characterizes this clade.

Darwin was aware of this phenomenon and called it "the tendency to vary in like manner" caused by a "community of descent" (Darwin, 1859). Later Darwin (1868) named this the "law of homologous variation" under which name it is often wrongly attributed to N.I. Vavilov who documented it extensively in cereals. Darwin illustrated the phenomenon by the independent origin of swollen "roots" (hypocotyl) in brassicas, particularly swede or ruta-baga (*Brassica napus* var. *napobrassica*) and turnip (*Brassica rapa* var. *rapa*). To these might be added radish (*Raphanus sativus*) in the closely related genus *Raphanus*. This latent homology (the competence to form swollen roots in *Brassica* and related genera) most probably has a molecular-developmental basis, but the details are unknown.

1.7 Processes of development

The main processes of development are given in Table 1.3. Highly important among these are the processes that give rise to the **position**, **identity** and **polarity** of organs. A phyllome arising as a primordium on the flank of an apical meristem is taken as an example. Its position will be set by phyllotaxis, a patterning mechanism based on the requirement for auxin for primordium development, and the auxin sink provided by the other primordia. The highest auxin concentration will therefore be at a distance from the other primordia, and this is where the new primordium forms (Chapter 3, Section 4.32).

The developing primordium also has what is termed an identity. For instance if B- and C-class MADS-box genes are expressed, downstream gene action ensues which ensures the development of that phyllome as a stamen. The MADS-box genes have conferred stamen identity on the developing primordium long before any stamen-like features are visible (Chapter 6, Section 6.11).

Furthermore, identity may be established as a gradient across an organ, and this is known as polarity. A good example of polarity is the abaxial–adaxial polarity that establishes across a leaf primordium that ensures that leaves develop as bifacial organs; in other words, the lower surface is different from the uppersurface (Chapter 4, Section 4.6).

These developmental concepts, position, identity and polarity as well as others, all have a molecular basis in gene expression and in the chemical signaling to which genes respond. Most of these concepts and their molecular basis have been worked out in model eudicots, especially *Arabidopsis*. However, these developmental

Table 1.3 A table showing the main processes of plant development.

	Process	Effect
I. Processes of quality and quantity	Identity	Nature of organs
	Polarity	Identity across organs (base to tip = proximodistal; relative to apex = adaxial/abaxial; side to side = mediolateral)
	Organ patterning	Identity within organs (e.g. stripes, spots)
	Modularity	Number of organs
	Growth	Size
	Relative growth	Shape
II. Spatiotemporal processes	Position	Spatial: positional signaling
	Timing or duration	Temporal

Note: Compare with Table 1.4.

Table 1.4 A summary of important evolutionary-developmental change.

	Evolutionary-developmental change	Effect
I. Changes of quality and quantity	Change in degree of polarity	Symmetry changes (more/less asymmetry)
	Homeosis	Change of identity
	Meristic or modular change	Number
	Change of size (absolute growth or heterometry)	Reduction/loss, or elaboration
	Shape change	Altered relative growth
II. Spatiotemporal changes	Heterochrony	Change of timing or duration
	Heterotopy	Change of position

Note: Compare with Table 1.3.

concepts are applicable to other groups of land plants, even though the details of the molecular mechanisms may not be. The elucidation of more general principles for the land plants as a whole is an important challenge for molecular-developmental biology.

1.8 Evolution of development (evo-devo)

The morphological phenotype is the result of developmental processes. In the same way, evolutionary changes in the phenotype are the result of evolutionary changes in developmental processes. The study of evolutionary changes in developmental processes is sometimes contracted to "evo-devo" and it represents a very fertile way of investigating evolution.

Developmental processes result from underlying gene action and so comparative molecular-developmental genetics is the most basic, powerful and satisfying means of testing evolutionary hypotheses and describing evolutionary change in the form of organisms. The developmental processes listed in Table 1.3 have counterparts as evolutionary processes, and a list of these corresponding evolutionary processes is given in Table 1.4. One of these is change in absolute growth or **heterometry** (Webster & Zelditch, 2005). For instance, *Isoetes* is the extant survivor of the clade that includes the extinct rhizomorphic lycopsids, yet *Isoetes* species generally have a cormoid stem often less than a centimetre high, whereas the Carboniferous *Lepidodendron*† may have grown up to 50 m or more in height. (The dagger represents extinct taxa.) The molecular-developmental changes underlying the heterometry in this lineage are not known. Another process, **heterochrony** or change of timing in development, may be involved: if the attainment of sexual maturity and the consequent cessation of growth is accelerated, absolute size will be decreased.

Heterochrony is without doubt a very important feature of evolutionary change (Bateman, 1994; Gould, 1979, 2000; Rudall & Bateman, 2004; Zelditch, Sheets & Fink, 2000). Developmental timing can be altered in several ways (Fig. 1.1), and these have been distinguished with different terminology (**paedomorphosis**, **gerontomorphosis,** etc.). Heterochrony overlaps with another

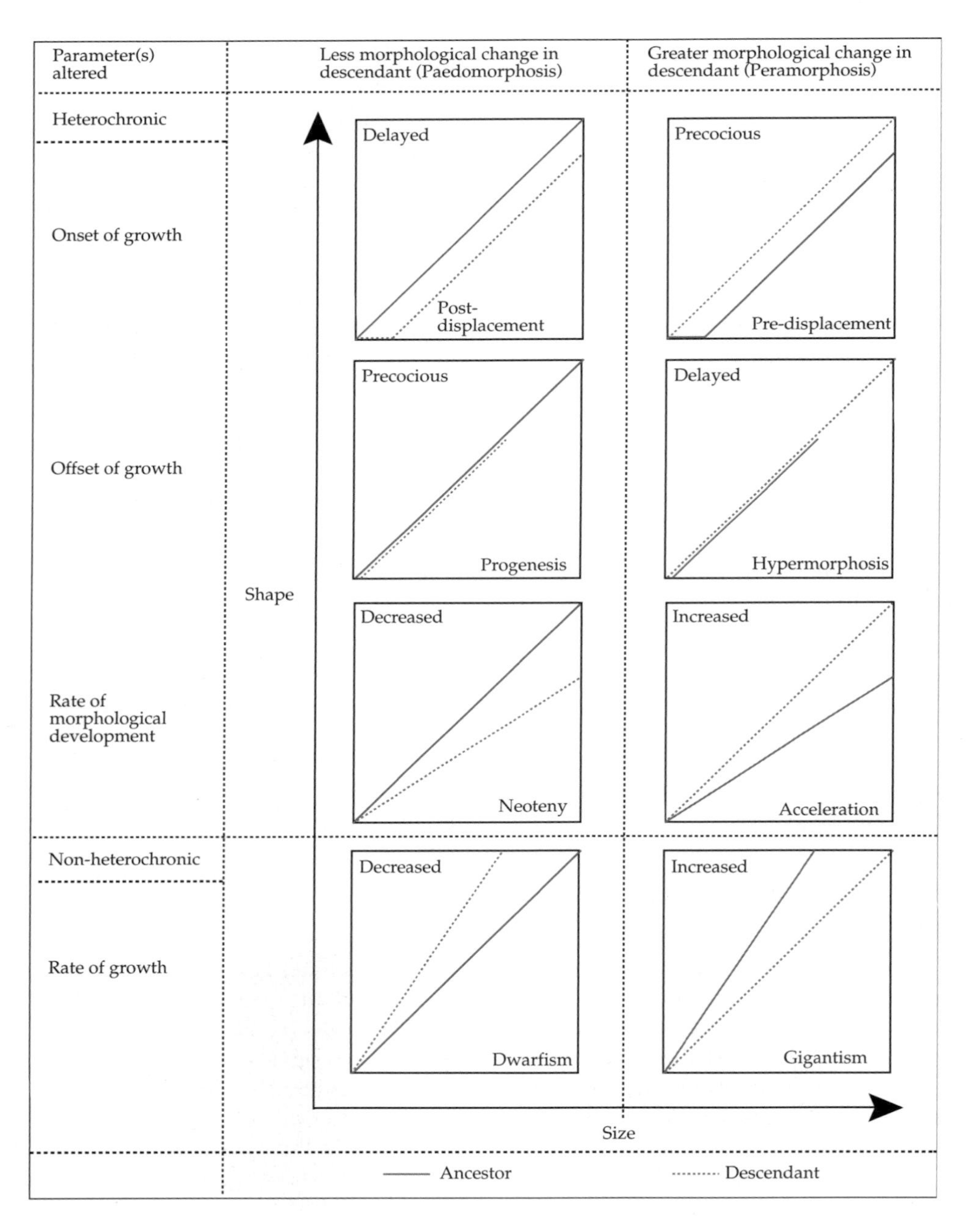

Fig. 1.1. The concept of heterochrony (change of timing in development), and its variants, illustrated by size versus shape graphs (after Bateman, 1994; Rudall & Bateman, 2004).

important process, **heterotopy** or change of position. A change of position in a developmental program often requires a change of timing (Baum & Donoghue, 2002), so it is quite possible for an evolutionary change to be both heterochronic and heterotopic. As Kellogg points out, it may depend on how one defines a developmental program:

If the program is defined as, for example, "make hairs on lemma," then the different hair patterns described [in the grass genus *Poa*] are produced by the amount of time the program is active (i.e. heterochrony). If the program is defined as "trichome development", then the program is activated sequentially in different sets of cells, resulting in a change in position of hairs (i.e. heterotopy). (Kellogg, 2002)

The process of heterotopy may involve that of **homeosis**, a change of identity. Indeed homeosis has been conceptually included within a broadened concept of heterotopy that includes **neoheterotopy** (the origin of a structure in a new location) and **homeoheterotopy** (the transfer of genetic identity to a recipient structure in a different location; Baum & Donoghue, 2002). If homeoheterotopy is complete it is known as homeosis (in the strict sense), as is found in the homeotic floral mutations of MADS-box genes in which complete transformations occur, such as the conversion of stamens into petals (Fig. 1.2). Partial homeoheterotopy leads to **partial homology** at the molecular level. This may be very important to functional evolution. Corner developed the concept of **transference of function**, exemplified by the hypothetical transfer of bird-attractive red pigmentation from the aril to the seed-coat in different species of the legume genus *Pithecellobium* (Corner, 1958). If this occurred by a change of expression pattern of the same pigment gene pathway, then aril and seed-coat share a homologous characteristic (red color). However, the aril and the seed-coat maintain their separate identity in other, morphological features. It is also possible that the similar pigmentation is due to a different pigmentation pathway (**nonhomologous transference of function**). Transference of function, via homeoheterotopy, has also been suggested for the evolutionary adjustments of the epicalyx, calyx, and ovary wall in relation to protection and dispersal that have occurred in the Dipsacales (Donoghue, Bell & Winkworth, 2003) and may be of wide applicability in the evolution of fruits, seeds, and accessory structures (Stebbins, 1970).

1.9 Evolutionary innovation: organs of novel form

Ganong's first principle of morphology, the principle of continuity of origin, states simply that

no functional structure ever arises *de novo*, but only from the modification of a pre-existing structure, which in turn arose from a still earlier, and so on backward through a longer or shorter chain ending only in the original protoplasmic variation, or in whatever it is which does lie at the beginning of specialization. (Ganong, 1901)

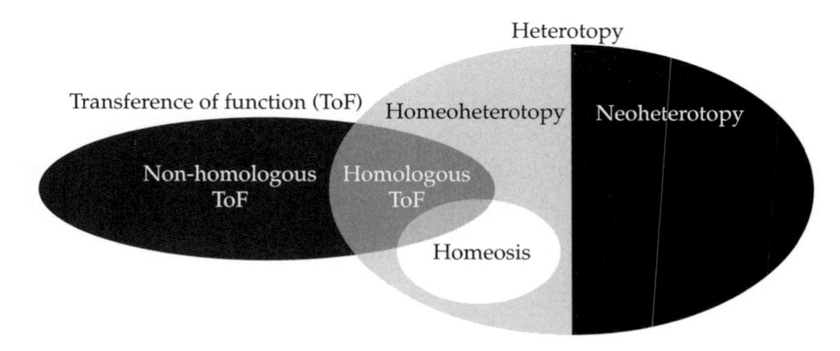

Fig. 1.2. The concept of heterotopy (change in position of an organ) and its relation to homeosis and transference of function (after Baum & Donoghue, 2002).

Such an axiom can also be couched in terms of gene networks or developmental pathways. A novel gene network does not spring into existence, *de novo*, but evolves from preexisting gene networks. The same applies to genes themselves. New genes do not arise *sui generis* within the genome but derive by mutation or duplication from preexisting genes.

Thus, in the context of land plants, the developmental process leading to axis formation was a basic one from which the developmental pathway to leaf, root and stem differentiated. Leaf, root and stem each in their turn differentiated into the large number of derivative organs, generically lumped (in Worsdell's terminology) under the terms **phyllome**, **rhizome**[2] and **caulome** (Worsdell, 1905).

The importance of gene regulation in evolution was pointed out in early work on the protein differences between humans and chimpanzees (King & Wilson, 1975). It is difficult to account for the large anatomical and behavioral differences between humans and chimpanzees by the rather small degree of molecular divergence. Instead regulatory changes controlling the expression of genes were postulated (Carroll, 2005). This idea continues to gather empirical support, particularly the notion that changes in *cis*-regulatory regions of transcription factors (i.e. regulation of regulators) are of key importance in morphological modification (Cronk, 2001; Doebley & Lukens, 1998). A good example is found in the evolution of maize (*Zea mays*). The maize allele of the gene *TEOSINTE-BRANCHED* 1 (*TB1*) confers the nontillering trait of domesticated maize, relative to its tillering progenitor (*Zea parviglumis*). There is evidence of a strong selective sweep of the 5-prime flanking regions in domesticated maize suggesting that the *cis*-regulatory regions of this gene have undergone intense selection for altered expression patterns of *TB1* (Clark et al., 2006; Wang et al., 1999).

[2] This usage, rhizome as a collective term for root-derived organs, used by Worsdell in what he calls the "original, literal sense," has for obvious reasons not been taken up, due to prior use in the sense of "underground stem." Radicome would be a better term if one were needed. It might be objected that it mixes Latin and Greek, but this is now common practice in scientific word formation.

An important mechanism for evolutionary innovation, although poorly documented in plants is **network capture**. Through changes of *cis*-regulation, a gene may be captured by an existing gene regulatory network, so resulting in **gene co-option** to a new function. A good example of this is the evolutionary gain of a novel wing pigmentation spot in *Drosophila biarmipes* (Carroll et al., 2005; Gompel et al., 2005). The general pigmentation gene, *YELLOW*, was captured by the regulatory network controlling wing development. The capture involved a mutation in *cis*-regulatory sequence that brought it under the control of the ENGRAILED regulatory protein. Such mutations tend to be genetically **dominant**, as they involve a **gain of function**.

Gene redundancy caused by gene duplication has long been recognized as an important raw material for evolution (Ohno, 1970). The proneness of plants to polyploidy, perhaps due in part to plant synthesis of spindle inhibiting alkaloids, such as nicotine and colchicine, makes duplication events particularly common in plants (Cronk, 2001). Also plants are often subjected to heat and cold shocks, having few means of temperature control, and these shocks are known to disrupt spindle formation. Even *Arabidopsis thaliana*, with just five chromosomes, has 60% of its genome segmentally duplicated by paleopolyploidization events (Blanc & Wolfe, 2004). Plant genomes undergo frenetic change, when viewed on a geological timescale, with genes being duplicated, destroyed and continuously repatterned on chromosomes.

Gene duplication can lead to loss of one copy by pseudogene formation and a return to a single gene state. Alternatively, it can lead to retention of both copies and divergence of gene function by subfunctionalization and neofunctionalization (Causier et al., 2005; Hileman & Baum, 2003; Lynch & Force, 2000). By this means, some ancient duplication events may have had a major effect on plant form. An ancient duplication separates the *KNOX* homeodomain-containing genes into two clades, class 1 and class 2 (Champagne & Ashton, 2001; Singer & Ashton, 2007). *KNOX* genes that are important in apical meristem determination and control, such as *SHOOTMERISTEMLESS* (*STM*), are class 1 *KNOX* genes. This duplication probably occurred c. 430–475 MyrBP, at around the time of the origin of the land flora. It therefore may have been important in allowing the evolution of a class of genes with meristem specific members, and in facilitating complex organization in land plants.

Repeated doubling of the plant genome has led to some very large gene families, and some of these gene expansions may be important in the evolution of morphology. In the human genome there are some 43 MYB transcription factors, yet in *Arabidopsis* this number is 198 (Chen et al., 2006). MYB transcription factors have a conserved DNA-binding domain consisting of up to three imperfect repeats of 51–53 amino acids (R1, R2 and R3). Most plant *MYB* genes contain only R2 and R3 (the so-called R1R2 MYBs of which there are 126). The existence of numerous *R1R2 MYB* genes is a plant specific trait. Several of these genes are known to be important regulators of morphogenesis, such as the *ARP* genes controlling leaf development (Chapter 4, Section 4.4). Through the evolutionary divergence of the MYB family (Zhang, Chopra & Peterson, 2000), and other similar gene families, genome duplication has had an enormously important impact on plant complexity.

1.10 Regressive evolution: the loss or reduction of organs and reversal of character states

In regressive evolution, a clear distinction needs to be made between the **loss of an entity** (such as an organ) and the **change of a quality** (i.e. the reversal of a character state transition; Mabee et al., 2007). The loss of leaves in *Cuscuta* (for instance) is an evolutionary innovation in terms of the leafy clade to which *Cuscuta* belongs. To see this as a reversal it is necessary to go back to leafless ancestors of the seed plants, which is not an appropriate comparison. The reverting of a character state, such as round leaves reverting to elliptic, may indeed be considered evolution in reverse. Another type of reversal is the **regain of a lost entity** (e.g. an organ type), such as the regain of inflorescence bracts in some mutants of *Arabidopsis*. These three types of regressive evolution are set out in Table 1.5.

Loss of organs may be by active suppression due to a gain-of-function (i.e. genetically dominant) mutation at a suppressor locus. An example of this appears to be the loss of bracts in the Brassicaceae, including *Arabidopsis*. In most eudicots the inflorescence branches are subtended by bracts. However, bracts are usually missing from the inflorescences of the Brassicaceae, although there is vestigial expression of bract-specific genes where bracts would be expected. These vestiges have been called "cryptic bracts." The leaf developmental gene *JAGGED* is required to form bracts (Dinneny et al., 2004) and this is excluded from the inflorescence of *Arabidopsis* by the action of the *BLADE ON PETIOLE* (*BOP*) genes acting in concert with the developmental gene *LEAFY* (Norberg, Holmlund & Nilsson, 2005). Thus bracts are actively suppressed in *Arabidopsis*. Loss-of-function mutations in the suppression gene network allows bracts to form, after having been evolutionarily lost, by **desuppression**. As the suppressed gene has an

Table 1.5 Genetic landscape of loss and reversal (see text for explanation).

	Dominant mutations Gain of function	Recessive mutations Reduction of function	Recessive mutations Loss of function in pleiotropic gene	Recessive mutations Loss of function in specific gene
Loss of organ	Suppression	Loss with vestige	Loss with possible neomorphism	Perfect loss
Reverting of character state	Cryptic innovation	Reversal with vestige	Reversal with possible neomorphism	Perfect reversal
Regain from loss	Recapture by conserved network	Partial desuppression	Desuppression with secondary phenotype	Desuppression

important role in leaf development elsewhere in the plant there is no likelihood that the *JAGGED* gene will degenerate making the reoccurrence of bracts impossible.

Alternatively, organ loss may be via loss or reduction of the function of a key gene required for organ formation. These loss-of-function mutations will be genetically recessive. If the loss of function is partial, **vestigial organs** will remain (**staminodes** are a common example of this). If the gene has pleiotropic effects then a **neomorphism** may accompany organ loss. An example of a neomorphism is the change in petal number accompanying a reversal to actinomorphy in *Antirrhinum* mutants. *Antirrhinum cycloidea/dichotoma* mutants commonly have six petals instead of the expected five. As well as being a key controller of floral symmetry, these genes apparently have an effect on primordium initiation, which leads to the neomorphic phenotype (six petals) that accompanies loss of radial symmetry. Perfect loss or perfect reversal, that is, loss without vestige or neomorphism, involving the clean deletion of an entity or quality, appears to be rare. However, it does occur. An example may be in lost petals of some legumes, in which there is no vestigial ontogeny whatever (in other legumes petal primordia are initiated but growth past a certain stage fails to occur; Tucker, 2001).

The apparent reversal of a character state transition may sometimes take place using a very different developmental mechanism to achieve the evolutionary prior condition. This is known as **cryptic innovation** (Citerne, Pennington & Cronk, 2006). An example is found in the bird-pollinated legume *Cadia*, which has reverted to an apparently primitive condition of radially symmetrical flowers instead of the strongly zygomorphic bee-pollinated flowers characteristic of other papilionoid legumes. However, this does not result from a loss-of-function mutation in the genes controlling floral symmetry. Instead the dorsally expressed *LEGUME CYCLOIDEA* (*LEGCYC*) gene controlling floral symmetry has expanded its expression pattern in *Cadia* to include all petals equally. All petals have therefore been homeotically converted into dorsal petals (Citerne, Pennington & Cronk, 2006). This is therefore a cryptic innovation that uses the machinery of zygomorphy to mimick an ancestral radial state.

The regain of lost organs is an interesting type of reversal. The easiest way to achieve this is if the original loss was by suppression. A loss-of-function mutation, to knock out the suppression, might reactivate the original organ, as in the *JAGGED/BOP* example above. However, it is probably necessary that the relevant gene pathway has a secondary function so it is not lost to deleterious mutation and pseudogene formation while inactive. Another means of organ reactivation is *cis*-regulatory **network recapture**, the reconnection of a conserved network to some key component. Network capture has been discussed in the previous section. The example given was the *ENGRAILED* network capturing the *YELLOW* gene by gain of *cis*-regulatory elements. One can also imagine this *cis*-regulatory element being lost, and regained, leading to evolutionary **flicker** (Marshall, Raff & Raff, 1994).

1.11 Fossils and phylogeny

Fossils and molecular phylogenetic trees of extant organisms provide the most direct, robust and reliable forms of evidence for the history of character evolution. Fossils present actual characters existing at a particular geological time. Phylogenetic trees, when well supported by a plethora of congruent characters, which molecular trees increasingly are, allow the reconstruction of character state evolution by character state mapping. New methods for incorporating fossils into joint morphological and molecular trees are therefore particularly exciting. Fossils, however, are patchy both in their occurrence and in the information that can be obtained from them. In the history of the early land plants, no deposit has been as important as the Rhynie Chert of Scotland, which holds exquisitely preserved petrifaction fossils, in which many details indicative of early plant biology can be seen (Edwards, 1986; Satterthwait & Schopf, 1972).

The early fossil history of flowering plants was based, until recently, largely on pollen, leaf and wood fossils, as few flower fossils, although useful for reconstructing angiosperm evolution, were known. Lately several excellent early flower fossil have been discovered (Friis, Pedersen & Crane, 2006) such as the Cretaceous *Archaeanthus* flower (Cretaceous) which has stalked follicles, emarginate leaves and scars of stamens, inner and outer tepals and bract scales (Dilcher & Crane, 1984). Another interesting fossil is the still controversial *Archaefructus* which has been interpreted variously as having elongated flowers in a "prefloral state" or an inflorescence of reduced flowers (Friis et al., 2003; Sun et al., 1998).

There is increasing consensus on the characteristics of the primitive flower, based on a combination of fossil evidence and phylogenetic trees (Doyle & Endress, 2000; Endress, 2004, 2006; Soltis & Soltis, 2004). A view of such a flower is emerging as radially symmetrical with numerous, separate, spirally arranged, poorly differentiated parts, simple pistils (composed of one carpel), a perianth of tepals (calyx and corolla not differentiated into petals and sepals), the stamen is somewhat leaf-like, without strongly differentiated filaments and ascidiate carpels (peltate tubular leaves; Table 1.6).

Molecular phylogenetic work has shed considerable light on the early evolutionary history of the land plants (Qiu et al., 2006, 2007). There are, however, still many uncertainties, such as the branching order of the three first divergent clades mosses, liverworts and hornworts (Fig. 1.3 suggests a likely solution). Recent phylogenies also strongly point to a monophyletic "monilophyte" group (ferns, horsetails, adder's-tongues and whisk-plants, see Figs 1.4 and 1.5; Des Marais et al., 2003; Pryer et al., 2001; Schneider et al., 2004). Furthermore, a monophyletic gymnosperm clade is increasingly strongly

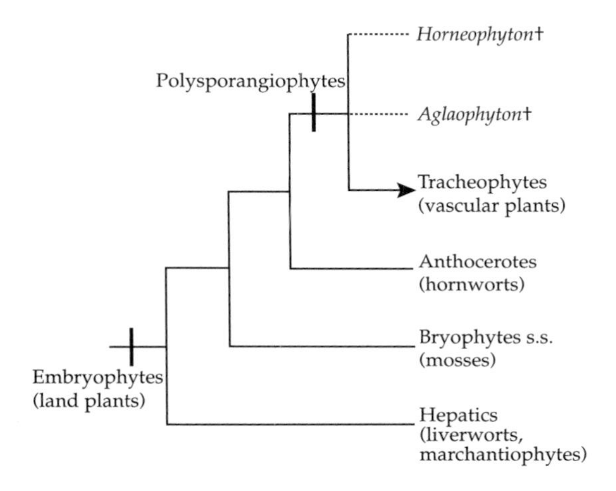

Fig. 1.3. Phylogenetic hypothesis for the embryophytes. There is still uncertainty over the branching order and relationships of the three earliest diverging groups.

Table 1.6 The three early divergent clades of extant angiosperms (from evidence of molecular phylogeny) with their putatively primitive and derived characters from a combination of fossil and phylogenetic (character state reconstruction) evidence.

	Primitive characters	Derived characters
Amborellales (*Amborella*)	Wood vesselless, perianth parts undifferentiated, spiral, numerous/indefinite in number, pollen inaperturate, pistils simple (single carpel)	Flowers unisexual, carpels in a single whorl
Nymphaeales (water lilies: Nymphaeaceae, Cabombaceae, Hydatellaceae)	Wood vesselless, perianth parts numerous and spirally arranged, separate, stamens numerous and spirally arranged, sometimes grading into petals, pollen monosulcate, pistils simple	Plants herbaceous, aquatic (in Hydatellaceae flowers reduced to single pistils or stamens)
Austrobaileyales (star anise and relatives: Illiciaceae s.l. [including Schisandraceae], Trimeniaceae, Austrobaileyaceae)	Perianth parts numerous and spirally arranged, separate; stamens numerous and spirally arranged; pistils simple	Vessels in wood; triaperturate pollen but not homologous with tricolpate pollen of eudicots; leaves with ethereal oil cells (single spherical cells filled with aromatic compounds)

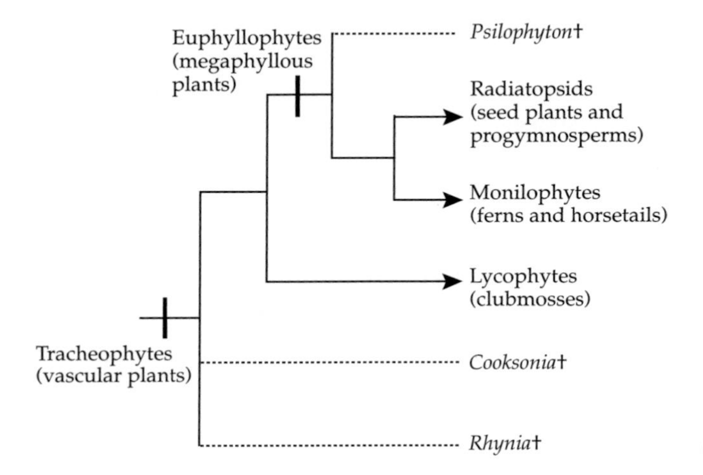

Fig. 1.4. Phylogenetic hypothesis for the tracheophytes.

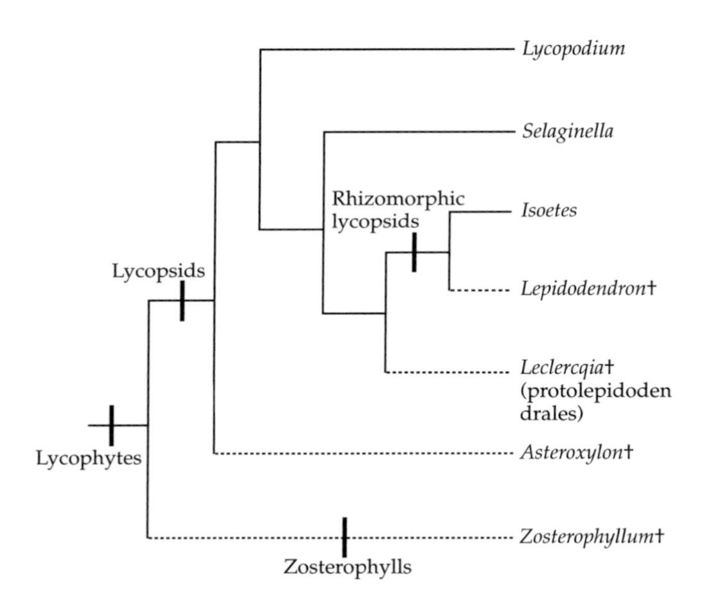

Fig. 1.5. Phylogenetic hypothesis for the lycophytes.

supported as sister group to the flowering plants (angiosperms, see Figs 1.6 and 1.7; Bowe, Coat & de Pamphilis, 2000; Chaw et al., 2000; Hajibabaei, Xia & Drouin, 2006; Nickerson & Drouin, 2004; Rydin, Kallersjo & Friist, 2002).

Recent phylogenetic studies on flowering plants are beginning to reach a consensus about the major features, although some problems still remain (Fig. 1.8; Chase, Fay & Savolainen, 2000; Qiu et al., 2005; Savolainen & Chase, 2003; Savolainen et al., 2000; Soltis, Gitzendanner & Soltis, 2007). The new information on angiosperm relationships has been summarized and systematized by the Angiosperm Phylogeny Group (APG), a group of molecular and morphological systematists who joined together to translate the results of molecular systematics of the 1990s into a classification in which all taxa are monophyletic. This led to the first published APG system (Bremer et al., 1998) and later a revised version, APG II (Bremer et al., 2003). APG uses informal names for higher taxa

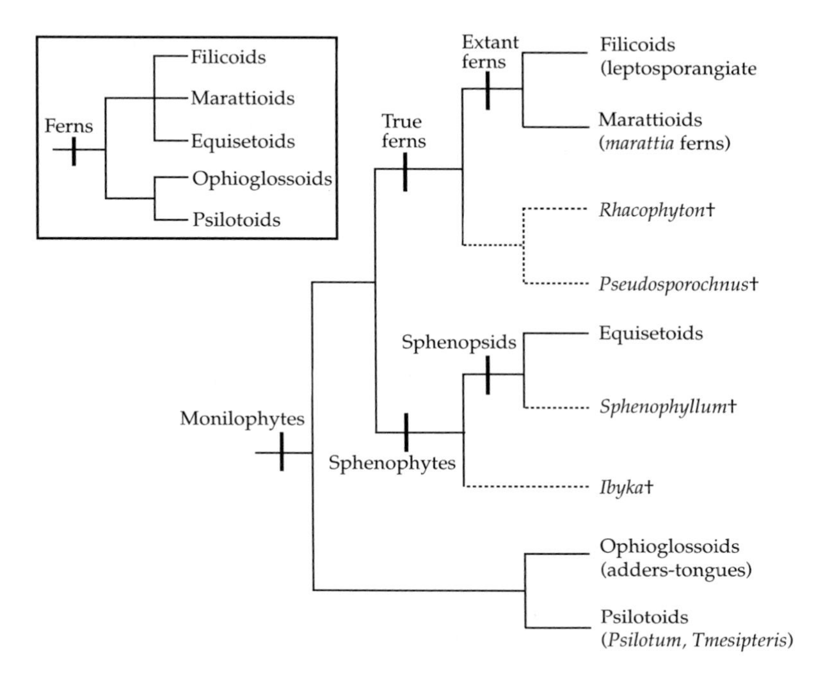

Fig. 1.6. Phylogenetic hypothesis for the monilophytes. The inset shows the uncertainty in the placement of the equisetoids and the broader usage of "ferns" adopted by some authors.

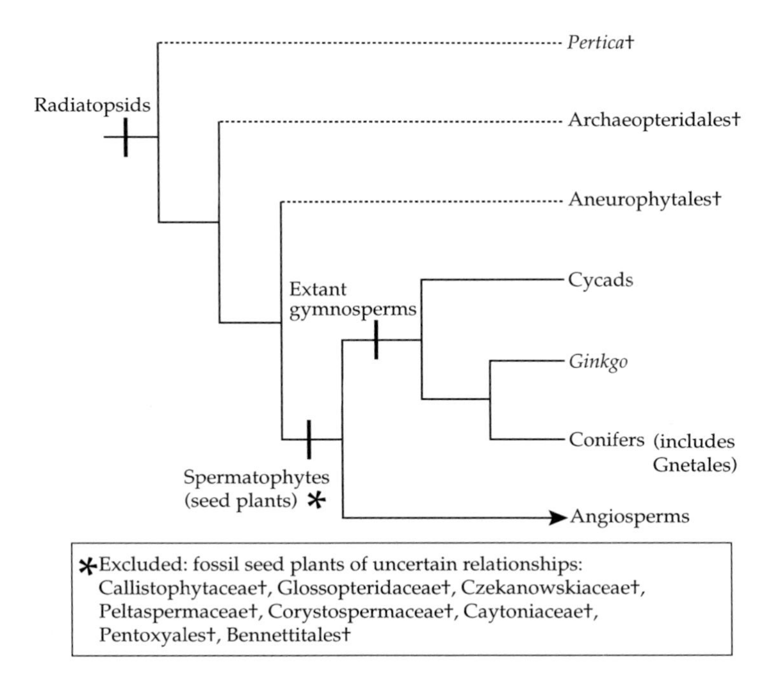

Fig. 1.7. Phylogenetic hypothesis for the radiatopsids (progymnosperms and seed plants). There is still some uncertainty over the relationships within the gymnosperms.

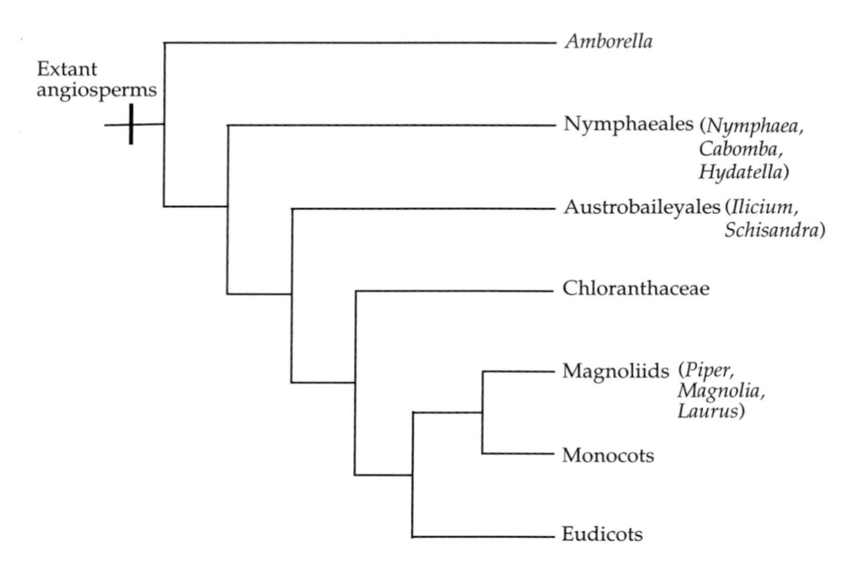

Fig. 1.8. Phylogenetic hypothesis for the angiosperms. There is still uncertainty about the relationships of *Amborella* to the Nymphaeales, Chloranthaceae to the magnoliids and monocots to the magnoliids.

Table 1.7 Classification of extant land plants.

Major clade	Subclade	Further subclade
1. Liverworts or hepatics (hepatophytes)		
2. Mosses (bryophytes s.s.)		
3. Anthocerotes (hornworts or anthocerophytes)		
4. Lycophytes (clubmosses and quillworts)		
5. Euphyllophytes	5.1 Lignophytes (radiatopsids)	5.1.1 Gymnosperms (conifers, gnetoids, ginkgoids, cycads)
		5.1.2 Angiosperms (flowering plants)
	5.2 Monilophytes	5.2.1 Equisetoids (horsetails)
		5.2.2 Ophioglossoids (adder's-tongues, including Psilotaceae or whisk-plants)
		5.2.3 Marattioids (eusporangaite ferns)
		5.2.4 Filicoids (leptosporangiate ferns)

and provides monophyletic but as far as possible traditional treatments for families and orders[3] (see Table 1.7).

[3] An easily accessible version of the APG classification, by P.F. Stevens, with a wealth of morphological and other information, is available on the web: http://www.mobot.org/MOBOT/research/APweb/.

1.12 Empirical morphology versus typology

A question can be asked whether the organ concepts described in this book, for instance leaves, stems and roots, are realities, or whether they are abstractions. Organs and morphological structures are treated here within the context of "empirical morphology." That is, organ concepts, based initially on morphological similarity, are treated as hypotheses until they are tested within a developmental, molecular and phylogenetic framework to determine that all instances of a particular organ or morphological structure are homologous.

Much miscommunication has occurred in the history of plant morphology because of the tension between two intellectual world-views: the German idealism deriving from Immannuel Kant (1724–1804) and English empiricism deriving from John Locke (1632–1704). In the nineteenth century, Friedrich Schelling (1775–1854), and others, developed an idealistic "Naturphilosophie," proposing that nature could be understood through a mixture of archetypal forms and vitalist essences. This was such a poor model for science that rather few German scientists after 1820 were much affected by idealistic philosophy. However, Alexander Braun (1805–1877), otherwise a brilliant plant morphologist, worked until near the end of his life in an idealistic framework. Darwin's (1859) *Origin of Species* (a tour-de-force of inductive empiricism which posited phylogeny rather than archetypes as the explanation for patterns of form in nature) as well as the emerging physiological viewpoint of Schleiden,[4] Sachs and others were powerful antidotes to idealism. The greatest German morphologists Wilhelm Hofmeister (1824–1877) and Karl von Goebel (1855–1932) were strictly empirical in outlook.

Nevertheless, a strand of idealism remained, sufficient for Wilhelm von Troll (1897–1978), despite having been a pupil of Goebel's, to return to an idealist stance. Troll saw morphology as a search for "types" that would intuitively appear in the mind of the investigator once the range of morphological variation had been examined. To quote Kaplan: "Like Goethe, Troll believed the types were real, not just abstractions, and that they stood behind the diversity one saw in the physical world. In many ways Troll held a platonic view of the biological world" (Kaplan, 2001). At its best, one might allow that idealistic morphology may provide a temporarily helpful framework for the ordering of information; in the spirit of Schopenhauer's dictum: "The task is not so much to see that which no one yet has seen, but to think what no one has yet thought, about that which everyone sees." Temporary, that is, until these concepts can be tested empirically in a phylogenetic or developmental framework. At its worst, idealistic morphology may be viewed as a junkyard of abstract ideas obscuring biological reality.

It is unsurprising that typological approaches have provoked strong antipathy. One reaction has been to downplay the existence of any organ concepts, even those that have strong empirical support, and this has led to what is variously called "process morphology", "continuum morphology" or "fuzzy morphology" (Rutishauser, 1995; Rutishauser & Isler, 2001; Sattler, 1996). Such a reaction is understandable but extreme, and not without critics (Weston, 1999).

Roots, stems and leaves do appear to have an empirical reality: with very little training it is possible for students to identify these organs correctly, without equivocation, in any of thousands of plants in the euphyllophyte clades. The general stability of these organs appears to result from the rather stable and conserved gene networks underpinning them. Exceptions that cause problems are rare. The most interesting of these, as in the bladderworts (*Utricularia*) and the podostems (Podostemaceae), are phylogenetically isolated examples that appear to be highly derived. They have not yet been fully examined in a molecular-developmental context. It is not a problem for empirical morphology that intermediate organs occur: indeed they are of great interest for the empirical morphologist. To use an analogy, the existence of a gasoline/electric hybrid car does not necessarily reveal the gasoline car and the electric car to be empty abstractions. Rather it indicates the possibilities that come from mixing two different things. The question of why organ identities are so rarely

[4] Schleiden, the father of cell theory, held an uncompromising reductionist view, expressed in his statement: "We should reject absolutely any hypothesis or induction that attempts to explain processes occurring at the level of the plant without taking into account changes at the level of individual cells" (Schleiden, 1842).

mixed is therefore an important one, and the answer is likely to involve developmental or functional constraints.

1.13 Plant teratology

Plant **teratology** is the study of aberrant phenotypes (**teratomorphs**), whether naturally occurring or induced, or whether genetically based or the result of developmental aberration. If used cautiously, that is, within a developmental and genetic framework, teratomorphs can be important sources of information about the developmental nature of plant traits. The angiosperm flower, in particular, has proved a rich source of these, often the result of the mutation of a critical gene after application of the chemical mutagen ethylmethane sulphonate (EMS; Meyer, 1966; Meyerowitz, Smyth & Bowman, 1989) or the insertion of a transposon, in lines with high transposon activity (Luo et al., 1991). These floral mutants have been pivotal in the advancement of plant molecular-developmental genetics.

Fascination with such phenotypes is longstanding. Linnaeus wrote extensively on the teratomorphic (peloric) phenotype of *Linaria* with five spurs instead of one (Cubas, Vincent & Coen, 1999; Linnaeus & Rudberg, 1749). More importantly, Goethe based his influential theory of plant metamorphosis on terata (Goethe, 1790). In the nineteenth and early twentieth centuries, compendia of abnormalities were published, notable among these being the *Vegetable Teratology* of Maxwell T. Masters (1833–1907; Masters, 1869). Sometime later attempts were also made to interpret these abnormalities in an evolutionary manner (Worsdell, 1916). Worsdell advanced the theory that terata represented atavisms, that is, throwbacks to a primitive evolutionary state. The interpretation by Saunders of *Matthiola* fruit terata provides an example of such thinking. She believed that aberrant fruits she discovered in *Matthiola* (with four rather than two valves) were atavistic: "in all these cases we are witnessing the reappearance of an ancestral character" (Saunders, 1923). Arber rightly gave a detailed rebuttal to both this particular interpretation and indeed to any interpretations that attempted to polarize character evolution using terata (Arber, 1931).

However, terata do have a legitimate use, and that is to reveal developmental possibilities. These possibilities can suggest explicit models of development, or provide tests of such models. Powerful insights can be so gained, as Goethe found from viewing plant terata. Guédès puts it thus: "terata are abnormal but no way haphazardly organized. Never will a stamen be transformed into a root or stem. Always it will become a leaf, a petal, a carpel or some other phyllome" (Guédès, 1979). This is highly suggestive that the stamen shares molecular-developmental features with other phyllomes, and if an evolutionary interpretation is required, we can say that it is homologous to other phyllomes.

These transformations can take two forms. First, a primordium occurring in a location (e.g. the third whorl of the flower) that is associated with a particular organ may develop as an organ associated with another location (e.g. the second whorl of the flower). This is **homeosis** (or **complete homeoheterotopy**). Alternatively, a primordium may develop as an organ intermediate between two organ types (**incomplete homeoheterotopy**). This latter case is similar to the intermediate organ types that occur in intermediate positions in normal development, for instance the intermediates between petals and stamens in many water lilies (*Nymphaea*).

A most striking modern use of teratology to develop an explicit developmental model is the use of homeotic terata of the flowers of *Arabidopsis* and *Antirrhinum* to develop the molecular-developmental ABC model of floral morphogenesis (Bowman, Smyth & Meyerowitz, 1991; Coen et al., 1991; Coen & Meyerowitz, 1991). One advantage of recent work over traditional teratology is the use of single gene mutants generated by artificial mutagenesis. This circumvents the complicating problem associated with naturally occurring terata: their cause may be due to viruses, environmental insult or a host of other factors.

Despite the simplicity of artificially induced terata resulting from mutagenesis, recently there has been considerable interest in naturally occurring terata (Cubas, Vincent & Coen, 1999; Hintz et al., 2006; Rudall & Bateman, 2003).

1.14 Corner and Sporne: a study in opposites

It is not possible here to attempt any sort of history of plant morphology. However, I would like to conclude this introduction with a reflection on the work of two botanists I knew personally and learnt much from, E.J.H. Corner (1906–1996) and K.R. Sporne (1917–1990). This, I hope, will give a little insight into some of the morphological preoccupations of the last century.

Corner is most widely known for his masterly survey of the plant world *The Life of Plants* (Corner, 1959) and Sporne for his three small but lapidary books on the morphology of pteridophytes, gymnosperms and angiosperms (Sporne, 1962, 1965, 1974). Both were interested in a problem that consumed much of plant morphology in the twentieth century: which of the characters possessed by flowering plants are primitive and which advanced? During much of the twentieth century the fossil record was not very helpful in resolving what Darwin had termed "an abominable mystery," the origin of flowering plants. Instead botanists sought to come at the problem by working back from neobotany.

The year 1939 finds Kenneth Sporne starting doctoral work in Cambridge on floral vascular systems, and Corner in Singapore establishing a mastery of tropical botany. When, during the early years of the Second World War, Sporne was asked by relatives what he was doing, he would reply, quite truthfully, that he was researching the question of which came first, willows or buttercups? It was generally assumed that he must be involved in war-work of the utmost importance and secrecy to require such an apparently ridiculous cover story. Actually, while Europe descended into chaos, he was seeking to resolve the opposing views attributed to Adolf Engler (1844–1930), the botanical colossus of Berlin, whose system implied that the first angiosperm had catkins, and Agnes Arber (1879–1960), the sage from Cambridge's Huntingdon Road, who, with her husband E.A. Newell Arber (1870–1918), favored an ancestor more akin to a buttercup or *Magnolia*.[5] It was assumed at the time that floral vascular systems were more conserved than other parts of the flower,[6] and thus might provide a window into the past.

In 1939 Sporne developed a method of clearing thick slices of flowers in lactic acid and examining them as whole mounts (Sporne, 1948). He assiduously taught this technique to all undergraduate students who passed under his tutelage, and I remember well mounding up the waxy putty of lanolin and resin to contain the pond of lactic acid necessary to examine the vasculature of the spike of *Ophioglossum*. However, Sporne soon lost faith in floral vasculature as a means of solving big questions in plant evolution, in part from his own experience and in part in response to the widespread criticism (Schmid, 1972) that the idea of vascular conservatism attracted.

An alternative, more physiological, view held that the vascular system of the flower was merely what was needed to irrigate the flower, and perhaps more importantly, irrigate the developing fruit. Years later, Corner would goad Sporne by presenting him with a loofah (the retted vascular system of the fruit of *Luffa cylindrica*, whose tangled and woody vascular bundles can be used as a back-scrub) with the rhetorical question: "tell me what that says about plant evolution!" Sporne took such goading in good part, and gave the loofah pride of place in his practical classes as an example of a complex floral vascular system.

In the rain forests of Malaya, Corner looked to the plant diversity of the tropics as a more likely place to find primitive characters than the

[5] The euanthial theory, as formulated by E.A.N. Arber and Parkin in 1907, was already implicit in the systems of Bessey (1897) and Hallier (1905). However, as the euanthial theory was most fully developed by Arber and Parkin, it is generally associated with those two names. Both the Bessey and Hallier systems, commencing the dicotyledons with *Magnolia*, were in direct opposition to Engler (commencing with *Casuarina*). As Hallier wryly noted: "The most striking point in the reception of my system is the silence of Professor Engler."

[6] The hypothesis of vascular conservatism was put forward by Henslow in 1890 but was mainly developed by Eames (see for instance Eames, 1931).

recently glaciated fields of Cambridgeshire. In the tropics Corner found many angiosperms of massive construction (similar to the cycads and the fossil Bennettitales) that he called "pachycaul." These plants have thick stems and short internodes, which are able to carry large flowers developing into large fruits bearing large seeds with arils. Here, Corner reasoned, was the primitive angiospermous syndrome, a syndrome he developed in a series of papers as the "Durian theory" after the massively constructed tropical fruit tree, *Durio zibethinus* (Corner, 1949, 1953, 1954a,b).

This combination of intuition and logic presented in dazzling style did not suit Sporne who desired a more rational means of identifying primitive characters. To this end he set to work, not making original observations on rainforest flowers, but in a dusty reprint collection, collecting data on the distribution of characters in angiosperm families in what would now be called a "meta-analysis." His reasoning was that while advanced character states should be scattered over the angiosperms, primitive character states should be correlated, as they once occurred together in a single lineage. This way, he felt, he was able to replace intuition by the cool judgement of statistics (Sporne, 1949, 1980, 1982). Despite the different methods, some of his conclusions were strikingly in agreement with Corner's, for instance in the primitiveness of the arillate seed.

Sporne, like many others, was deeply skeptical of Corner's intuition and liked to cite their disagreement over the yew (*Taxus*) as an example. The question at issue was whether *Taxus* was a conifer. Sporne followed Florin's seemingly unassailable logic that *Taxus* had no close relationship with modern conifers but represented a separate line of gymnosperm evolution. The reasoning went like this: Florin had shown that the conifer cone derives from a compound polyaxial structure. *Taxus* with its single terminal ovule could only derive from the complex female reproductive structures of conifers if it is the result of extreme reduction and the very end of a long reduction series. However, as the fossil history of *Taxus* is apparently very ancient (the fossil *Palaeotaxus*†, with an ovule like a yew but a different cuticle, comes from the lowermost Jurassic), it cannot be the end point of a reduction series (Florin, 1948).

This logical edifice held no sway with Corner for whom yew was clearly a conifer. Sporne told the story of Corner coming to him brandishing a branch of yew and saying "Just look at it, you can tell it's a conifer!" Sporne liked to tell that story as an example of how intuition can lead one into error. However, it can no longer be seen that way. Modern molecular phylogenetics unequivocally places *Taxus* right in the middle of the extant conifers, just as Corner maintained. Logic can fail as surely as intuition if the assumptions on which it is based are flawed.

What mattered to Corner was "thought", and by thought he meant not mere logic-chopping but the joining of reason with observation and inspiration as the only way to drag ourselves out of the librarian-guarded mire of the existing corpus. He made no secret to me that this was a mire he regarded Sporne as inhabiting. As for Sporne, he concluded his lecture series with the observation that the most important qualities in any science are humility and skepticism, humility deriving from how little we know and skepticism deriving from how little other people know. And when he spoke of skepticism I fancy he may have had Corner's ideas in the back of his mind.

How have Sporne and Corner's ideas of primitive angiosperm characters fared in the modern age? Toward the close of the twentieth century a number of exquisitely preserved early angiosperm flowers were discovered (Friis, Pedersen & Crane, 2006), greatly improving the contribution of paleobotany to the problem. Furthermore, the advent of fairly reliable molecular phylogenies for the angiosperms allowed characters to be directly optimized on trees and primitive character states to be determined much more directly and reliably (Endress, 2001b; Friis, Pedersen & Crane, 2005; Soltis & Soltis, 2004). As might be expected, it seems that neither Sporne nor Corner were wholly wrong or wholly right. Some of their character states appear to be primitive in at least some clades of angiosperms although surprisingly few can be regarded as primitive in the context of the angiosperms as a whole. Sporne's identification of the unisexual flower as primitive is a notable failure, likely due to an artefact in his statistical method (Thompson, 1986). However, it is evident that unisexual flowers, by reduction, evolved early in a number of lineages. Similarly, there is little evidence that early angiosperms were pachycauls, as Corner suggested, although the pachycaul habit does appear early in some clades and there is even a member of an early diverging angiosperm clade, the water lily *Nymphaea*, which has the massive stems with short internodes, the large flowers and fruit, and seeds with arils, just as required by the Durian theory.

CHAPTER 2

The organography of stems

2.1 Origin and homology of stems

Of the three fundamental units of the plant body (stem, leaf and root), the **stem or AXIS** is the oldest and arguably the most interesting. In **gametophyte,** the stem is a **unifacial** development of the **bifacial** and often irregularly growing **thallus**. The first land plants seem to have had a prostrate thalloid "stem" in the gametophyte rather like today's thalloid liverworts. This has evolved into axial stem structures in both liverworts and mosses. It should be noted that the word "stem" is sometimes used in a narrow sense to denote only leafy axes with a stem/node architecture. However, making a distinction between leafy and leafless stems merely obscures the basic developmental commonality of all radial subaerial axes.

In the **sporophyte**, the stem originally functioned as the stalk of the **sporangium**. In seed plants, stems are now hugely elaborated, distinguished from roots by their **exogenous** rather than **endogenous** branching. Before the end of the Silurian period, stems circular in cross section, like those of *Cooksonia*, which may well have been erect, had evolved (Fig. 2.1). The terrestrial environment of the Silurian was not a highly competitive environment so it is unlikely that these early stems were reaching for the light. Rather, the erect stems were sporophytes raising sporangia above the ground for more efficient dispersal. Every centimeter of increased height helps to escape the "boundary layer" of still air near the ground. Later, after the evolution of leaves, the primary function of stems became one of supporting the leaf canopy to optimize photosynthesis.

It is interesting to consider the homology of the gametophytic bryophyte stem and the sporophytic vascular plant stem. In mosses and leafy liverworts stems occur mainly in the gametophyte (although the sporangium is usually raised on a stem-like **seta**). In vascular plants, by contrast, the stem is characteristically elaborated mainly by the sporophyte, and gametophytes are often not at all stem-like (as in many fern **prothalli**). However, there are exceptions and in lycophytes, although the gametophytes are underground, they are often somewhat stem-like (**cormoid**) in morphology. The cormoid gametophyte is horizontal (**diageotropic**) rather than upright (**negatively geotropic**), but that is a property of many underground sporophyte stems.

It is hypothesized that the first land plants had thalloid gametophytes with sessile sporophytes and that this gametophyte thallus then evolved into a stem-like form. There are two plausible mechanisms for the gametophyte stem: the **thalloid hypothesis** and **sporangiophoral hypothesis**. In a thallus, the functions of growth and support are combined with that of photosynthesis. The thalloid hypothesis suggests that stems evolved as a result of increasing division of function between a central core for growth and support and lateral outgrowths (**phyllidia**) for photosynthesis. Modern day thalloid liverworts, such as *Riccardia* (with a narrow, much-lobed thallus), serve as analogues for possible intermediate states.

The alternative hypothesis, sporangiophoral origin of stems, states that stems evolved as specialized organs (**sporangiophores**) of thalloid liverworts: stalks to raise the sporophyte into the air. The sporangiophore then came to dominate over the thallus by **heterochronic development**, specifically **gerontomorphosis** (i.e. the encroachment of juvenile stages by mature stages). An analogue for a possible intermediate stage is found in the

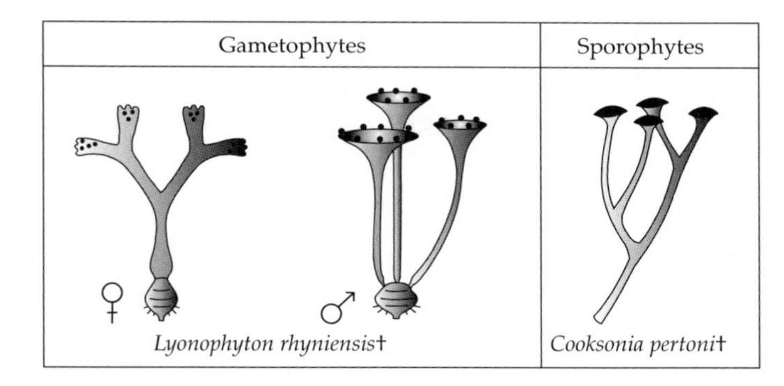

Fig. 2.1. The stems of gametophytes and sporophytes of the early land plant *Cooksonia* (after Taylor, Kerp & Hass, 2005).

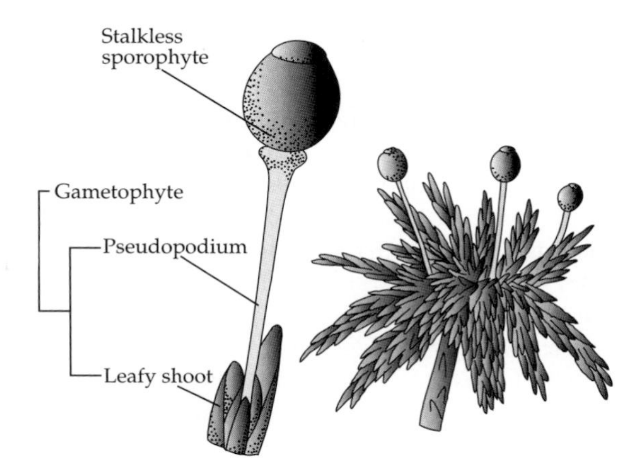

Fig. 2.2. The pseudopodium of *Sphagnum*, a leafless sporophyte stem.

thalloid liverwort *Marchantia*, which has strongly developed sporangiophores that are erect and circular in cross section. Developmentally however, the *Marchantia* sporangiophore is very different from a stem, being tubular and deriving from an invagination of the thallus.

If the **sporangiophoral hypothesis** is correct, there has been transference of function from the gametophyte sporangiophore to the sporophyte seta in most mosses and leafy liverworts. However, in some mosses (e.g. *Sphagnum*) the gametophyte produces a stem-like (but leafless) organ called a **pseudopodium** that raises the stalkless sporophyte (Fig. 2.2).

Turning now to the sporophyte stem, it is probable that this has evolved from the progressive elaboration of a stalk subtending the sporangium. This is the **interpolation hypothesis** (Blackwell, 2003; Kato & Akiyama, 2005). This suggests that a *de novo* structure is interpolated as a developmental stage between the sporophyte **foot** (the transfer tissue at the base of the sporangium) and the sporangium itself. This interpolation hypothesis was first proposed by Bower[1] as the **antithetic theory** (Blackwell, 2003; Bower, 1908, 1935). Bower was challenging Pringsheim's alternative theory that the ancestors of land plants had **isomorphic alternation of generations** like certain seaweeds, and that on land the two generations diverged from their originally identical state. Modern phylogenetic knowledge has shown that the **isomorphic theory** (or **homologous theory** as it is traditionally called) is incorrect. The closest relatives of the land plants, *Chara* and *Coleochaete,* are **haplontic** (i.e. with haploid phase of the life cycle dominant) and there is no evidence for isomorphic ancestry. On the other hand, the probable haplontic ancestry of the land plants is consistent with an interpolated origin in which the complex sporophyte represents a *de novo,* interpolated structure (Fig. 2.3).

It should be noted, however, that the **seta** subtending the sporangium of present day mosses and liverworts is unlikely to be homologous to vascular plant sporophyte stems, as it was once thought. The seta is best interpreted as an enlarged foot (Kato & Akiyama, 2005). Its

[1] Although Bower proposed the theory under the name "antithetic theory" in 1908, he later (Bower, 1935) came to prefer the term "interpolation theory."

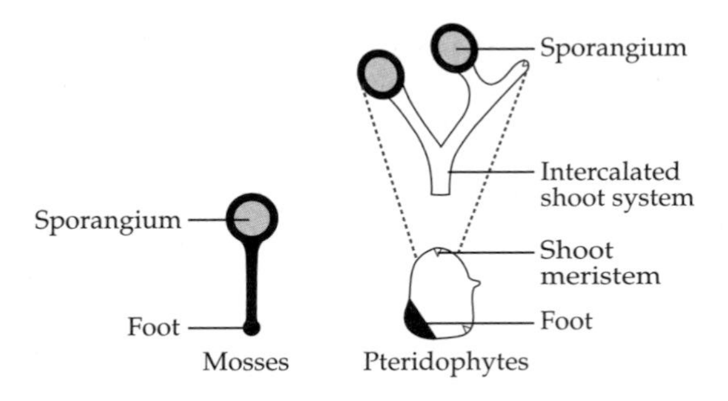

Fig. 2.3. The interpolation of the sporophyte generation (after Kato & Akiyama, 2005).

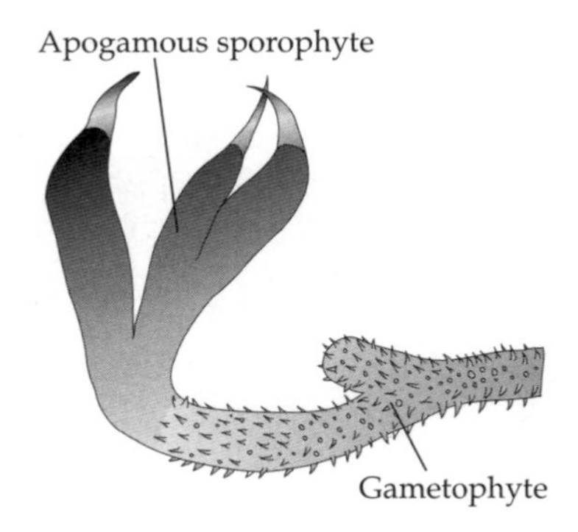

Fig. 2.4. Gametophyte sporophyte transition in apogamous *Tmesipteris*, showing seamless transition from gametophytic to sporophytic morphogenesis (after Whittier, 2004).

mechanism of growth is different from that of vascular plant stems, being precocious and without a persistent apical cell. This is particularly clear in the seta of leafy liverworts in which most of the elongation growth is by cell expansion and not apical growth. Furthermore, vascular plant stems are characterized by the possession of certain derived biochemical properties, which are probably important in their robustness and complexity. An example is the possession of certain xylan types, universal in vascular plant secondary cell walls, but in bryophytes only found in hornwort pseudoelaters and spores (Carafa et al., 2005). At present it seems that the bryophyte sporophyte is therefore best regarded as a **sporogonium** (a footed sporangium) rather than as a stemmed sporangium.

Although the sporophytic stem of vascular plants may be regarded as a *de novo* interpolation, there may have been transfer of developmental programs between the sporophyte and gametophyte phases of the life cycle. Initially, this was probably transfer of developmental capacity from stem-like gametophyte organs such as the sporophore or radially organized thallus to the interpolated sporophyte stem. Subsequently, there may have been transfer of stem developmental pathways to the gametophyte. The similarity between gametophyte and sporophyte stems may be illustrated by apogamous psilotoid *Tmesipteris* (Fig. 2.4). The *Tmesipteris* gametophyte takes the form of a mycotrophic subterranean stem, radial in cross section and with persistent apical growth. If cultured in the correct conditions (Whittier, 2004), they convert from producing gametophyte tissue to producing sporophyte tissue without a change in ploidy level (apogamy). When this happens, there is a seamless transition from gametophyte stem to sporophyte stem indicating the fundamental developmental unity of the two organ types.

Stem formation in the gametophyte of mosses is an orchestrated developmental process. In mosses a spore germinates to form a filamentous (and unstemlike) **protonema**. In response to hormone signals, specific cells of the green protonema (**chloronema**) develop as a new filament type called **caulonema**. Subapical cells of the caulonema produce a lateral outgrowth at their apex, and this becomes a separate **initial cell**. The initial cell divides to form an undifferentiated mass of cells that eventually grows into the leafy shoot (Schumaker & Dietrich, 1998).

2.2 General characteristics of stems

The stem presents a way of organization almost unique to plants. The only animal analogues are found in the growth of monopodial thecate hydrozoa such as *Dynamena pumila*. These anomalous hydrozoa have a stem-tip which grows to produce an elongated stem mass, circular in cross section like a typical plant stem, which produces polyps by lateral budding (Berking, Hesse & Herrmann, 2002). Terminal growth is provided in these hydrozoa by a terminal polyp primordium, which remains undifferentiated in a persistent growth state. In plants the terminal growth persists through the maintenance of a population of one or more **stem cells**. Continued division of cell lineages resulting from these stem cells causes the stem cells to rise on a mound of derivatives, creating a self-extruding cylinder. The evolution of variations on this simple basic principle has given rise to myriad life forms in plants.

The term "stem cell" is originally a zoological term for an undifferentiated cell that is competent to be maintained in a state of cell division or to differentiate into other cells. It derives from "stem" only in the genealogical sense of a fundamental stock or line. However, it is highly suitable when adopted into botany as stems themselves are the creation of stem cells. In vascular plants the stem cell region at the apex of stems is called the **shoot apical meristem** (**SAM**). There is scope for further confusion here as the term **meristem** carried no original connotation with stems or with stem cells in its etymology but instead derives from the Greek term *meristos* (divided) and refers to a place in a plant where cell division occurs in an organized fashion.

Stems take many different forms, from typical stems to flat disc structures (in bulbous angiosperms), leaf-like structures and thorns. Nevertheless, stems show a variety of unifying characteristics, and the term **caulome** can be used for the generalized and collective stem-system of a plant, as opposed to a specific stem.

All types develop from a SAM, even though many stems can form **secondary meristems** to complete their growth. In vascular plants width growth can be completed by a peripheral meristem called a **cambium** (this is a feature shared with roots). Stems can also elongate from meristematic tissue below the apex. These meristems, which are inserted (intercalated) into the mature stem, are called **intercalary meristems**. This allows stems to grow from the base. This feature is a means by which a rapidly elevated canopy is produced by many grasses. For instance, the phenomenally high growth rate of some tropical bamboos is contributed to by active intercalary meristems in the internodes. An intercalary meristem is also found in some leaves, which can grow by means of a **basal meristem** (at the base of the leaf blade).

A characteristic feature of stem is that they may bear **leaves** (or **phyllidia** in the case of bryophytes). Indeed leaves are such a universal feature of vascular plant stems, that the term **shoot**, meaning a combined organ (stem with leaves), is often used as the fundamental unit of plant construction along with the root. However, the association with leaves, although widespread, should not be used to define stems as many early land plant fossils are known without leaves. Once gained, very rarely has the ability to produce leaves been lost but there are occasions. An example is the parasite *Cuscuta,* which consists of stems, haustoria (possibly modified roots) and only a few small scale leaves.

Leaves are borne on the stem at **nodes**, giving stems their characteristic alternating **node** and **internode** structure, as stem vasculature is perturbed at nodes by the presence of vascular connections to leaves (**leaf traces**), often to the extent of leaving gaps in the stem vasculature (**leaf gaps**).

Stems are generally persistent organs in contrast to leaves, which are expendable. Leaves have a lifespan that is generally between 1 and 10 years, whereas stems can last hundreds or even thousands of years. However, there are circumstances in certain plants under which stem units form an **abscission layer** and are shed in a manner analogous to that of leaves. This process is known as **branch drop** (**cladoptosis**) and is characteristic of the response to drought in conifers such as *Thuja plicata.*

Although stems derive from a bifacial thallus they generally have **radial symmetry** in cross section and are unifacial (having only one morphological surface). This is in contrast to the generally bifacial nature of leaves. The internal architecture of stems is usually radial too, being dominated by the vascular cylinder or stele. One of the prime functions of stems is to transport water from the ground to the dry aerial environment and therefore stems are often highly vascularized. In young stems there is often a thick central **medulla** (**pith**) surrounded by the **xylem** tissue. In species with strong secondary thickening this subsequently gets to be squeezed out of existence by continuing production of xylem tissue. In contrast to the initially thick central medulla, the **cortex** or **rind** of the stem is often a relatively thin layer encasing the vascular tissue. In this there is a marked contrast to **roots** in which there is little or no medulla and a thick cortex—a consequence of the centralized position of the root vasculature.

2.3 The structure of the seed plant SAM

The seed plant SAM has a complex organization whose spatial structure can be broken down in two ways depending on where the focus is on (1) the plane of cell division (tunica-corpus model) or on (2) the organizational control (zonal model). The tunica-corpus model (Schmidt, 1924) reflects the fact that apical meristems have a skin caused by cells in the outermost layers dividing periclinally (around the slope), that is, in the surface plane, therefore producing discrete cell layers. Although these layers can vary in number from one to five (in *Hippuris*), there are often only two tunic layers, designated L1 (the epidermis) and L2 (the subepidermis). Below these two layers is a cell mass in which anticlinal (against the slope) divisions are much more common, so no layers are obviously distinguishable. This is the corpus (sometimes called the L3). The lack of cell exchange between these skins and the inner stem tissues received great support from classic work on periclinal chimeras (Satina, Blakeslee & Avery, 1940) using histological markers (in this case artificially induced polyploid nuclei) to follow cell lineages.

The zonal model takes an alternative view. This recognizes four main zones (Foster, 1941): the zone of apical initials (AI), the organizing center (OC), the peripheral zone (PZ) and the rib zone (RZ) (Fig. 2.5). The AI is the zone that contains the stem cells and so ultimately is the source of all cells of the apex. In contrast, the OC is the zone of very slowly dividing cells whose function is to maintain the AI in stem cell state. The AI and the OC are sometimes lumped into a single zone: the central zone (CZ).[2] However, apart from the fact that their functions are interconnected (see next section) there is little to connect the two zones, which are maintained in very different physiological states. OC cells are often large and vacuolated, whereas the stem cells of the AI are small and fast dividing. The PZ contains cells that are fast dividing and contribute to the lateral organs and outer parts of the stem. The RZ contains the cells that divide to contribute to the central core of stem.

Foster referred to the OC cells as "central mother cells." This is a misnomer as they are not mother cells in the sense that they give rise to other cells. Rather they function to provide a micro-environment (the stem cell niche) that permits a stem cell population to be continuously maintained. They are thus best thought of as accessory cells that maintain the correct molecular signaling environment for the stem cells. Interestingly, Buvat (1952) recognized the quiescent nature of the OC, and called it the *"meristème d'attente"* or waiting meristem. Waiting for what? The *meristème d'attente* was thought, wrongly, to be waiting, Cinderella-like, to be turned into a flower.

Both views of the SAM (tunica-corpus and zonal) are useful but it should be noted that as they are overlapping rather than complementary concepts, they are difficult to combine

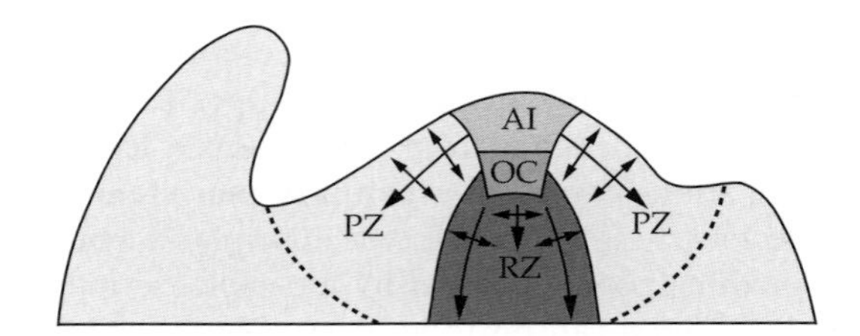

Fig. 2.5. The organization of the seed plant meristem showing the different regions. AI= apical initials, PZ = peripheral zone, OC = organizing center and RZ = rib zone. The lengths of the arrows give an approximate indication of the amount of cell divisions orientated in that direction.

[2] The CZ apparently appears to be derived from Foster's (1939) use of the term "central tissue" for the RZ and OC together in *Cycas revoluta* (the apical initials were excluded from this region).

into a single view. It is possible to designate the upper cells of the AI zone as **dermatogen** cells. This is because they are distinct in only dividing periclinally, and so only give rise to the tunic layers. This is not always so simple. Some conifers have **open dermatogen** (e.g. *Taxus*) in which there is frequent exchange of cell lineages between inner and outer layers of the AIs by means of anticlinal divisions. In contrast, other conifers (the Cupressaceae are examples) have **closed dermatogen**: the tunic-forming outer AIs are positionally very stable and thus chimeras formed by mutations in the dermatogen are very stable (Guédès, 1979).

2.4 Variation in SAM morphology

The SAM is the engine of stem production. In its simplest form it consists of a single tetrahedral **apical cell** and the surrounding cells, as was well documented 150 years ago by Hofmeister (1857). The tetrahedral apical cell cuts off derivatives from its three internal faces. However, despite the simplicity of this system, we do not know what signaling processes maintain the apical cell in a stem cell state, and most of our knowledge of SAMs comes from angiosperms, which, like gymnosperms, have a population of multiple progenitor cells, whose organization is tightly controlled. A recent study of the apical meristem of *Selaginella* has shown that the two AI cells are transient implying a position-specific rather than a lineage-specific mode of stem cell maintenance (Harrison, Rezvani & Langdale, 2007).

Apical meristems vary from 50 μm across to 3 mm across. Size of apical meristem is correlated with size of plant stem. Therefore plants of massive construction such as cycads, palms and barrel cacti have apical meristems typically in the millimeter range. Rapidity of stem elongation also affects meristem size. Twining plants with rapidly elongating long internodes tend to have very small apical meristems. In these plants the leaf primordia that would contribute to meristem bulk are quickly carried away from the meristem by elongation.

Most angiosperm SAMs are in the range between 0.1 and 0.3 mm. Meristem size varies in those species with episodic leaf production (e.g. **decussate** species). During the part of the **plastochron** (developmental time separating leaf initiation) when leaf primordia are absent, the apex is relatively broad. It then becomes narrow while leaves are initiated as much of the PZ is appropriated by the leaf primordia.

The shape of the apical dome varies too. Some are almost flat while others, such as *Hypericum*, are steeply domed. The shape of the SAM appears to have important effects on leaf production with the enlarged meristems of the *abphyll* mutants of maize leading to a **decussate phyllotaxy** (Jackson & Hake, 1999) and the flat SAMs of the *sho* mutants of rice (Itoh et al., 2000) leading to irregular phyllotaxy. The *sho* mutants suggest that domed meristems appear to have a greater number of stem cells in proportion to nonstem cells. In plants with long internode length, the apical meristem is relatively protuberant. In plants with short internodes, such as rosette ferns and cycads, the SAM is often deeply sunk into a pit formed by the enlarging primordia surrounding the apex, sometimes to the extent of making the apex difficult to dissect out.

The single apical cell as the unitary stem cell appears to be the norm in bryophytes, lycophytes and monilophytes (White & Turner, 1995). Mostly these cells are tetrahedral, each of the three internal faces cutting off blocks of cells (merophytes). The merophytes may form helices implying a regular pattern of divisions from the three faces. In *Salvinia* and *Azolla* (Croxdale, 1978) the apical cell is reported to have only two faces, the two merophytes correlating with the abaxial–adaxial patterning of the horizontal shoot system.

In *Lycopodium* and the marattioid ferns, a single apical cell is difficult to discern and these groups may represent the ones that have undergone a transition to a more diffuse stem cell population. What is not yet clear is whether there is a population of cells responsible for maintaining the stem cell state of the apical cell. If the apical cell is homologous to the AIs of seed plants, then patterns of expression of *CLAVATA3* (*CLV3*) and *WUSCHEL* (*WUS*) homologues in ferns would be very interesting to study. These might

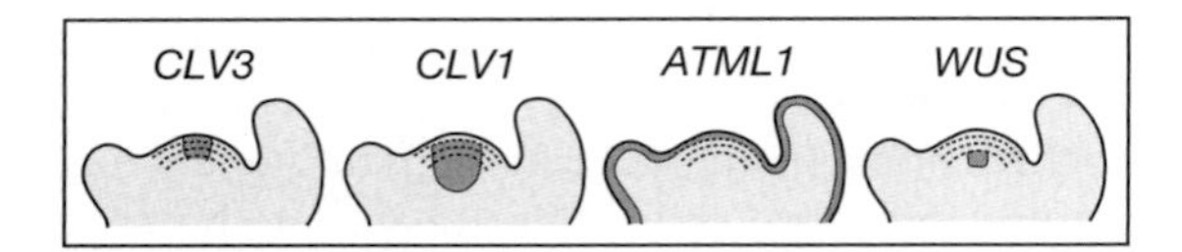

Fig. 2.6. Expression patterns of some important genes in the SAM. *CLV1/3 = CLAVATA1/3*, *WUS = WUSCHEL*. The three dotted lines mark the approximate boundaries of the outer tunica L1/L2 and the central corpus. This layering is caused by a predominance of periclinal rather than anticlinal divisions in the outer layers that is characteristic of angiosperms (after Wurschum et al., 2006).

answer both the question of whether there are pteridophytes with multiple stem cells (*CLV3* is a stem cell marker) and where the OC lies in ferns (*WUS* is an OC marker; Fig. 2.6). There is some indication that the monilophyte apex is zoned and in some species "central cells" have been described, which could possibly be the seat of the OC. Alternatively, the apical cell might be self-organizing, remaining in a stem cell state through endogenous processes. In the past it has even been suggested that the apical cell itself might be the OC, in relative stasis, stimulating adjacent cells to gain stem cell status. However, recent histological evidence is against this latter hypothesis (White & Turner, 1995).

Mention should be made here of the highly abnormal teratological forms of the apical meristem called fasciations. The most common type is the banded fasciation. In this it appears that the SAM divides into several in one plane, without separating. This creates a stem in the form of a flattened strap with ridges corresponding to each of the meristemoids (hence fasciation, from the Latin *fascis*, a bundle). Some plants, such as *Linaria* and *Forsythia*, seem more prone to fasciation than others, possibly due to susceptibility to causal pathogens. One pathogen known to cause fasciation is the bacterium *Rhodococcus* (*Corynebacterium*) *fascians*. *Rhodococcus* secretes signal molecules that disrupt meristem function (Crespi et al., 1992; Vereecke et al., 2002). Fasciation can also be caused by mutation. The cockscomb celosia (*Celosia argentea* var. *cristata*) is a horticultural fasciation mutant that is caused by a single gene (recessive) mutation. The gene responsible is not known but *Arabidopsis* mutations in genes important for meristem function can cause fasciation-like phenotypes. Examples are *CLAVATA1* (*CLV1*), *FASCIATA1* (*FAS1*) and *FAS2* (Leyser & Furner, 1992). The *FAS* genes correspond to subunits of the chromatin assembly factor protein (CAF1), which is necessary to maintain the correct epigenetic state for meristem function (Kaya et al., 2001; Ono et al., 2006).

Another type of fasciation is the **ring fasciation** often found in the inflorescences (**capitula**) of composites. In this the inflorescence apex assumes the form of a ring. In species with ray and disc flowers, ray florets form both on the inside and outside of the ring with disc florets in between.

2.5 Molecular signaling in the maintenance of the SAM

A central problem is how the stem cell population is maintained and regulated. During the life of a plant, the apical meristems vary somewhat in size, in accordance with the size of the branch or main stem being produced. However, the size of the apical meristem must be tightly regulated as runaway expansion or contraction would be disastrous for the plant.

At the core of the regulation is the *WUSCHEL–CLAVATA* regulatory feedback loop (Brand et al., 2000; Schoof et al., 2000). In the center of a SAM there is an OC that is the equivalent of the quiescent center (QC) of the root. Above this is the stem cell zone, and *WUSCHEL* (*WUS*) expression in the OC induces stem cell fate in the overlying cells. Cells of the OC can therefore be thought of as providing a suitable microenvironment within the stem cell niche for the maintenance of stem cells. However, the signal from the OC might progressively accumulate causing the stem cell zone to expand. This is prevented by the stem cells expressing CLAVATA3 (CLV3) a small mobile protein that activates the CLAVATA1–CLAVATA2 (CLV1–CLV2) receptor complex, which in turn leads to suppression of *WUS*. This negative feedback loop therefore checks any tendency toward WUS overactivity, hence regulating meristem size (Fig. 2.7).

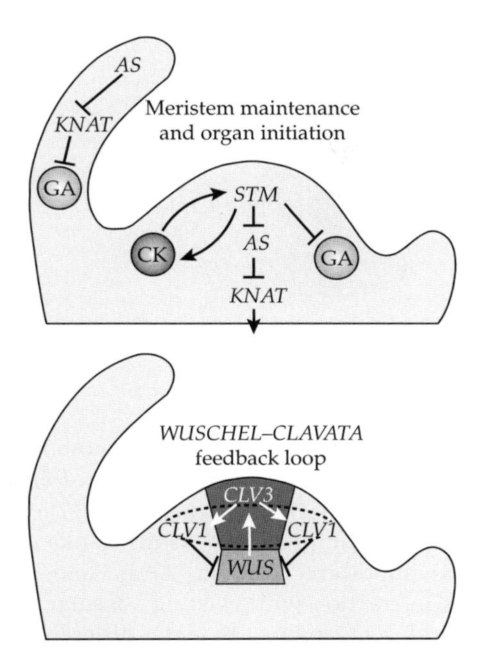

Fig. 2.7. Gene networks responsible for meristem maintenance (top) and meristem size (bottom). *STM* = *SHOOTMERISTEMLESS*, *AS* = *ASYMMETRIC LEAVES* 1/2, *CLV* = *CLAVATA* 1/3, *KNAT* = *ARABIDOPSIS KNOTTED-LIKE*, *WUS* = *WUSCHEL*, *CK* = *CYTOKININ*, *GA* = *GIBBERELLIN* (after Bäurle & Laux, 2003).

WUS acts, at least partially, by repressing the negative regulation of cytokinin (CK) response. High CK levels are necessary for the stem cell niche microenvironment and are promoted by *KNOTTED HOMEOBOX* (*KNOX*) genes (Jasinski et al., 2005; Yanai et al., 2005). Normally when CK levels are locally elevated *ARABIDOPSIS RESPONSE REGULATOR* (*ARR*) genes are induced to moderate the positive effects of CK on cell division. In the SAM, no such moderation is appropriate and WUS knocks down negative regulation by *ARR*s.

In the same way that CK are positive contributors to stem cell fate, gibberellins (GA) have adverse effect. *KNOX* genes (particularly *SHOOTMERISTEMLESS, STM*), in addition to promoting CK, also inhibit GA synthesis (Chen, Banerjee & Hannapel, 2004; Hay et al., 2002; Sakamoto et al., 2001). *STM* expression is therefore vital to maintain stem cell fate and to protect stem cells and their derivatives from premature differentiation. STM expression is eliminated from the developing leaf, and this control of expression is now known to be due to the K box, an upstream conserved noncoding sequence (Uchida et al., 2007; Fig. 2.8).

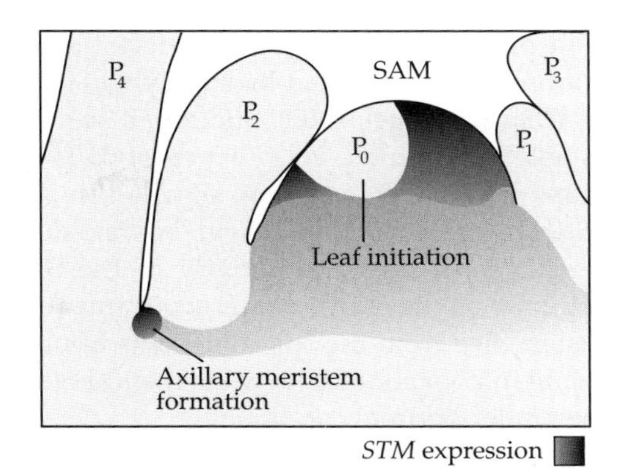

Fig. 2.8. STM expression in the SAM showing the elimination of STM expression from the site of leaf initiation but possible retention of expression in axillary positions (after Grbić & Bleecker, 2000).

Plant growth by SAMs is rather like having numerous small tightly controlled cancers at the extremities. The result of control failures (often leading to a type of developmental abnormality called **fasciation**) is serious for the plant. Perhaps for this reason the *CLV3* negative feedback loop is not the only negative regulator of *WUS*. Other genes such as *HANABA TARANU* (*HAN*) are also involved. *HAN* is a strong negative regulator of *WUS* that is expressed at the SAM margin and appears to limit the spatial extent of the OC (Zhao et al., 2004).

Part of *WUS* regulation is through positive regulators, which are required for correct meristem function. One of these mechanisms is chromatin regulation. The chromatin remodeling factor *SPLAYED* (*SYD*) is a gene which directly targets WUS and presumably facilitates transcription by allowed the transcriptional machinery better access to the DNA (Kwon, Chen & Wagner, 2005).

Another type of positive regulation is provided by *STIMPY/WOX9* (*STIP*). This gene is a *WUS*-like homeobox gene that appears to stimulate WUS production and is, like *WUS*, negatively regulated by *CLV3* adding an extra layer of control on the system. Fascinatingly, the *STIP* phenotype can be rescued by supplying sucrose (Wu, Dabi & Weigel, 2005). Sucrose, it seems, is thus able to stimulate WUS expression. This is perhaps an indication of how signals about the nutritional status of the plant can be transduced to regulate the size and activity of the SAM. Nutritional status clearly has a major impact on meristem size from the spindly shoots of plants growing in poor conditions to the thick shoots of plants under optimal conditions.

What is clear from this recent work is that the meristem is subject to tight controls, many of which act through *WUS*. *WUS* is therefore a key node in meristem regulatory network. It has been shown that the positioning and size of the *WUS* expression domain (in *Arabidopsis*) can be modeled (Jonsson et al., 2005) in terms of the Turing reaction–diffusion field that emerges from a correctly tuned interaction between a *WUS* activation signal and a WUS inhibitor (plausibly *CLV3*). It will be interesting to take this to a comparative level and examine whether such a mechanism works with meristems of different sizes and shapes.

2.6 Development of stem structure

2.6.1 Development of procambium

The SAM rises higher carried up on the pile of cells it produces. Some of these cells become elongated and mark the beginning of what will become, in seed plants, **vascular bundles**. These files of precursor cells are called **procambial strands**. The first sign that cells are fated to become procambial (at least in the leaves) is the expression of At *HOMEOBOX GENE8* (*HB8*; Scarpella, Francis & Berleth, 2004). Developmentally we can take either a forward or a back view of the differentiation of procambial strands. The process can be envisaged as being controlled by the SAM, which stacks back procambial strands on top of existing vasculature. Alternatively one can see this development as being driven by the existing vascular bundles, which project themselves forward by causing cells above them to become vascular.

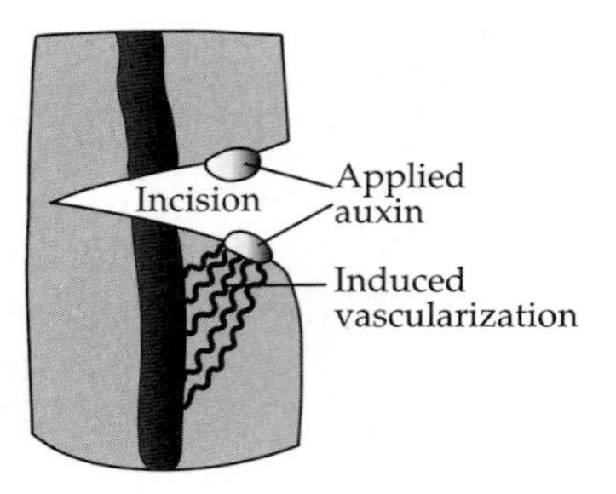

Fig. 2.9. Auxin-induced vascularization showing polar effect (after Berleth et al., 2007).

Both ways of seeing are valid, as both signaling from the apex and signaling from the stem are involved. Polar auxin transport (PAT) from the lateral organ primordia at the flanks of the apex is important (Fukuda, 2004; Stein, 1993), but its interaction with existing vascular tissue is critical. PAT ensures that the differentiation of procambial strands is highly controlled and the integrity of the vascular system, as a single system, is maintained (Fig. 2.9). When PAT is disrupted, for instance by a bud or a wound, PAT "whirlpools" can be formed. Normal vascular differentiation in long strands is prevented and instead vascular circles form (Sachs & Cohen, 1982). Interestingly, the progymnosperm *Archaeopteris*†, the first fossil to have trunks like those of modern gymnosperm and angiosperm trees, also shows these vascular circles (Rothwell & Lev-Yadun, 2005). This is a rare case where there is direct evidence for a cellular process in the fossil record (Fig. 2.10).

A great step forward would be a complete understanding of this process at the molecular level, and information about the genes involved in this process is now emerging.

First, the gene network responsible for PAT is clearly of critical importance. Auxin, produced in lateral organ primordia, travels down the shoot by PAT and induces cells at the head of existing procambial strands to differentiate as more procambium. The canalization hypothesis (Sachs, 1981, 2000) has been put forward to explain this, whereby the procambial strand acts as an efficient drain for auxin exiting the shoot apex, creating local auxin concentrations around

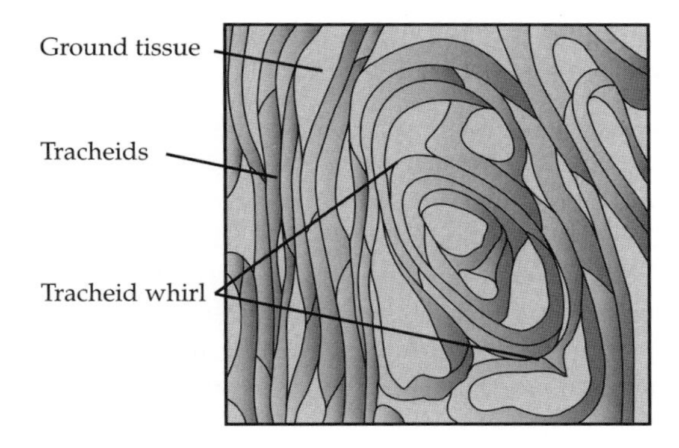

Fig. 2.10. Vascular whirls in a fossil induced by circular patterns of auxin flux. An example of the fossilization of physiology (after Rothwell & Lev-Yadun, 2005).

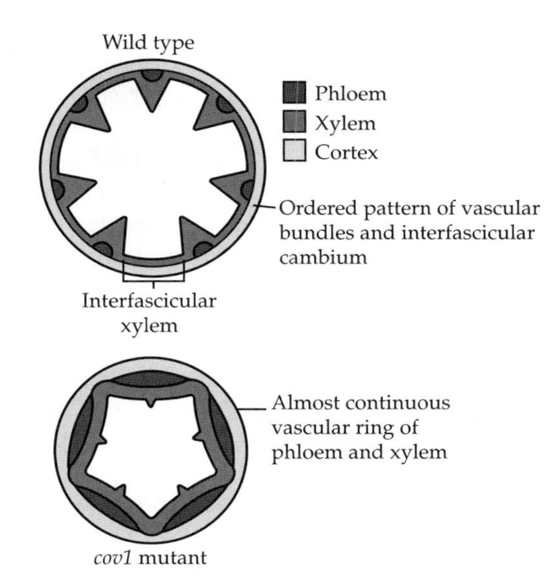

Fig. 2.11. The *COV* mutant of *Arabidopsis* (after Parker et al., 2003)

procambial strands. Auxin regulates PINOID (PID), which in turn spatially positions the auxin efflux carrier PIN-FORMED (PIN) at the base of cells so establishing polarity (Friml et al., 2004). This is potentially a self-organizing process whereby auxin establishes polar flow, which in turn draws more auxin to the region.

The auxin flowing through the procambial region stimulates auxin-dependent differentiation pathways. For instance, *MONOPTEROS* is a gene specifically expressed in the procambium that is necessary for the correct differentiation of vascular tissue (Fukuda, 2004). It is an auxin response factor (*ARF*) gene and activates transcription of those genes with the appropriate auxin response elements in their control regions. This is presumably the mechanism by which auxin exerts direct effects on vascular differentiation.

As well as an auxin-dependent mechanism of vascular patterning, there appears to be an auxin-independent mechanism. Two *Arabidopsis* mutants are known that have precocious differentiation of **interfascicular cambium**, the cambium that forms between the vascular bundles to create a complete ring of vascular tissue during secondary growth.

In wild-type *Arabidopsis* the stems exhibit a pattern of discrete bundles, and the interfascicular cells differentiate as fibers. In the loss-of-function mutant *high cambial activity* (*hca*), the interfascicular cells become cambial and it is suggested that this is due to increased sensitization to CK (Pineau et al., 2005). The *HCA* gene is therefore a negative regulator of vascularization perhaps by blocking the CK responsivity of the interfascicular cells. The continuous vascular ring mutant (*cov1*) similarly develops interfascicular vascular tissue, except in this case there is no strong cambial activity (Parker et al., 2003). The gene *COV1* is thus also a negative regulator of vascular development (Fig. 2.11).

As negative regulators of woodiness, these genes are therefore, at least partly, responsible for maintaining *Arabidopsis* as a herbaceous plant. It may therefore be interesting to examine these genes in woody plants that have evolved from herbaceous ancestors, as appears to have happened on islands (the phenomenon of insular woodiness).

2.6.2 Primary xylem and phloem

The procambial strands are composed of meristematic cells, which divide to produce cells that differentiate into **xylem** and **phloem** while the stem is still elongating. This is the **primary xylem** and **primary phloem**. Bizarrely, primary xylem can be divided into two distinct types, depending on when it forms. The xylem that differentiates first has narrow tracheids (or vessels in the case of angiosperms) and practically no **xylem pith** (**parenchyma**) or **fibers** (**sclerenchyma**). This is dubbed **protoxylem**. Later, so-called **metaxylem** forms with larger tracheary elements. Metaxylem may also contain **sclerenchyma** cells and the cells may be arranged in rows. In all these characters the metaxylem is transitional to secondary xylem. It is possible to make a similar distinction between **protophloem** and **metaphloem** but as the differences are usually slight this is scarcely worth doing.

This division of the primary xylem into two types is very ancient within vascular plants. It can be plainly seen (as small versus large lumen widths of tracheary elements) in cross sections of petrified fossil stems, and it is a great aid to paleobotanists in providing extra characters. The spatial relations between protoxylem and metaxylem also provide characters for the classification of fossils. When the protoxylem is external (toward the outside of the stem) to the metaxylem it is denoted **exarch**, when internal **endarch**. A third pattern, when the protoxylem is surrounded by metaxylem, is **mesarch**. The endarch pattern in protosteles (which have the vascular tissue in the center of the stem) means that the protoxylem is exactly in the center of the stem and this pattern is sometime called **centrarch**.

Why should there be two clearly identifiable types of primary xylem? It is possible that there is a physiological reason and that the protoxylem morphology is more suitable, in some ways, for very young stems. Alternatively it may be an inevitable consequence of the vascular developmental process and that conditions in the very young stem make the formation of highly differentiated metaxylem impossible. At the moment the answer is not known. However, metaxylem does appear to be an evolutionary innovation, as the size distribution of all tracheary elements in Devonian fossils like *Rhynia* is consistent with protoxylem, as is the protoxylem-like helical and annular thickening.

2.6.3 Molecular basis of primary vascular patterning

In searching for the molecular basis to primary vascular patterning, two processes have to be considered. The first is the provision of positional information along the radial axis. Much of the patterning appears to be along this inside–outside gradient. For instance, phloem is usually on the outside and xylem on the inside, and in endarch plants protoxylem is on the inside relative to the metaxylem on the outside.

Second, differentiation of vascular tissue appears to follow a "decision tree" in responding to positioning signals. The first fork in the decision tree is xylem versus phloem and the second, within xylem-fated cells, is metaxylem versus protoxylem. We expect the existence of molecular switches to inform these decisions following the transduction of appropriate positional information.

Types of positional information along the radial axis include (1) cell–cell signaling between successive cell layers, (2) a radial gradient of biochemical signals such as phytohormones or (3) position-specific expression of transcription factors, induced by biochemical gradients.

An example of cell–cell signaling is provided by the secreted proteoglycan, and arabinogalactan/lipid transfer protein dubbed xylogen (Motose, Sugiyama & Fukuda, 2004). Evidence for xylogen came from the *Zinnia* system, in which *Zinnia* leaf mesophyll cells can be

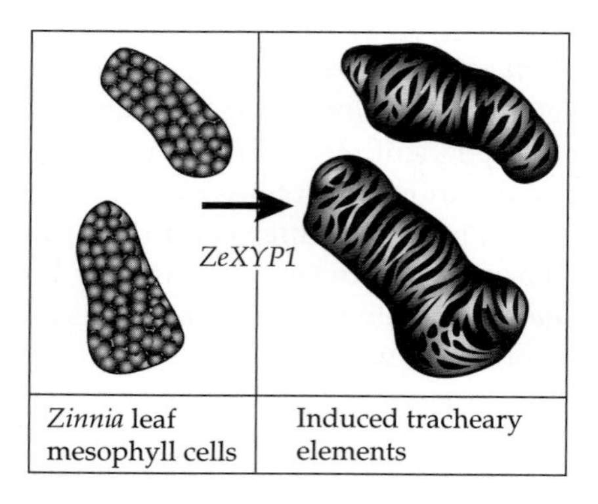

Fig. 2.12. Tracheary induction in the *Zinnia* mesophyll system mediated by the xylogen protein *ZeXYP1* (after Fukuda, 2004).

induced to differentiate into tracheary elements in culture (Fig. 2.12). Under certain conditions, differentiation is highly clumped implying cell–cell inductive reinforcement of tracheary element differentiation. The underlying gene, *Zinnia elegans XYLOGEN PROTEIN1* (ZeXYP1), has orthologues in *Arabidopsis* (*AtXYP1* and *AtXYP2*). Double knockout of *XYP1* and *XYP2* in *Arabidopsis* give rise to mutant phenotypes with defective vasculature.

Broader biochemical gradients across stems may potentially be established by molecules produced peripherally in the epidermis or cambium versus signals produced centrally in the inner parenchyma. For instance, there is evidence that a brassinosteroid ligand may be produced in the center of stems, which potentially activates a pathway leading to *HD-ZIPIII* gene expression in specific domains (Cano-Delgado et al., 2004; Emery et al., 2003; Fig. 2.13).

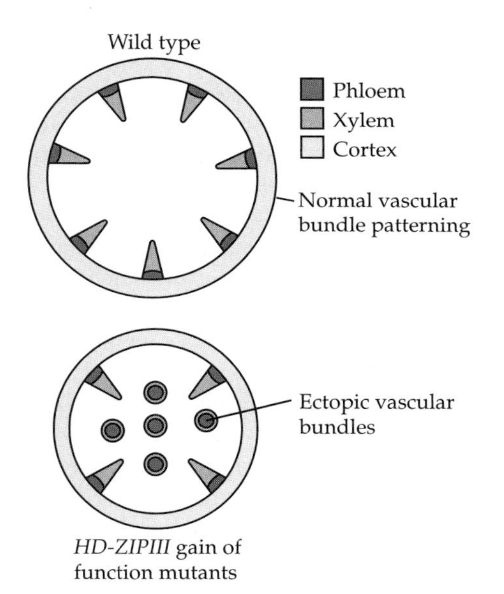

Fig. 2.13. *HD-ZipIII* mutants in *Arabidopsis* (after Sieburth & Deyholos, 2005). The lower part shows the amphivasal vascular bundles, characteristic of HD-ZIP III overexpression.

It should be noted that the inside of stems has some developmental equivalence to the adaxial surface of lateral organs, and the outer surface of stems is equivalent to the abaxial surface of lateral organs (Chapter 4; Fig. 2.14). Thus genes conferring adaxial identity are expressed in the center of stems and those conferring abaxial identity are expressed to the outside of stems. This is of considerable importance to xylem versus phloem differentiation. For instance, the abaxial identity gene *KANADI* is expressed in developing phloem, while the adaxial identity *HD-ZIPIII* gene, *REVOLUTA* (*REV*), is expressed in developing xylem (Emery et al., 2003; Fig. 2.13).

After the transduction of positional information comes cell-fate switching. *ALTERED PHLOEM DEVELOPMENT* (*APL*) is a transcription factor of the MYB coiled-coil-type that confers phloem identity (Bonke et al., 2003). It thus appears to be part of the machinery required for the xylem versus phloem switch. Eliminating *APL* expression causes the cells to develop xylem fates, implying that *APL* may be antagonistic to an equivalent xylem identity gene.

Possible components of the molecular switch for protoxylem versus metaxylem have been identified in *Arabidopsis* and *Populus* (Kubo et al., 2005). These are the NAC-domain transcription factors, *VASCULAR-RELATED NAC-DOMAIN6* (*VND6*) and *VASCULAR-RELATED NAC-DOMAIN6* (*VND7*). Repression of *VND6* inhibits metaxylem while repression of *VND7* inhibits protoxylem. These NAC-domain genes belong to a family of related vascular-expressed genes. Timing the initial divergence of this gene family would be of great interest as it potentially codiversified with land plant vascular systems.

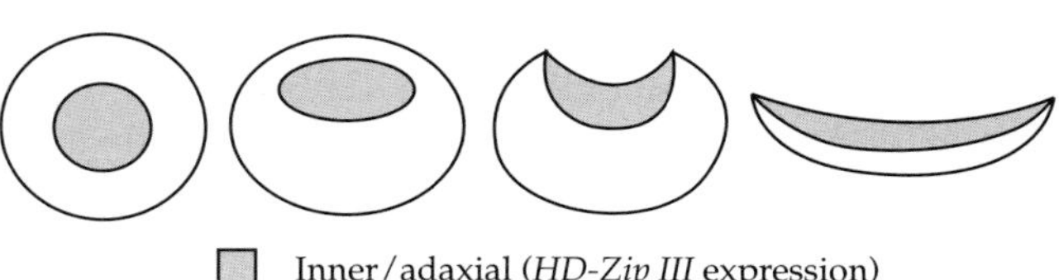

Fig. 2.14. Scheme illustrating the evolution of a bifacial leaf (right) from a radial stem (left).

2.6.4 Secondary growth

Later, after the stem has stopped lengthening, at least some secondary growth is likely to occur, increasing stem width. Some thickening of the primary tissues may occur bulking up the parenchyma ground tissue and increasing stem width. An extreme example of this is found in the swollen stems of kohlrabi (*Brassica oleracea* var. *caulorapa*). These are massive with storage parenchyma derived from primary thickening.

In most seed plants the increase in stem girth is due mainly to **secondary thickening**. In this process, **secondary xylem** and **secondary phloem** form within the vascular bundle from the **bundle cambium** (**fascicular cambium**). Furthermore, a **between-bundle cambium** (**interfascicular cambium**) forms, and secondary xylem is produced all around the stem. The formation of an interfascicular cambium is a key stage in the development from a primary eustele (a network of vascular bundles) to the secondary hyperstele (a woody trunk).

Monocots have lost the ability to make an interfascicular cambium and instead have anomalous secondary thickening. The secondary thickening (such as it is) results from a diffuse differentiation of fibro-vascular bundles in a massive parenchyma produced by prolonged primary thickening. Thus palm trees can have sturdy and thick trunks due to the density of vascular bundles each associated with large number of heavily lignified fibers (sclerenchyma). In dicotyledonous angiosperms there are relatively few vessels but a large proportion of **sclerenchyma fibers**. It is these fibers that make hardwoods hard. In conifer softwoods there are relatively more thin-walled tracheids than fibers. It is the water-conducting cells that make softwood soft. Sclerenchyma fibers are of immense economic importance in the forestry industry. In contrast, secondary thickening of the swede or rutabaga (*Brassica napus* var. *napobrassica*) produces little sclerenchyma but much unlignified secondary wood parenchyma. The vegetable is a tuber formed from the secondary hypocotyl and if it was extensively sclerified it would be inedible. Cultivated beetroot (*Beta vulgaris*) is also an edible swollen hypocotyl, but one with highly anomalous secondary thickening. Continued growth takes place through the formation of successive cambia.

2.7 Organization of the stele

In archaeology a stele refers to a standing block of wood or stone that was erected for commemorative purposes by ancient civilizations. In plants it refers to the "block of wood" (vascular system) inside a plant, comprising the **xylem**, **phloem** and their accessory tissues. These comprise all the developmental derivatives of the **procambium**. It can be thought of as both the plumbing and the support system of the plant as it contains the bulk of both the conducting tissue and the lignified tissues of the plant. As plants have diversified, different evolutionary solutions to the problems of support and conduction have led to an at first bewildering diversity of solutions, manifested as different stele types (Beck, Schmid & Rothwell, 1982). Botanists have named these stele types by prefixing a suitable word to stele, as in dictyostele, plectostele, actinostele, and so forth (see Box 2.1).

There are three basic types of stele (Fig. 2.15). First there is the **protostele**, in which the stele occupies the central position in the stem. This is characteristic of mosses and lycophytes in which the xylem (or **hydroids** in mosses) is at the very center and the phloem (or **leptoids** in mosses) is around this.

Second there is the **siphonostele**, in which the stele forms a tube around the center of the stem. This tube is not usually complete but has holes in it where vascular connections leave to supply the leaves. The siphonostele is characteristic of the monilophytes. A siphonostele may have phloem on the outside of the tube only, in which case it is called **ectophloic** (e.g. *Osmunda*). Alternatively, the phloem may be on the outside and the inside of the tube, in which case it is known as amphiphloic (e.g. *Dicksonia*).

Finally there is the **eustele**, in this the stele forms as a ring of discrete strands, called **vascular bundles**, with phloem on the outside and xylem on the inside. This is characteristic of all

Box 2.1 Synopsis of stelar types.

1 **Protostele**: stele a central mass
  1.1 **Haplostele**: stele a simple rod (e.g. mosses—with endarch differentiation of hydrome; **Selaginella**—often dorsoventrally flattened; *Gleichenia*; *Rhynia*†)
  1.2 **Actinostele**: stele star shaped (e.g. *Psilotum*, *Asteroxylon*†).
  1.3 **Plectostele**: stele a series of xylem plates surrounded by phloem (e.g. *Lycopodium*).
2 **Siphonostele**: stele a cylinder
  2.1 **Solenostele**: a stele consisting of an obvious cylinder with few holes caused by leaf gaps (e.g. *Adiantum*)
  2.2 **Dictyostele**: a stele consisting of a highly dissected cylinder by virtue of numerous leaf gaps (e.g. *Dryopteris*).
3 **Eustele**: stele (at least at first) a ring of discrete elements each with phloem on the outside and xylem on the inside (e.g. *Archaeopteris*†, *Pinus*, *Quercus*)
  3.1 **Primary eustele**: a stele composed of a ring of vascular bundles (which may branch and interconnect to varying degrees) (e.g. *Quercus*, actively growing young shoot of oak)
  3.2 **Hyperstele**: a eustele after extensive secondary thickening, forming the entirety of the plant stem (a trunk) (e.g. *Quercus*, oak trunk).

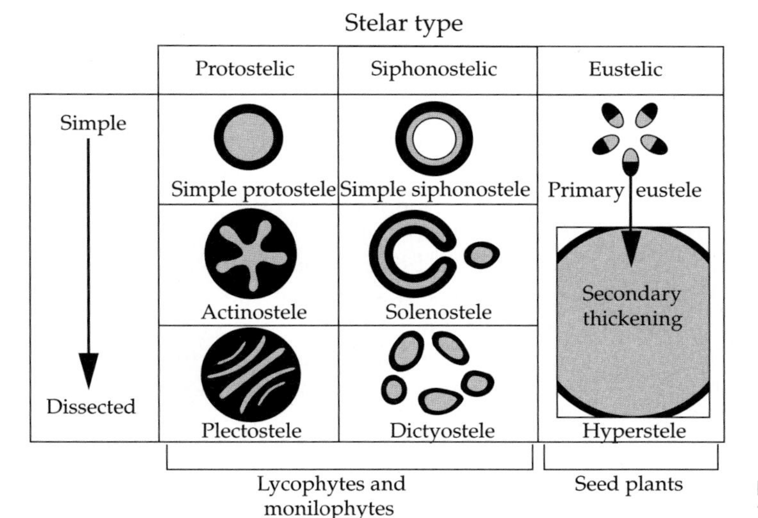

Fig. 2.15. Different types of stelar organization. Xylem = gray, phloem = black.

seed plants. Later in development an interfascicular cambium may form between the bundles of a eustele, joining the bundles up into a tube, in which case the vascular tissue comes to resemble an ectophloic siphonostele. It is not a siphonostele, however, as its origin is completely different. It is merely a eustele with secondary development.

Many eustelic plants (i.e. trees) have the ability to engage in massive and prolonged secondary growth. This destroys, by compression, all the tissues in the center of the stem (trunk). Furthermore, the cortex is often exfoliated and replaced by phloem fibers (bark). Thus, the stele comes to occupy every bit of the stem volume. Such massive eusteles, representing a stelar takeover of the whole stem, are termed **hypersteles**. The origin of hyperstele was a momentous event in earth history as it enabled plants to rise to heights of up to 100 m, carried up on their massive hypersteles. The first plant in the fossil record to exhibit a hyperstele is the progymnosperm *Archaeopteris*†.

The evolution of stele types may be interpreted biomechanically (Roth & Mosbrugger, 1996; Speck & Vogellehner, 1988). The earliest vascularized land plants such as *Cooksonia* (Edwards, 2003) were protostelic. Subsequent evolution of siphonosteles and eusteles therefore involves moving the vascular and supporting tissue from the center to the periphery of stems and dispersing it among the ground tissue. This has biomechanical benefits relative to both (1) rigidity (weight for weight, a tube is more rigid than a rod) and (2) the efficient irrigation of the parenchyma tissues and the transpiring plant surface.

2.8 Oriented development in stems

Stems are generally **erect** (**orthotropic**) as they are negatively geotropic. They thus grow straight up against the force of gravity. This particularly applies to the **primary shoot** or **first-order branches** (i.e. those expressing **apical dominance**). There is often only one shoot of this type, called the **leading shoot** or **leader**. Second-order shoots (i.e. lateral branches from the primary stem) are rarely completely orthotropic but instead have a characteristic branch angle. Third-order shoots (lateral shoots from lateral shoots) may have yet another characteristic branch angle. A good example is furnished by the silver birch (*Betula pendula*). The first-order branches are vertical, second-order approximately horizontal and third-order pendulous. The branch angle changes by around 90° at each order of branching.

Branch angle (also called the **setpoint angle**) is thought to be the result of a balance between **orthogravitropism** (the tendency for organs to align themselves parallel to the direction of gravity) and **epinasty** (the tendency of lateral organs to bend away from the main axis; De Vries, 1872). Epinasty is an active developmental process that counteracts the orthotropic response to gravity (**negative geotropism**). Very strong epinasty results in the **horizontal** (**plagiotropic**) branching seen in many trees including cedars of Lebanon (*Cedrus*), common beech (*Fagus silvatica*) and *Pinus excelsa*. Loss of epinasty gives rise to **fastigiate** branching in which all branches are erect. A large number of trees have fastigiate mutants including *Populus nigra*, the European black poplar. The fastigiate Lombardy form (*P. nigra* var. *italica*) is planted throughout temperate regions. Loss of negative geotropism gives rise to **pendulous** branching. The weeping willow (*Salix chrysocoma* or *Salix babylonica*) is a natural example. However, there are many horticultural mutants that show this characteristic. An example is the weeping form of the common ash (*Fraxinus excelsior* var. *pendula*). The most common mutants are these **loss-of-function mutations**, either loss of epinasty or loss of negative geotropism. Mutations that change branch angle quantitively (i.e. change the outcome of the antagonism between epinasty and negative geotropism) are subtler and are either less common or more often overlooked. There are, however, plants with plagiotropic growth-forms such as *Juniperus horizontalis*.

The silver birch, *B. pendula*, is an interesting example. The first-order shoots are erect (orthotropic). The second-order shoots (main branches) are horizontal (plagiotropic) and the third-order shoots (minor branches) hang down vertically. Thus through the development of the tree, shoot orientation goes through three perpendicular switches.

As mentioned above, loss of negative geotropy in second-order branches when the primary stem is erect allows for branches that hang down. However, if no shoots are erect, there is nothing to hang down from, and all the stems lie flat along the ground. Such stems are termed **procumbent** or **prostrate**.

Climbing plants also appear to have reduced negative geotropism. They tend to follow a support; however, that support might be inclined. Thigmotropism therefore trumps geotropism.

An exception to this orientation model that balances epinasty against orthogravitropy may well apply to horizontally growing shoots like rhizomes. An alternative model suggests that these may be directly **diagravitropic** and thus respond to gravity in a direction perpendicular to orthogravitropic shoots (LaMotte & Pickard, 2004a,b).

2.9 Molecular control of stem orientation

The orientation of stems is due to a combination of tropic responses and **epinasty**. **Phototropy** and **gravitropy** are mechanisms for sensing the external environment and, in most stems, for growing toward light or away from gravitational pull (**negative geotropism**). There are examples of stems that are **negatively phototropic**, such as the **pedicels** of *Cymbalaria muralis* fruits, which function to bury the fruits in crevices. These pedicels are positively phototropic when young so there is a developmental switch that reverses the polarity of the process after fertilization.

Auxin appears to be the key regulator of all tropic responses, including epinasty (Blancaflor & Masson, 2003; Esmon, Pedmale & Liscum, 2005; Esmon et al., 2006; Romano, Cooper & Klee, 1993). This is an old discovery: an auxin gradient across the stem is the basis of the 1928 Cholodny–Went hypothesis (Went & Thimann, 1937) for plant tropisms. However, it is only recently that the molecular basis of the response has been elucidated. In *Arabidopsis*, relocalization of the auxin efflux transporter *PIN3* occurs as a response to **gravistimulation** (Friml, 2003; Friml et al., 2002), leading in turn to an asymmetric distribution of auxin across the stem and asymmetric (corrective) growth. Auxin has a direct effect on growth by derepressing *AUXIN RESPONSE FACTOR* (*ARF*) genes that in turn promote transcription of those growth genes with auxin response elements (AREs; Esmon et al., 2006).

The gravity signal appears to be picked up initially by sedimentation of **amyloplasts** (**starch grains**) in the shoot **endodermis**. The endodermis is a tissue of small starchy cells, which forms the innermost cell layer of the cortex. When stems are bent sideways the amyloplasts sediment against the cell side walls. It is thought that this triggers a signal cascade from membrane bound receptor proteins that is eventually transduced into the auxin transport system (Fig. 2.16).

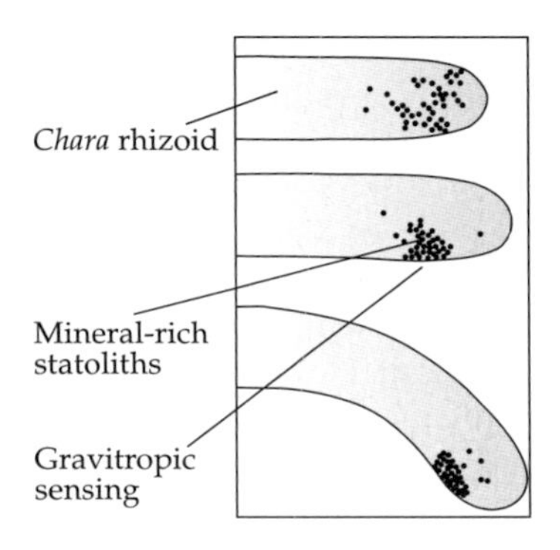

Fig. 2.16. Gravitropic sensing in the *Chara* rhizoid mediated by the sedimentation of cytoplasmic particles (after Blancaflor & Masson, 2003).

In the phototropic response of *Arabidopsis PHOTOTROPIN* genes (*PHOT1* and *PHOT2*) act as the light receptors, interacting with members of the *NRL* (*NPH3/RPT2-like*) gene family, such as *NONPHOTOTROPIC HYPOCOTYL3* (*NPH3*) and root *PHOTOTROPISM2* (*RPT2*), to form photosignaling complexes (Inada et al., 2004; Motchoulski & Liscum, 1999). Blue light then causes this photosignaling complex to change phosphorylation state. This difference is then transduced into changed auxin concentrations between the light and dark side of the stem.

2.10 Change of directionality with stem age

A common feature of stems is a change in growth directionality with age (Fig. 2.17). **Decumbent** stems are prostrate but rise up at the tips, as in *Sagina decumbens*. **Ascending** plants are a version of this, with stems erect, although curved or horizontal at the base. **Drooping** refers to a shoot that is erect but with the terminal part horizontal. **Nodding** refers to a more extreme type of droop, when the stem is erect except for the youngest part, which hangs down. Many floral stems nod, and when this occurs in the mature flower it is for the obvious reason of inclining the flowers downward as a likely adaptation to pollinators or other feature of floral biology which may be as simple as keeping out the rain. Nodding flowers are well exemplified by fritillaries (e.g. *Fritillaria meleagris*). However, for reasons that are obscure, many plants nod in bud only. Examples include poppies (*Papaver*) when in bud. In the most extreme example of this behavior, *Euphorbia characias* has vegetative stems that nod when they are forming floral buds. By this means it is possible to determine which stems of *E. characias* will flower, before any floral parts are visible.

Nodding buds in species with erect flowers may be a **vestigial** feature, at least in groups where nodding flowers are an ancestral feature. Alternatively nodding buds may have evolved before nodding flowers, for reasons unknown. This feature would then facilitate the transition to nodding flowers. In poppies bud nodding may have permitted the evolution of species that nod in flower, such as in *Meconopsis punicea*, by a process of **heterochrony** (a change of developmental timing). If the flower opening is brought forward into the nodding phase, or if the nodding phase is extended to flower opening, a nodding flower will result. Nodding or drooping flowers may be random with respect to external direction as in *Narcissus*. Alternatively they may be directionally oriented as in the sunflower, *Helianthus* (in *Helianthus* the direction of nod is southerly, that is, toward the sun). As erect stems are radially symmetrical there do not appear to be any internal asymmetries to determine the direction of nod.

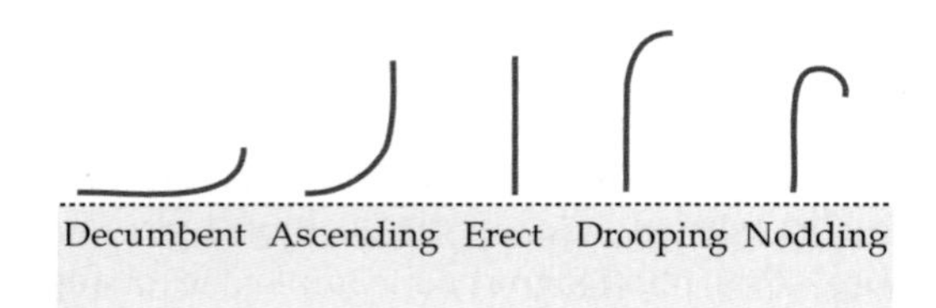

Fig. 2.17. Main forms of directionality change during plant development.

2.11 Chirality in stem morphology

The **helical stem** is an important morphological innovation as it allows **twining** as a mechanism of **climbing**. In twining plants the main stem wraps around a support and hence twiners are also referred to as **voluble plants**, from the Latin *volvere*, to turn. Initial contact with a support is by a phenomenon called **circumnutation** (Fig. 2.18). Darwin used ingenious experiments to study this process in real time. The stem apex describes an arcing sweep through space caused by asymmetrical growth pulses. On contacting a support the asymmetrical growth allows helical twining, the **winding response**. In addition to the winding response (exaggerated circumnutation), changes in stem anatomy occur to fix the helical growth. This is known as **thigmomorphogenesis**, or morphogenesis resulting from touch stimuli.

Circumnutation and the winding response require the presence of a functional endodermis (Kitazawa et al., 2005), which is the seat of

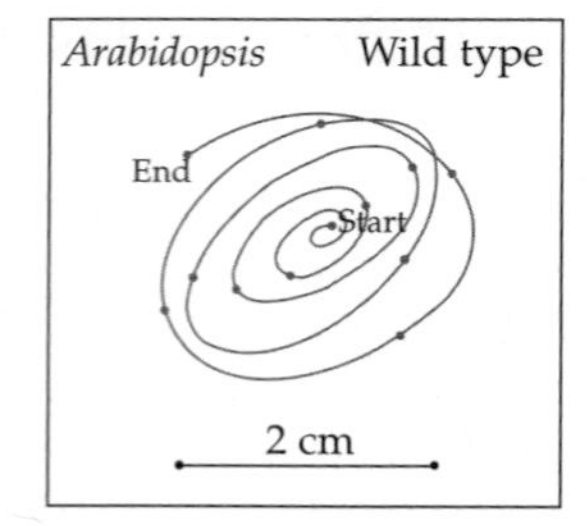

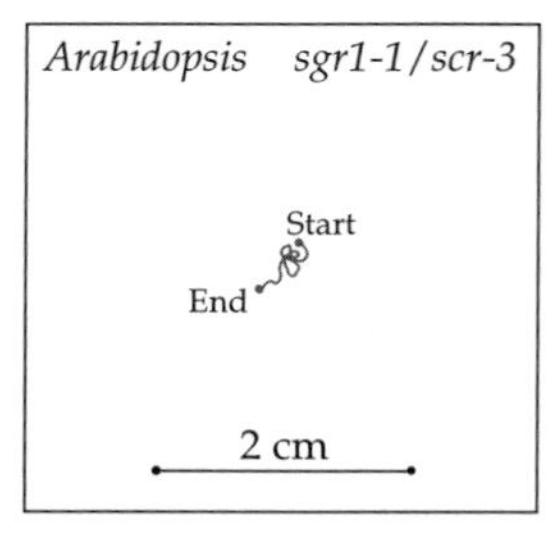

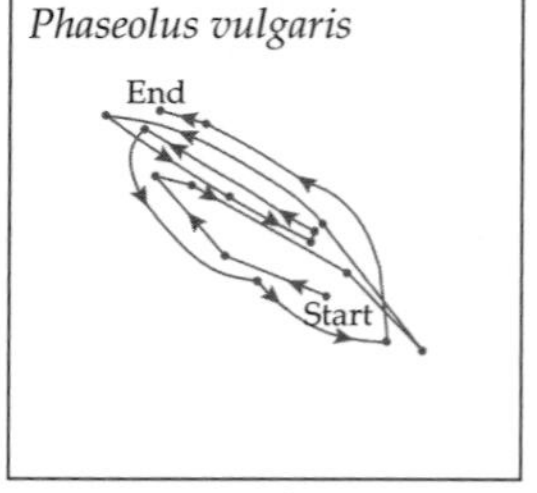

Darwin (1875)

Fig. 2.18. Circular movement in stems (circumnutation), showing normal circumnutation in the *shoot gravity response* (*sgr*) and *Scarecrow* (*scr*) double mutant (after Kitazawa et al., 2005–left and centre).

the gravitropism response in stems. The endodermis contains a sensitive detection system for starch grain (**amyloplast**) sedimentation within the cells. The amyloplasts act as the plant's "plumb line" in gravitropism.

In *Ipomoea* (*Pharbitis*) *nil*, the Japanese morning glory, the variety "Weeping" lacks an endodermis and fails to circumnutate or wind (Kitazawa et al., 2005). Weeping's thigmomorphogenesis is unimpaired. In this mutant it has been shown that *SCARECROW*, a GRAS transcription factor important in root and shoot development, is defective. *SCR* and the related GRAS gene *SHORTROOT* (*SHR*) are responsible for aspects of tissue patterning in developing roots and stems including the formation of the endodermis. SHR controls SCR expression and so acts upstream of SCR in the same pathway. This is an elegant demonstration that a functional gravitropic response is required for circumnutation, but it is not yet clear why, as circumnutation is a helical and not up or down growth.

The helix described by twining plants may be **left-handed twining** (**sinistrorse**) or **right-handed twining** (**dextrorse**). A left-handed, or anticlockwise, helix ascends to the left from the perspective of an observer within the spiral looking toward the growing point and is commoner that right-handed twining. An example of sinistrorse twining is *Calystegia sepium*, or bellbine. Dextrorse plants include *Humulus* and *Lonicera*. The existence of both dextrorse and sinistrorse twining has even been celebrated in song.[3]

The systematic distribution of dextrorse and sinistrorse individuals is interesting. In most twining genera all the species twine the same way. However, in *Dioscorea* both left- and right-handed individuals are known. In *Solanum dulcamara* there is variation at the level of individuals. So, in an analogous way to how handedness is distributed in humans, some individual plants are left-handed, some right-handed. Even more extreme is *Loasa aurantiaca* which may have left- and right-handed stems on the same plant.

[3] A comic song by Michael Flanders and Donald Swann entitled "Misalliance" is an allegorical tale concerning the dalliance of a honeysuckle (*Lonicera* sp.) and a bindweed (*Calystegia* sp.), see Hussey (2002).

Chirality in plant stem is an interesting phenomenon because there are no obvious cues in the external environment that could impose left- or right-handedness, and neither, in a radially symmetrical stem, is there any obvious anatomical cues for handedness. However, chirality is a natural property of certain molecules, which could therefore potentially supply handedness cues at the biochemical level.

The orientation of cell expansion is maintained by the **cortical microtubule cytoskeleton** (i.e. the pattern of microtubules underlying the cell-wall associated plasma membrane, transverse to the direction of cell elongation). There are numerous genes responsible for keeping cell orientation straight, through the cortical microtubule cytoskeleton. These include *SPIRAL1* (*SPR1*) and *OTHER SPIRAL1-LIKE* (*SP1L*) genes. A loss-of-function mutation in *SPR1* results in dextrorse helical growth (Nakajima et al., 2004).

Furthermore, mutations that alter the nature of the cortical cytoskeletal tubulins themselves also affect chirality (Fig. 2.19). In *Arabidopsis*, mutant forms of the tubulin genes *TUBULIN ALPHA4* (*TUA4*) and *TUBULIN ALPHA6* (*TUA6*) have been found that have sinistrorsely twisted growth phenotypes, *lefty2* and *lefty1* (Thitamadee, Tsuchihara & Hashimoto, 2002). These phenotypes may result from the mutant tubulins having a right-handed oblique orientation within the cortical microtubule cytoskeleton. What is not yet clear is the extent to which these helical

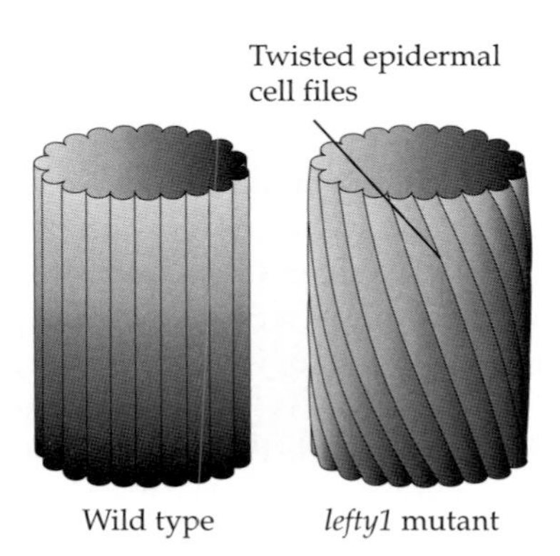

Fig. 2.19. The *lefty* mutant in *Arabidopsis* showing the helically oriented epidermal cell files (after Thitamadee et al., 2002).

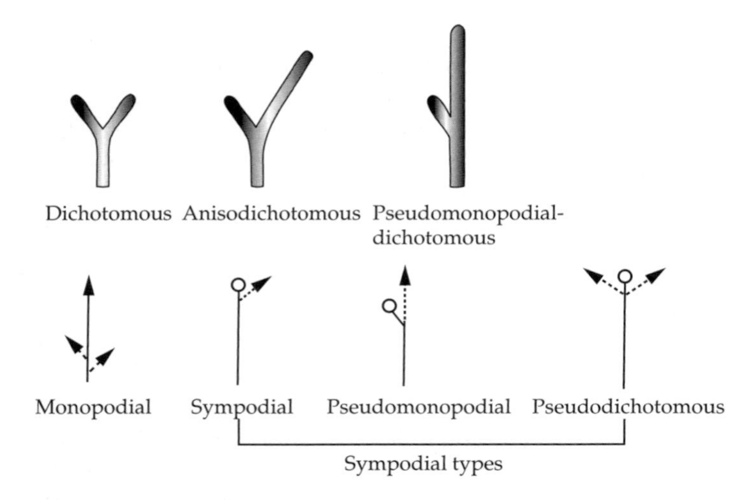

Fig. 2.20. Branching types. Branching may be either dichotomous (top row), monopodial or sympodial. The circle indicates cessation of growth through determinacy (usually flowering).

cytoskeletal mutants have any bearing on the growth of naturally helical twining plants.

Occasionally helical stems are found in non-climbing plants, usually as mutants. An example is the spiral rush, *Juncus effusus* f. *spiralis* in which internode growth of the **calamus** is spiral. More disorganized helical growth occurs in the "contorted" mutants of trees. Probably the best known of these is *Corylus avellana* var. *contorta*. Many of these appear to be single gene mutants, although as they are generally sterile this has not been formally demonstrated. The asymmetrical growth is possibly due to a problem with the hormonal regulation. The mutation may be a loss-of-function mutation in a gene that normally downregulates the sensitivity to hormones of plant tissues in particular places.

2.12 Architecture of the stem

Growth may be sympodial, monopodial or dichotomous, and these concepts are extremely important for understanding plant architecture. Most plant stems have a SAM that is capable of indefinitely prolonged or **indeterminate** growth. Growth of this sort is termed **monopodial**. However, in some plants the SAM ceases growth in a regular pattern, either because it aborts or because it converts to a structure with **determinate**, that is, limited growth. The commonest determinate structures terminating a SAM are reproductive structures such as a flower or an inflorescence. When this happens the stem can only continue to grow by the activation of lateral buds below the terminated apex. This form of growth is called **sympodial**. Trees can thus usefully be distinguished as sympodial, with terminal inflorescences (*Aesculus, Magnolia*) or monopodial, with lateral inflorescences (*Acer, Fraxinus*). The same distinction applies to bulbs. Tulips (*Tulipa*) and hyacinths (*Hyacinthus*) are sympodial, with a terminal inflorescence and subsequent growth being axillary. *Ornithogalum caudatum*, the false sea onion, is an example of a monopodial buld in which the inflorescences are axillary.

Another possibility is that the SAM may divide into two daughter SAMs (**dichotomous** growth). This is exceedingly rare in seed plants but common in nonseed plants. Dichotomous growth can also mimic monopodial growth if one meristem of the two daughter meristems is dominant (i.e. **anisodichotomous**), and there thus appears to be a strong continuity of a single meristem. This type of growth is therefore **pseudomonopodial-dichotomous** (Fig. 2.20).

In a similar way, a surprisingly large number of flowering plants are sympodial but mimic monopodial growth. These are termed **pseudomonopodial**. In this case the axillary bud which takes over growth assumes a terminal position on the stem, so to the casual observer, growth appears as a prolongation of a single stem. Almost all the temperate deciduous trees with alternate leaves are pseudomonopodial, a trait that is well illustrated by elm (*Ulmus*).[4] The apical bud aborts in the autumn and a lateral bud at the tip of the stem (subterminal) resumes growth in the spring assuming a terminal position on the twig.

In contrast, most temperate trees with opposite leaves, such as ash (*Fraxinus*), are truly monopodial. The association of growth habit with leaf arrangement is curious and remains unexplained. However, as opposite-leaved trees have two axillary buds at the same level, one would have to grow, and not the other, if growth is to be pseudomonopodial. It may be that the difficulties in regulating this contribute to the infrequency of the pseudomonopodial habit in opposite-leaved trees. In the opposite-leaved *Syringa* and *Staphylea*, however, the terminal bud aborts and both subterminal axillary buds develop (they are **mixed buds**, producing both vegetative shoots and inflorescences) and thus growth is **dichasial** (**pseudodichotomous**), and sympodial by abortion rather than by flowering.

Another plant that is pseudomonopodial is the grapevine (*Vitis*). The SAM terminates in a determinate structure (a tendril or an inflorescence: for the switching between them see section below) at almost every node. Growth is taken over by the axillary bud of the nearest leaf, but this bud assumes a terminal position so a grapevine stem looks superficially like an unbranched monopodial structure, when it is in fact sympodial.

2.13 Molecular control of determinacy and phase change in stems

Stems are **determinate** if the SAM arrests through abortion, permanent cessation of growth or by becoming a flower or determinate inflorescence. Growth can then only continue by means of lateral branches (**sympodial growth**).

[4] This is very common in temperate trees: other examples include *Tilia, Robinia, Corylus, Betula, Salix,* some *Quercus* and *Fagus.*

Determinacy is often associated with a phase change from vegetative to reproductive, as most (although not all) inflorescences are determinate structures. A **racemose inflorescence** is usually determinate by petering out after producing a number of lateral flowers. A **cymose inflorescence** is determinate by virtue of producing a terminal flower.

The gene *CENTRORADIALIS* (*CEN*) in *Antirrhinum* is required to maintain prolonged **indeterminacy** in the inflorescence, which is a long spike that continues growing for some time (Bradley et al., 1996, 1997). The loss-of-function mutation of *CEN* has a more determinate phenotype with a **terminal flower**. Conversely, *CEN* overexpression, for instance in tobacco (*Nicotiana*), delays both flowering and the cessation of shoot growth. The highly indeterminate plants that result grow for much longer without flowering (Amaya, Ratcliffe & Bradley, 1999). The promotion of indeterminacy by *CEN* and its homologues is very widespread in the angiosperms. Similar effects have been reported for *Arabidopsis* (*TERMINAL FLOWER1, TFL1*), tomato (*SELF-PRUNING, SP*) and grapevine (*VvTFL1*) with minor differences due to specific plant architecture (Boss, Sreekantan & Thomas, 2006; Bradley et al., 1997; Pnueli et al., 1998).

In *Arabidopsis* when a phase change to inflorescence occurs, the effects of *TFL1* are overridden by transcriptions factors including *LEAFY* (*LFY*; Parcy, Bomblies & Weigel, 2002) and such transcriptional repression of the indeterminate **vegetative phase** is required for the maintenance of the determinate **floral phase**. Reversals are rare. However, in *Impatiens balsamina*, developed as an important model by Battey and coworkers (Tooke et al., 2005), reversals from flower to vegetative phase are the norm under certain conditions. *I. balsamina* flowers under short-day conditions, and if flowering plants are transferred to long-day conditions the shoot apex reverts. It seems that in this species a signal produced in the leaves is required to maintain the determinate flowering phase (Pouteau et al., 1997; Tooke & Battey, 2000).

The phase change from vegetative to floral phase is initiated through the transduction of external environmental and internal physiological signals, which have been the subject of intensive recent study in *Arabidopsis*. For instance, in long days the phloem-expressed gene *CONSTANS* (*CO*) activates transcription of *FLOWERING LOCUS* T (*FT*), a RAF-kinase-inhibitor-like protein (An et al., 2004). FT mRNA moves to the SAM as a flowering induction signal (Huang et al., 2005). An important integrator of these signals, and an important gene for the maintenance of the flowering phase, is the gene *LFY* (*FLORICAULA* (*FLO*), is the homologue in *Antirrhinum*; Blazquez et al., 1997). Transition to the flowering phase is marked by *LFY* upregulation and artificially induced constitutive expression of *LFY* causes premature flowering (Weigel & Nilsson, 1995).

Flowering is not the only phase change during the life cycle of the plant. Trees and shrubs often have a phase change between juvenile and adult, which may affect the SAM profoundly. In the tree composite *Lachanodes arborea* (Asteraceae), the **juvenile phase** has a tall, unbranched woody stem with a large SAM making large leaves (Cronk, 1981). In the **adult phase** the plant is much branched with thin twigs and small leaves, indicating a major change in the regulation of the SAM. Plants such as these would be interesting models for studies of phase change regulation.

2.14 The thickness of stems and its consequences

Another useful pair of terms to describe plant architecture is **pachycaul** and **leptocaul** coined by E. J. H. Corner. Pachycaul, or thick stemmed, plants have a massive SAM producing a massive stem. However, this comes with a syndrome of associated characters and the term pachycaul refers to the syndrome and not just to the thickness of the stem (Niklas, Cobb & Marler, 2006). The internodes of pachycaul plants are short and therefore the leaves tend to be clustered in terminal rosettes. They also tend to produce few branches. The big stems can also support big leaves, big flowers and big fruit. An extreme example is the palm family (Arecaceae) in which the majority of species have a massive unbranched stem with a rosette of leaves at the top. Some palms such as the coconut (*Cocos*) and the double coconut (*Lodoicea*) use this massive construction to support massive fruit.

The opposite is leptocaul (thin-stemmed). In leptocaul plants the SAM is small, the stems are in consequence narrow. The internodes tend to be large and the leaves are in consequence widely distributed. Leptocaul plants produce bulk not because of the size of their individual stems, but rather by the number of them. They are well branched. In consequence of the slender but numerous stems, the leaves, flowers and fruit tend to be small but numerous, following Corner's rules (White, 1983a,b). An example of a highly leptocaul tree is willow (*Salix*), whose slender branches form elegant trees. There is a continuum of form between the extremes of leptocauly and pachycauly but the terms are useful to express an axis of tree architectural variation.

Pachycauly, branching pattern and the occurrence of **sympodial** versus **monopodial** branching were features originally used by Hallé and Oldeman (Hallé, Oldeman & Tomlinson, 1978) to classify tree architecture into a series of architectural models or types. These range from **Holttum's model** (the palm tree with its simple unbranched pachycaul stem) to complex models like **Troll's model**, to which many temperate deciduous trees belong. Hallé and Oldeman based their work on the tree diversity of the tropical rain forest. It is here that the majority of the models occur, and there is a progressive reduction of architectural diversity toward the poles.

2.15 Origin of new stems through dichotomous or lateral branching

Shortly after a **phyllome** forms on the flank of the SAM, a new but very small SAM forms in association with it (the axillary bud). The tightness of the developmental linkage between leaf and axillary bud in seed plants is extraordinary and unparalleled. In seed plants the bud is in the **axil** of the phyllome but in monilophytes it may be below the phyllome and this association appears to have been independently evolved. In bryophytes there is no such association between **phyllidia** and lateral meristems. Neither do lycophyte **microphylls** have axillary buds (although they may have **axillary sporangia** and axillary **ligules** in the case of Sellaginellaceae and Isoetaceae).

Instead, bryophytes and lycopods characteristically branch by dichotomous division of the SAM. In **dichotomous branching**, which appears to be the ancestral method of producing new SAMs, the single apical cell cuts off a cell, which, instead of being nonstem cell, is also a stem cell. Thus two apical cells then each produce clusters of daughter cells around them and soon are pushed apart and establish separate stems. Dichotomous branching also occurs, but more rarely, in seed plants. A good example of this is the palm *Hyphaene*, and it is also known in *Chamaedorea* and *Strelitzia* (Fisher, 1974, 1976).

There is a major divide between dichotomous branching and **lateral branching** through lateral meristems. Some bryophytes do produce lateral buds and shoots but these are sporadic rather than associated with every leaf.

The nearly constant association between the axils of leaves and lateral buds, which is only found in seed plants, is a mysterious evolutionary feature. It is true that floral phyllomes in angiosperms lack **axillary buds**, and that some gymnosperms sometimes produce leaves with **empty axils**. Nevertheless, the association between leaves and buds is striking. Even cotyledons have axillary buds, and in some plants the epicotyl dies and the adult plant is produced solely from stems arising in the cotyledonary axils, as is common in *Euphorbia* section *Anisophyllum*.

The axillary bud is likely to have first evolved in the progymnosperms (*Progymnospermopsida*†) but it is difficult to know, as paleobotanists pay little attention to axillary regions in their reconstructions. The best reconstructed progymnosperm, *Archaeopteris*†, still lacks a reliable reconstruction of the axil to know whether it had axillary buds. It was certainly branched and this branching appears to be lateral rather than dichotomous. However, its lateral branching might be from adventitious buds or from lateral meristems produced instead of leaves at the SAM.

It is questionable whether these lateral SAMs (axillary meristems) are formed *de novo* from cells that have already left the meristematic state, as in adventitious buds, or whether they represent fragments of the apical meristem which become marooned on the flanks of the original SAM.

Stems may form *de novo*, as **adventitious buds**, from epidermal or other tissues of a variety of organs, including leaves and roots. Adventitious buds on the roots of trees (**root suckers**) are an important means of regeneration in aspens (*Populus* spp.), in which the stems form **endogenously**. Adventitious buds also form commonly on leaves. In ferns this is an important means of vegetative reproduction. In bracken (*Pteridium aquilinum*), buds form at the base of the petiole. In *Asplenium* they either form on the lamina (*Asplenium bulbiferum*) or at the leaf tip (*Asplenium erectum*). In flowering plants this means of reproduction is less frequent, perhaps because of greater leaf determinacy. However, *Bryophyllum* (*Bryophyllum daigremontianum* and other species) is well known to form rows of "plantlets" at the margin of the lamina.

Also, although stems are well provided with axillary buds in the axils of leaves, adventitious buds form readily on stems too. In annual species of *Linaria* the main stem in often supplemented by stems that form from adventitious buds on the **hypocotyl**. In woody plants new buds form from the **cambium**, mainly when stimulated by **mechanical damage**.

2.16 Stem integration and the shoot

In extant plants the stem rarely occurs in isolation. Instead, leaf and parent stem are tightly integrated into a single functional unit, the **shoot**, in many plants. Furthermore, in many extant plants there is tight integration between a leaf and its associated **axillary meristem**, even to the extent that the axillary stem can be subsumed into the leaf in some species, leading to shoots appearing to issue from the leaf, a phenomenon known as **epiphylly** (Dickinson, 1978). An example of this is the epiphyllous inflorescence of *Helwingia japonica*, in which an **inflorescence primordium** that formed in proximity to a leaf primordium gets transported up the leaf as the leaf primordium expands (Dickinson & Sattler, 1975).

Despite the tight integration that has evolved through the modern SAM, there are examples of stem existing without leaves. *Cuscuta* (dodder) never forms any leaves other than minute scale leaves, and does not even form cotyledons (Truscott, 1966). Leafless plants also occur in the fossil record and thus it is convenient to consider the stem as a more basic component of plant organization than the shoot.

2.17 Forms adopted by the stem

The stem, considered as the all nonleaf product of the SAM, is referred to generally as the **caulome** in distinction to the **phyllome** (leaf-organs). It is an evolutionarily plastic organ taking many forms in response to ecological conditions and developmental position. Many of these stem units have been given special names.

One stem unit in seed plants that is considered morphologically special enough to be given a separate name is the **hypocotyl**. It is the first stem of the new **sporophyte**. Whereas most stem units lie between two nodes, the hypocotyl is unique in being between a node and the root. At its base is the **cross-zone**, the region of conversion between the vascular systems of the root and the stem. It is also the part of the stem that is formed as part of the embryo and is contained within the seed. It is therefore the stem counterpart of the **cotyledons** (leaves which are also given a special name by the slender virtue of their being present in the seed). Similarly, the stem between the cotyledons and the first subsequent leaves is called the **epicotyl**. Very rarely the two cotyledons become separated due to asymmetrically produced stem tissue, called **mesocotyl**, between them. This state is found in *Streptocarpus* (Möller & Cronk, 2001) and certain other Gesneriaceae. However, the term mesocotyl is also used for the stem between the cotyledon and coleoptile in grasses.

Another type of stem that is sufficiently distinctive to be given a special name is the **pseudopodium**. The pseudopodium is found only in bryophytes such as *Sphagnum* and *Andreaea*; it is a leafless stem that bears the sporogonium on its swollen tip. The pseudopodium occurs in a number of different groups of bryophyte and appears to be independently derived. However, the striking similarities in form indicate that similar developmental pathways are involved in all cases. The **sporophore** of *Marchantia* can be thought of as an extreme form of pseudopodium. The name pseudopodium is not a very good one as it is also used for the protoplasmic extensions of amoeboid protozoa.

Other forms adopted by the stem are usually associated with ecology. The amazing evolutionary plasticity of the caulome has allowed numerous **life forms** to evolve in flowering plants. These life forms are the basis of much of the structure and function of modern terrestrial ecosystems.

There is thus a direct link between the morphogenetic potential of the stem and the Earth's major biomes. Arguably the role of the stem is even more telling than that of the root and the leaf. Part of this potential is due to the evolutionary plasticity of the SAM, and part due to a multiplicity of types of secondary growth. Between them, the two major meristems of the stem, the SAM and the cambium, have allowed plants to rise 100 m into the air or to retreat belowground to escape drought, cold and fire.

2.18 Stem morphology associated with life form

2.18.1 Aerial stems

When secondary thickening is extensive, a **trunk** is formed as in the **tree** life form. In forming a tree the stele expands by secondary thickening to form the entirety of the stem as the cortex and pith are eventually lost (a **hyperstele**). An alternative to this is the evolutionary expansion of the cortex to form fleshy tissue, perhaps by

neoteny (Altesor, Silva & Ezcurra, 1994). These **stem succulents**, or **sarcocaulescents**, include many members of the Cactaceae, Apocynaceae (Asclepiadaceae) and species of *Euphorbia* (Euphorbiaceae).

Alternatively, in flooded conditions, the cortex and pith (medulla) may develop large intercellular air spaces to promote long distance transport of gases to and from submerged stems or roots. Tissue with a large amount of intercellular air space is called **aerenchyma**. The mangroves such as *Rhizophora* and *Avicennia* are examples of aerenchymatous plants. Gas exchange between the atmosphere and stem aerenchyma may be either through stomata via the leaf spongy mesophyll or more directly through lenticels (pores) on the stem surface.

Grasses have a specialized type of stem called a **culm** or **straw**. Culms can grow rapidly and have little secondary thickening. They are unbranched stems with long internodes and distinct, often thickened, nodes. Culms grow not only by means of a SAM but also by means of an intercalary meristem. This is a zone of meristematic cells in the mature internode that can lengthen the culm by growth from near the base. This makes grasses an effective life form in the presence of large herbivores because the loss of the SAM does not necessarily stop culm growth. The small cells or the intercalary meristem are well provided with sugars and are the pale sweet stem bases that are often chewed by country people. More importantly, most of the world's food is produced on culms, notably on flowering culms of wheat (*Triticum*), maize (*Zea*), barley (*Hordeum*), oats (*Avena*), rye (*Secale*), sorghum (*Sorghum*), millet (*Panicum*) and rice (*Oryza*). Woody grasses such as bamboos (Bambusoideae) have culms with extensive **sclerenchyma**, allowing them to be used in the construction of buildings and for sundry other uses.

Length of culm is an important agronomic trait. Long culms are difficult to harvest and prone to **lodging** (collapse). The "green revolution" is the name given to a fruitful period in plant breeding in which rice and wheat varieties were bred that had short culms. The genes underlying this trait are now known to be mutations in growth inhibiting DELLA transcription factors which make them resistant to the growth derepressing effects of giberellin (Peng et al., 1999). Culms are often unbranched above, but produce many branches (called **tillers**) at ground level. Tillering allows grasses to form clumps. The absence of tillering is one of the main differences between cultivated maize and its direct ancestor, teosinte (*Zea mays* ssp. *parviglumis*), from which it was domesticated in southern Mexico not more than 9000 years ago. The control of tillering is largely due to the TCP transcription factor, *TEOSINTE BRANCHED1* (*TB1*; Wang et al., 1999; Fig. 2.21).

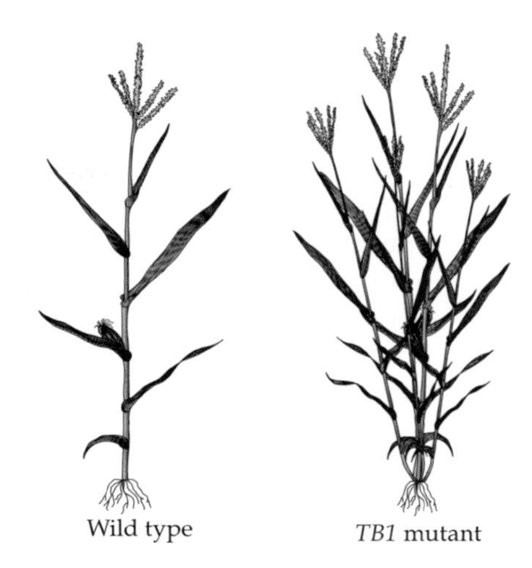

Fig. 2.21. The role of the gene TEOSINTE BRANCHED1 (*TB1*) in controlling tillering in domesticated maize. Tillering is disadvantageous under cultivation as it prevents the formation of a large single stem capable of bearing large ears.

An extreme form of culm is distinguished as a **calamus** or **rush**. A calamus is a stem formed of a single very long internode, and is therefore leafless. Calami are green and alone provide for the photosynthetic needs of the plant. Examples include rushes (*Juncus*) and bullrushes (*Holoschoenus*). The cattail (*Typha*), often mistakenly called bullrush, is most certainly not a rush, as it has true leaves. The stem morphology of rushes has an economic consequence as dried calami of rushes may be woven into mats and baskets with the minimum of preparation, as there are no leaves to remove from the long stems.

2.18.2 Stems at or below the ground

There are many interesting morphological modifications of stems due to their being on or under the ground. The **stolon**, or **runner**, is a horizontal stem with long internodes that is efficient at vegetative spread and in foraging for favorable environmental conditions. Some stolons creep along the ground typically rooting at the nodes as in the strawberry (*Fragaria*), others form arching stems which root at the tip, as in the blackberry (*Rubus*). A horizontal (i.e. **diageotropic**) stem underground is termed a **rhizome**. These may be primarily storage organs, as in *Iris*, or primarily for propagating and foraging as in the sand sedge (*Carex arenaria*). The rhizomes of *C. arenaria* can cover huge distances in straight lines under sand, producing two types of node, which alternate regularly. Some nodes bear only **scale leaves** with undeveloped axillary buds, other nodes develop aboveground shoots. Rhizomes usually produce **adventitious roots** in abundance, and often the leaves produced by the rhizome develop into normal aboveground foliage leaves. In contrast, the **tuber** is primarily a storage organ. It is a swollen, and apparently leafless, caulome, although it may have scale leaves. In the potato (*Solanum tuberosum*) the tubers are produced at the end of rhizomes and consist of many internodes (the "eyes" of a potato are the **axillary buds**). On the other hand, in horsetail (*Equisetum arvense*), each tuber is a single internode. The diminutive Australian lycopsid, *Phylloglossum drummondii* is remarkable in producing the so-called "marsupial tuber", a leafless geotropic branch.

Rosette herbs have a **stock**, or **caudex** as it is termed in some literature. The stock is a short, stout, vertical underground stem. It has very short internodes (leaves of rosette herbs being tightly packed together). The stock is the only, or at least the main, persistent stem. The only aboveground stems are generally inflorescence stalks. Good familiar examples of a rosette herb are *Hypochaeris radicata* and *Taraxacum officinale*, two Eurasian weeds that are now cosmopolitan. The stock may be directly contiguous with a persistent **primary root** as in *T. officinale*. Others plants have only adventitious roots produced by the stock after the primary root aborts, as in *Valeriana dioica*. The stock may branch, in which case a tight **clump** of shoots is formed (as opposed to a **patch** of shoots formed by a horizontal underground rhizome). A swollen stock is known as a **corm** (as in *Musa, Cyclamen*)

Finally, **bulbs** exhibit a very distinctive underground stem. Bulbs have a small, flattened stem at the very base of the bulb called a **disc stem** (not to be confused with a floral disc). The **disc** bears the swollen storage leaves of the bulb and has a SAM in its center from which arise aboveground leaves and inflorescences. The existence of the disc indicates the enormous flexibility of the SAM. In **thorns**, the SAM produces a stem that is narrowed to a sharp point and precociously lignified, while in bulb discs the SAM produces a very obtuse (sometimes almost flat) structure (Fig. 2.22).

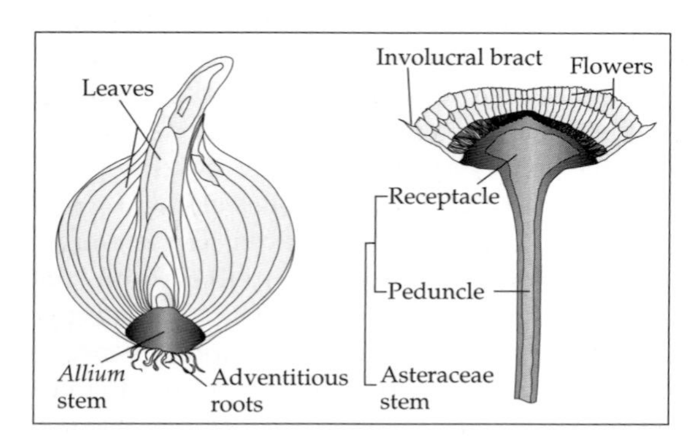

Fig. 2.22. Specialized stem shapes: the stem of a bulb and the receptacle of a compositae capitulum.

2.19 Buds as specialized stems: survival and vegetative reproduction

A specialized form of stem is the resting bud. These are stems formed in miniature and held under hormonal arrest. Axillary buds, formed in the axils of leaves, are such buds. They are often arrested by the apical dominance of the SAM of the parent axis. However, even the SAM of the main stem can go into arrest on the onset of unfavorable seasons and these are called terminal buds. The formation of a resting bud from a continuously growing axis is accompanied by major morphogenetic changes in the apex. Apical growth slows, internodes shorten and, usually, minute scale leaves (prophylls) form, instead of normal leaves, to protect the resting bud (almost all temperate trees are of this type). Buds invested with a covering or tegument of scales is called a scaly bud. This contrasts with the naked bud as in *Frangula alnus*. In naked buds the SAM is covered only by undeveloped, but otherwise normal, foliage leaves.

The terminal bud is the form in which most temperate trees survive summer drought and winter cold. They then produce a flush of growth when winter ends and temperatures increase. This new growth is called spring, as it literally springs from the old stems. In the English language, the first season of the year is named after these shoots. By early summer the growth of temperate trees is usually completed and new terminal buds are formed. However, in wet years, a second flush of growth can occur around the month of August and these shoots so produced are called **lammas shoots** after the pagan summer festival of Lammas.[5] In turn these lammas shoots will form a terminal bud to overwinter the SAM.

Shoots that grow in flushes between periods of growth arrest, marked by the formation of resting buds, are termed **sylleptic**. Shoots that continuously grow without the formation of a resting bud are termed **proleptic**. This is a more common state in the wet tropics where equable climate permits continuous, or at least a seasonal growth.

Buds, often with rudiments of leaves or even inflorescences can lie dormant for some time before becoming active. This is known as **preformation**. Many temperate trees form vegetative or inflorescence buds the summer before flowering. These buds overwinter in a dormant state until resuming growth in the spring. Such preformed buds may contain just inflorescences or just leaves (e.g *Ulmus*, *Fagus*), or may be mixed, with leaves and flowers (e.g. *Syringa*, *Aesculus*). Preformation may be extreme, particularly in arctic environments. The arctic-alpine herb *Acomastylis* (*Geum*) *rossii* (alpine avens) requires three years for development of its buds (Meloche & Diggle, 2001).

The Danish botanist Raunkiær (Raunkiær, 1934) used the position of the winter resting buds to classify temperate plants into life forms. Many of his terms have become common in botany. Thus many herbaceous plants are **hemicryptophytes** defined as having perennating buds at soil level (hemicryptos means literally "half-hidden"). **Phanerophytes** include trees and have winter buds on stems that are raised aboveground while geophytes perennate with buds belowground. Water plants produce winter buds that may detach from the main stem which are called **turions** (or **hibernacula**). An example is given by the turions of *Hydrocharis* and *Stratiotes*.

In some plants, buds are modified for dispersal and drop from the plant. They often have nutritive reserves, and if these are in the leaf part they are called **bulbils** and if in the stem part, **tuber buds**. Tuber buds occur in *Ranunculus ficaria* and as "air potatoes" in *Dioscorea*. Bulbils occur in ferns and angiosperms, as buds on phyllomes (e.g. *Cystopteris bulbifera*), in axils of phyllomes (*Lilium bulbiferum*) or in place of flowers, as in *Allium vineale* and *Titanotrichum* (Wang & Cronk, 2003). In *Titanotrichum* the transition from floral to bulbiliferous inflorescences is marked with a pronounced downregulation of the transcription factor *GESNERIACEAE FLORICAULA* (*GFLO*; Wang, Moller & Cronk, 2004a). This is the Geseneriaceae homologue of the *Antirrhinum* gene *FLORICAULA* (*FLO*) and the *Arabidopsis* gene *LEAFY* (*LFY*).

Mosses and liverworts often produce reproductive buds called **gemmae**. These are small amorphous pieces of thallus (caulome) capable of developing into stems. They may develop from rhizoids as in some species of *Tortula* and *Barbula* (**rhizoidal gemmae**). Alternatively, they may develop from phyllidia, thallus or stems. They may be produced in specialized **gemmae cups** as in *Tetraphis and Marchantia* or

[5] Lammas is held on August 1. It is one of the quarter year fire festivals in the Celtic calendar, after Beltane (May 1) and before Samhain (October 31).

alternatively directly from other organs, such as stems, as in the **cauline gemmae** of *Aulacomnium androgynum*.

2.20 Surface form

Although stems often form simple cylinders, they may take a variety of other forms. If they are circular in cross section with no obvious surface features they are called **terete**. However, they may be angular in cross section. Stems of monocots are commonly **triangular** in cross section, as in the triangular inflorescence stalk of sedges (*Carex*). Stems of the Lamiaceae are often **quadrangular** in cross section.

Surface features also vary. The surface may be uneven with longitudinal **ribs** and **furrows**. The relative sizes of the ribs and furrows determine whether the stems are described as ribbed or furrowed. Ribs that are taller than they are broad are called wings. **Winged** (**alate**) stems are common in the Asteraceae as in yellow starthistle, *Centaurea solstitialis*. Some of the patterning of stems may derive from the leaves. Internodes form from the growth of plates of cells between the leaves. If this growth includes the leaf bases then features of petioles can extend down the stem (i.e. they are **decurrent**). Thus, if the petiole is winged, these wings are frequently decurrent down the stem. These prolongations are called **leaf tails.** It has even been suggested that the ribs at the angles of labiate stems (formed of sclerified tissue called **collenchyma**) can be interpreted as prolongations of leaf bases (Guédès, 1979).

Further oddities can form by the mixing of leaf and stem tissue in internode elongation. In **concaulescence** the internode elongates between a leaf and its axillary bud, thus carrying the leaf away from it bud, as in *Symphytum*. The axillary bud can thus appear to have formed just under the leaf above instead of in the axil of its parent leaf. **Recaulescence** is the elongation of the lateral tissue below an axillary bud so that leaf and axillary bud appear to be displaced up a lateral stem away from the main stem, a phenomenon often seen in the bracts of solanaceous plants including *Lycopersicum* (Weberling, 1989).

Stems may bear processes such as **hairs** (**trichomes**), **prickles** or **warts** (**tubercles**). **Lenticels** (pores functioning for gas exchange) are a characteristic feature of young woody stems, often distinguished from the surrounding epidermal cells by cork or wax production. In the Cactaceae the stem is swollen into a **leaf cushion** at the point of leaf (i.e. spine) insertion. The genus *Mamillaria* gets its name from its pronounced leaf cushions that resemble teats (from the Latin *mamilla*).

Stems after secondary thickening may have a bark. In corky barks this may form from a separate **cork cambium** (**phellogen**) producing corky tissue known as **phellem**. More usually, however, bark is fibrous and is formed from phloem fiber cells, ultimately produced from the cambium of the stele. Either way, when a bark forms, the original cells of the epidermis and cortex are lost. Bark morphology can be quite variable depending on the developmental and anatomical feature of the underlying cambium and phloem (Whitmore, 1962). Tropical foresters utilize these characters (slash characters) to identify tropical trees by slashing the bark with a machete.

2.21 Stems associated with flowers

2.21.1 The receptacle

The floral parts are borne on a very interesting stem called a **receptacle**. It is a greatly modified stem and the modifications are very important in the evolution of flowers. The receptacle may be elongated and cylindrical or compressed and discoid. It may be cone shaped or hollow, like a cup, or even as a deep pitcher that completely encloses the **gynoecium** as in a rosehip (fruit of *Rosa* sp.). The receptacles often have terminal outgrowths or plates, which are called **discs**, often in the shape of a ring (**annular discs**). These rings may be outside the stamen whorl (**extrastaminal discs**) or inside the stamen whorl (**intrastaminal discs**). These often function as nectaries by having either a nectar-bearing surface or specific nectarial **discal glands** (as in *Reseda* and *Cadaba*). The disc may even bear discal scales (as in *Passiflora* and *Bridelia*).

The great plasticity of the receptacular stem has contributed greatly to the diversity of the flower. Most importantly the capacity of the receptacle to undergo **invagination** and enclose the **carpels** during evolution has permitted the evolution of the **inferior ovary (epigyny)**. The inferior ovary is one of the most important evolutionary trends and it has arisen many times in evolution (Rudall, 2002; Soltis, Fishbein & Kuzoff, 2003) possibly as a protection from insect predation on ovules. In doing so, the receptacle has often taken on a dispersal function, taken in other plants by the fruit wall. In apples, the receptacle has become fleshy, anthocyanin pigmented and delicious. The transfer of characteristics from the fruit wall (a leaf derived organ) to the receptacle (a stem derived organ) is an example of what E. J. H. Corner calls **transference of function** (Baum & Donoghue, 2002).

A word of caution is needed here about the role of the receptacle in the inferior ovary. It is exceedingly difficult to distinguish between tissue derived from the receptacle and tissue derived from **congenital fusion (adnation)** of the bases of the sepals and petals. In the last century this caused both a lot of anguish, and a lot of ink to be expended on the matter. Payer, Hofmeister and Sachs suggested a receptacular origin and Van Tieghem and Velenovsky promoted the adnation theory. Gene expression studies could potentially resolve this issue. If genes characteristic of the receptacle are expressed in the perigynous tissue it would confirm the receptacular theory, while the contrary would be true if genes characteristic of petals or sepals were expressed (Gustafsson & Albert, 1999). The problem with such studies is that as the nature of the organ has changed the gene expression patterns are likely to have changed too. Like many controversies in science, it is probable that both views are right depending on the particular example.

Given the importance of the receptacular inferior ovary in plant evolution, it would be very interesting to know the developmental genetic basis of this trait. We may assume that it involves a repatterning of the SAM of the flower so that instead of growth being even, there is a cessation of stem growth in the middle and a transfer of growth to the flanks of the meristem so that the sepal, petal and stamen primordia are carried up well above the level of the carpel primordia.

It would be of great interest to study how the genes important in controlling the SAM behave during the development of an inferior ovary. It is important to realize that the receptacle of the flower is the seat of the SAM that produces the floral phyllomes. It is **determinate**, generally being used up in the production of the last floral phyllomes, that is, the carpels. However, it is not always determinate. Environmental conditions can reactivate the floral SAM and it can revert to a vegetative apex, as in the terminal flowers of some cultivated *Impatiens* and in Goethe's rose. Such cases are always striking, as a vegetative stem growing out of the center of a flower is always unexpected.

2.21.2 Pedicel, peduncle and scape

Other stems associated with flowers are given special names not because of any intrinsic qualities but because of their topographical relations with other parts. They are morphological way-markers in a morphologically complex region: the inflorescence. The **pedicel** is defined as the stem of a single flower. The **peduncle** is defined as all other stems of an inflorescence that are not pedicels. The peduncle may be a much-branched system in a complex inflorescence, and we may recognize a **primary peduncle** as the stalk at the base of the entire inflorescence. Remaining portions of the peduncle may be designated as first-order, second-order, third-order **peduncle branches,** and so on.

The peduncle and pedicel have the important function of supplying water and nutrients to the flower and fruit. They can also form **abscission layers** to control reproduction, as in the "June drop" of apples (natural self-thinning of the apple crop, when the fruits are young, prevents an overburden of mature fruit).

Some plants have **sessile** flowers, in which the pedicel and peduncle are so reduced as to be almost undetectable. An example is provided by the Asteraceae, in which the flowers are sessile in the head (**capitulum**). Where inflorescences have been reduced to a single flower a node can usually be detected between the peduncle and pedicel; so although the peduncle subtends a single flower, a clear distinction can still be made between it and the pedicel.

The **scape** is a type of peduncle characteristic of two life forms, **rosette herbs** and **cormophytes** (especially bulbs). It is a largely historical contingency that this type of peduncle, rather than others, was singled out to be dignified as a separate morphological concept. However, the term is widely used and so merits discussion here. A scape is an inflorescence stalk which arises directly from the ground and is generally the only aboveground stem of the plant. It may be naked and consist of a single internode only as in daffodils (*Narcissus*) and snowdrops (*Galanthus*), or it may have a few scale leaves as in turmeric (*Curcuma longa*). It may bear a single flower (*Narcissus*) or many, as in lawn daisies (*Bellis*) or dandelions (*Taraxacum*), but is never branched.

2.22 Special modifications of stems

2.22.1 Thorns and hooks

Stems have proved versatile in producing morphological innovations. A strikingly modified stem is the **thorn**. A thorn is a **short shoot** (i.e. a lateral stem of **determinate growth**) that functions as defence against vertebrate herbivores. It is a stem in which the SAM progressively narrows the width of the stem as it grows until the stem ceases growth in a sharp point. The molecular developmental control of this progressive narrowing of the SAM would be interesting to study. For instance, the *CLAVATA* gene network is known to negatively regulate meristem size (see above).

The thorn is also unusual in that it lignifies rapidly right to a hardened (**indurated**) tip. Thorns may either be naked or have a few scale leaves. Sometimes the scale leaves bear axillary buds, which in *Pyracantha* (Rosaceae) even develop into inflorescences. Such complex thorns are best termed **thorn-shoots**.

Thorns can usually be recognized as modified stems by the presence of a subtending leaf.

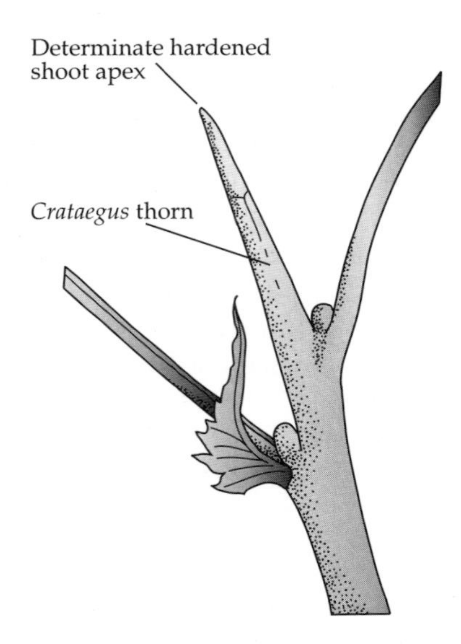

Fig. 2.23. A thorn form from the terminal potion of a shoot in *Crataegus*.

A thorn uses up an axillary bud, but **serial buds** may form in the same axil and these may develop into ordinary vegetative shoots.

Thorns should not be confused with **spines** (modified leaves, as in Cactaceae) or **prickles** (outgrowths of the epidermis). Prickles, as developments of the stem epidermis, are familiar in roses (*Rosa*) and cane fruit (*Rubus*). In both cases the degree of production of prickles is a target of breeding, as thornless varieties make for easier cultivation. The number of prickles is related to the degree to which localized meristematic activity occurs in the epidermis. Quantitative trait loci have been identified for prickliness (Crespel et al., 2002) and although the underlying genes have not yet been identified, meristematic control genes are obvious candidates.

True thorns are widely dispersed in the angiosperms and are well known in the Leguminosae (e.g. *Gleditschia*), Rosaceae (e.g. *Crataegus, Pyracantha*), Rhamnaceae (e.g. *Zizyphus*), Eleagnaceae (e.g. *Hippophae*), Punicaceae (*Punica*) and Rutaceae (e.g. *Citrus, Zanthoxylum*). In herbivore defence there may be advantages to producing thorns quickly on new shoots (Fig. 2.23). Thus, in *Punica* the axillary buds develop into thorns in the same year that the parent shoot is produced.

Such early produced branches are known as **anticipated shoots**.

Similar in development to the thorn but more complex, because of the asymmetrical growth necessary to produce curvature, is the **hook** or **uncus**, used for climbing. In *Uncaria* these are formed in the inflorescence from sterile pedicels. In *Ancistrocladus* they are formed from stem apices, and in *Lavunga* from **short shoots**, in a manner similar to thorns. However, it should be noted that hooks may also be formed from prickles and leaves, and that even **trichomes** may curve into **hooklets**.

2.22.2 Tendrils

Tendrils are specialized twining organs. Some tendrils are shoot-derived while others are leaf-derived. A good example of the shoot-derived tendril is in the family Vitaceae in which the tendril is thought to have originated from the primary shoot axis, which is converted to a tendril immediately after forming a node.

The main evidence for this is that the tendril is not axillary but is on the opposite side of the stem to the leaf (i.e. it is **leaf-opposed**). Under this model, shoot growth continues only from the activation of the axillary bud. The center of the old shoot is **usurped** by the axillary bud and the tendril is displaced to become leaf-opposed. Because of this, it appears to the casual observer that the tendril (a terminal structure) is an adventitious lateral organ on the opposite side of the stem to the leaf. However, although this sympodial process is almost certainly the origin of the leaf-opposed tendril, the system has evolved from historically sympodial to functionally and developmentally monopodial, with a continuity of meristematic growth.

In the Vitaceae, the tendrils may be further interpreted as modified terminal inflorescences. Tendrils and inflorescences are therefore both part of the fundamentally sympodial organization of the family. There are several lines of evidence for the derivation of tendrils from inflorescences. First, the genus *Ampelopsis* does not produce tendrils, but does have tendrillate inflorescences. Second, when inflorescences are

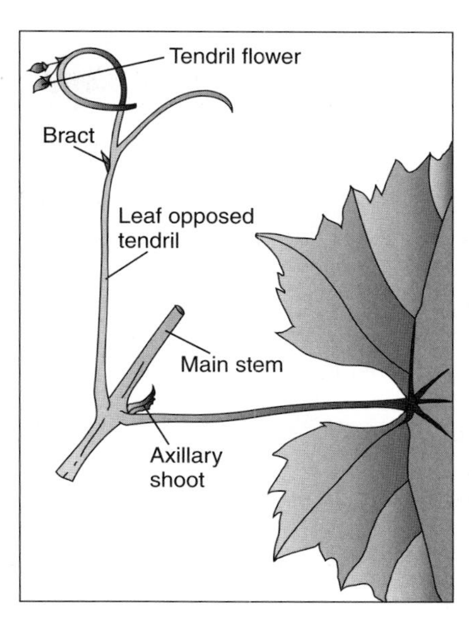

Fig. 2.24. Tendrils in *Vitis* as modified terminal inflorescences (1). Some tendrils bear a few flowers.

produced in *Vitis*, they are produced instead of tendrils (Fig. 2.24; 2.25). Third, in *Vitis* there is a tendency for otherwise normal tendrils to produce a few flowers. Fourth, floral genes are expressed in developing tendrils (Fig. 2.26).

Meristems of *Vitis* differentiating at a node can thus adopt either tendril fate or inflorescence identity. Interestingly, tendril meristems, as well as inflorescence meristems, express the floral genes *Vitis vinifera FRUITFULL* (Vv*FUL*) and *V. vinifera APETALA1* (Vv*AP1*), further exemplifying the inflorescence nature of tendrils (Calonje et al., 2004). However, this does not explain how tendril versus inflorescence identity is induced.

It appears that the tendril/inflorescence developmental switching involves GA. The evidence for this comes from a mutant of the grape variety Pinot Noir called Pinot Meunier. Pinot Meunier is thought to be genetically identical to Pinot Noir except for a somatic mutation in the outer layer of the plants (i.e. the cells deriving from the L1 layer of the apical meristem). When whole plants were produced from the L1 layer by tissue culture they were dwarf and produced inflorescences in place of tendrils.

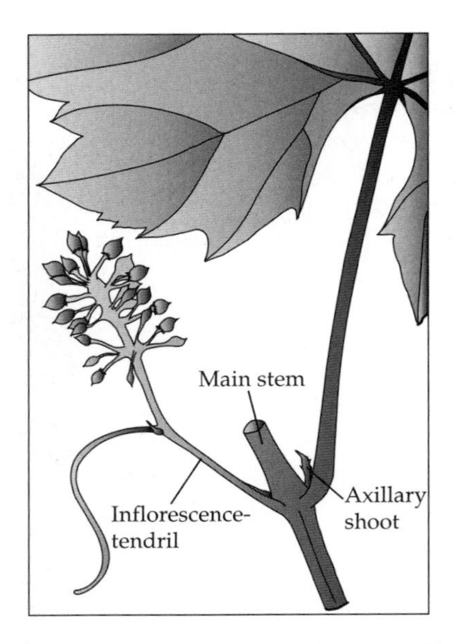

Fig. 2.25. Tendrils as modified terminal inflorescences (2). Some inflorescences bear tendrils.

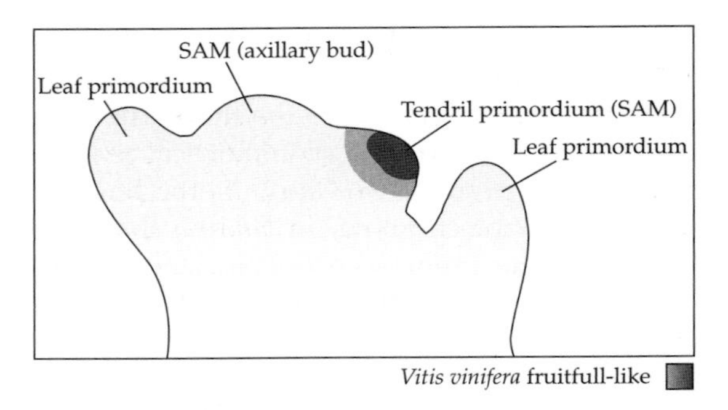

Fig. 2.26. Tendrils as modified terminal inflorescences (3). Tendril primordium showing expression of floral gene, *VvFul* (after Calonje et al., 2004).

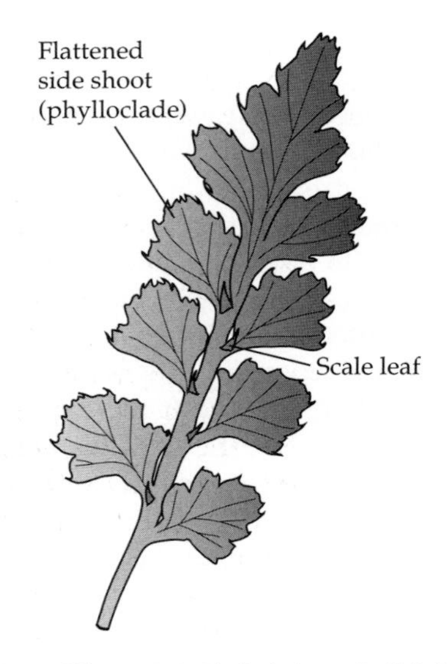

Fig. 2.27. Leaf-like stems in *Phyllocladus asplenifolia* bearing cladodes (phylloclades) in the axils of scale leaves.

This phenotype is due to a loss of function of the grapevine homologue (Vv*GAI*) of the *Arabidopsis* gene *GIBBERELLIN INSENSITIVE* (At*GAI*). GAI is a member of the DELLA family of transcription factors and is responsible for various types of growth suppression. The application of GA represses GAI allowing stem elongation and other GA responses. The *gai* mutant is resistant to repression by GA. The lack of a GA response in the grapevine mutant explains the phenotype of dwarfism (short internodes). However, GA (in *Arabidopsis* at least) is known to stimulate inflorescence production by upregulating the gene *LEAFY* (*LFY*), and so the promoting of flowering by the loss of the GA response is puzzling. Two possibilities present themselves. First, in contrast to *Arabidopsis*, GA may be an inhibitor of inflorescence production in grapevine (Boss & Thomas, 2002). Second, there may be a *GAI* independent pathway for GA to stimulate LFY production. If this is the case, then the mass production of inflorescences is explained, as the grapevine *gai* mutant has more endogenous GA than the wild type (Boss & Thomas, 2002).

The leafy (*LFY*) homologue in *Vitis*, *VITIS LEAFY* (*VFL*) is certainly a candidate to be the regulator of the inflorescence/tendril switch. In *Arabidopsis*, a threshold of LFY expression is required for floral initiation (Blazquez et al., 1997). However, *VFL* is expressed in both developing tendril and developing inflorescence meristems, as well as in leaf primordia. At first glance therefore, *VFL* seems to have a general role in promoting organ indeterminacy at young stages (Carmona, Cubas & Martinez-Zapater, 2002). However, in tendril meristems, *VFL* is only transiently expressed, perhaps so placing the tendril meristem below the necessary threshold of *VFL* expression for required for inflorescence development (Calonje et al., 2004; Fig. 2.26).

Some tendrils in the Vitaceae are further modified for clinging. The tendrils of *Parthenocissus trifoliata* (Boston ivy) terminate in **adhesive discs**, or **hapteres**, permitting them to cling tightly to the walls of ancient Universities. Hapteres are defined as organs that anchor without absorbing nutrients. The derivation of the tendril from the inflorescence gives rise to the intriguing possibility that the adhesive discs of *Parthenocissus* and *Cissus* are derived from modified floral buds (Gerrath & Posluszny, 1989).

In *Passiflora* the tendrils are modified axillary shoots. However, as in *Vitis*, they can also be interpreted as modified inflorescences. In *Passiflora* inflorescences, flowers are produced laterally to the tendril meristem (Shah & Dave, 1971), or put another way the inflorescence shoot has a terminal tendril.

2.22.3 Cladodes

Stems can also simulate leaves. Such simulated leaves are called **cladodes**. Cladodes should not be confused with the similar **phyllodes**. Phyllodes are transformations of the most stem-like portion of the leaf, the **petiole**, into simulated **leaf blades**. To add to the confusion cladodes are sometimes called **phylloclades** or **cladophylls**, terms that should be dropped to avoid confusion.

Cladodes are stems of **determinate growth**, flattened and leaf-like. They are often produced in the axil of a true leaf, which may be reduced to a **scale leaf**. Cladodes may consist of one internode or several.

If cladodes comprise more than one internode they may have scale leaves of their own. There are many well-known examples such as the gymnosperm *Phyllocladus* (Fig. 2.27) and the monocots *Ruscus* and *Asparagus*. The prickly pear cactus, *Opuntia*, has complex cladodes formed of large pad-like stems of very many internodes, and which bear scale leaves converted into spines. Although stems never simulate leaves very convincingly, it should be remembered that leaves evolved from stems in the first place. The existence of cladodes show that stems have a capacity for laminar growth and the evolution of cladodes may possibly repeat some of the processes involved in the original evolution of the megaphyllous leaf. It is not known, but would be interesting to know, to what extent cladodes borrow regulatory genes involved in true leaf formation, by means of **network capture**.

CHAPTER 3

The organography of the root

3.1 The era of the root

The root is a vascular organ, so the first land plants, being nonvascular, cannot have had roots. Vascular plants radiated between c. 410 and 395 million years ago and the first vascular plant fossils have well-developed axes ("stems") but no roots.[1] Subsequently, rooting structures appear, at first seeming like modified stems (rhizomes), then unequivocal roots. Thus it appears that the **root** evolved as a *de novo* structure perhaps some 15 million years after the first appearance of vascular plants (Raven & Edwards, 2001). The root is therefore entirely a feature of the land flora and the **holdfast** of seaweeds represents a non-homologous structure.

Roots are now ubiquitous in vascular plants, with the exception of a few plants that have secondarily lost them. This ubiquity of the root has had a major impact on terrestrial ecosystems. The evolution of the root was largely responsible for the formation of soil by promoting weathering of rock and the incorporation of organic matter into sediment. Roots altered biogeochemistry by injecting carbon dioxide belowground by root respiration, thereby promoting the conversion of silicates into carbonates (Raven & Edwards, 2001). This process sequestered huge amounts of atmospheric carbon in mineral form in the earth. Root turnover, the birth and death of roots, further injected large quantities of carbon into the earth, this time in organic form. The root may thus be considered as a one-way air to earth carbon pump. It is thought that there may have been a 10-fold decrease in atmospheric carbon dioxide in the Devonian period, 400–350 million years ago (Raven & Edwards, 2001). If this is correct, the root is the prime candidate as the effector of this massive drawdown, as the Devonian saw a progressive increase in the size and complexity of roots.

Today roots are important determinants of ecosystem function. Almost all mineral ions entering terrestrial ecosystems do so either by crossing a cell membrane in the root (directly or through mycorrhizal associates) or by crossing a cell membrane of a bacterium. Of the two pathways, the root pathway is much more important. The fine roots responsible for this uptake are calculated to have a surface area greater than leaf surface area and to represent 5% of atmospheric carbon (Jackson, Mooney & Schulze, 1997). Furthermore, taking into account rates of root growth, it has been suggested that fine roots represent a third of global net primary productivity (Jackson, Mooney & Schulze, 1997).

In most ecosystems over half of this root biomass is within 10–30 cm of the surface (Schenk & Jackson, 2002). However, roots are known to penetrate to great depths. Mesquite roots (*Prosopis*) have been reported to penetrate to 175 ft (Phillips, 1963). Cave systems can provide natural root observation chambers (rhizotrons) to observe root growth at depth. Studies of cave systems in Texas have revealed numerous forest trees that root to 5 m or more, with *Quercus fusiformis* rooting down to 25 m

[1] It should be noted that the use of the term "stems" for early land plants (before the root/shoot dichotomy was established) is inappropriate if stems are considered antithetic to roots in vascular plants. "Telomes" or "subaerial axes" are more general terms. However, the sub-aerial axes of early vascular plants and the stems of later land plants share functional and morphological characteristics (and probably developmental mechanisms). Using a different term obscures this basic commonality of radial sub-aerial axes (stems in the broad sense).

(Jackson et al., 1999). These deep roots probably have an important role in "hydraulic redistribution." This is the extraction of water at depth to recharge the water status of the bulk root biomass occurring in the dry surface layers of the soil. Hydraulic redistribution from depth allows plants to maintain functional surface root systems, even in drought, so maintaining evapotranspiration and mineral uptake. The efficient propulsion of water from great depths into the atmosphere appears to have a significant effect on global climate (Lee et al., 2005c).

3.2 The origin of roots

Extant nonvascular plants do not have roots but have analogous absorptive structures called rhizoids. These are outgrowths of the surface layers of the gametophytes and usually consist of a single cell file. However, they can become more complex structures and even regenerate stems as in the rhizoidal gemmae of mosses. Rhizoidal gemmae are cell clumps that form on rhizoids and act as vegetative propagules. Whole gametophytes can regenerate from them. Interestingly the rhizoids of moss gametophytes share at least some of their molecular developmental machinery with the root hairs of flowering plant sporophytes. The development of both rhizoids and root hairs is controlled by basic helix–loop–helix transcription factors, genes of the ROOT HAIR DEFICIENT6-LIKE (RHD6-like) group (Menand et al., 2007).

The nearest a bryophyte comes to producing an organ with the essential characters of a root is in the specialized leafless stems of the peculiar mosses, *Takakia* and *Haplomitrium* (Grubb, 1970). However, these structures are not homologous to vascular plant roots and are more stem-like than root-like in many of their features. Nevertheless, they illustrate how stems can become dimorphic and evolve different developmental trajectories, just as we assume happened in the evolution of the root.

Early vascular plants appear to have had no roots but dimorphic subaerial axes ("stems" in the broad sense). Some of these were upright (probably negatively geotropic and positively phototropic) and bore sporangia and, in some cases, leaves. Others were probably horizontal in growth and performed an anchoring and nutritive function by bearing rhizoids or hosting vesicular–arbuscular (VA) fungal association, similar to modern mycorrhizae. These latter stems or "rhizomes," as present in early fossils like *Aglaophyton*†, are good candidates for evolution into roots by progressive divergence of their developmental trajectory. The presence of fossil rhizomes with VA fungi (Phipps & Taylor, 1996; Taylor et al., 1995) is of particular interest as it indicates that the root was possibly a mycorrhizal organ from its inception. Mycorrhiza-like associations are rare in modern stems, but are a nearly universal feature of roots.

It has been suggested that an important driver of the evolution of the root was the selective advantage inherent in providing a more suitable habitat for VA fungi (Brundrett, 2002). It is even possible to speculate that some of the features of the root may have originally been plastic responses of the phenotype to fungal endophyte invasion that were later hard-wired into development, as endophyte invasion became ubiquitous. Endophyte invasion of plant tissues are extremely common, whether mycorrhizal or not, and are common in nonvascular plants. The evolution of the land flora therefore cannot be divorced from their coevolving fungal associates. The *Glomales*, the fungal group now commonly responsible for VA mycorrhizae, have a fossil history as long as land plants (Redecker, Kodner & Graham, 2000).

Rhizomes (but apparently not roots) have been described in the protracheophytes, *Horneophyton*† and *Aglaophyton*†, all plants with little vascularization, as well as in the fully vascular plant, *Rhynia*†. Rhizomes are also present in early lycophytes such as *Asteroxylon*† and *Zosterophyllum*†, and in early euphyllophytes such as *Psilophyton*† and *Protopteridophyton*†. The earliest convincing roots appear in Lycophytes during the early Devonian, as in the case of *Drepanophycus thuringensis*† (Li & Edwards, 1995). However, in the euphyllophyte clade the first really convincing roots do not occur until the middle Devonian, as in the cladoxylalean

Lorophyton goense† (Fairon-Demaret & Li, 1993; Raven & Edwards, 2001).

3.3 Homology of roots

The root is recognized by a suite of characters, the possession of all or most of which serves to distinguish it from the stem. However, in the transformation of stems to roots, it follows that organs of intermediate and mixed character are likely to have existed. This makes it problematic to pinpoint the origin of the root. This uncertainty about what constitutes a root also bedevils attempts to say whether the root has one evolutionary origin or many.

One important question is whether roots are homologous between extant lycophytes and other vascular plants (euphyllophytes). The lycophytes are considered by some to have diverged from euphyllophytes at a time when plants were still rootless, implying independent derivation of roots in both lineages. The placement of apparently rootless plants like *Zosterophyllum*† and *Psilophyton*† is crucial in this regard. If these plants are correctly inserted into the phylogenetic tree above the divergence between lycophytes and the other vascular plants, it implies that vascular plants were rootless at the split, and roots must have been gained twice. *Zosterophyllum*† is usually placed with the lycophyte clade and *Psilophyton*† with the euphyllophyte clade, supporting a dual origin of roots. However, it should be borne in mind that the phylogeny of these fossils may be incorrectly reconstructed. There is nothing obvious in the root morphologies of lycophytes versus other vascular plants to suggest that they must have separate origins.

The zosterophylls may hold the key to the evolution of roots in lycophytes (Gensel, Kotyk & Basinger, 2001). Zosterophylls show **H-branching**, which consists of very divergent dichotomies, which have the obvious potential to be distinguished into root-like and stem-like portions. This indeed happens in the notable **K-branching** of *Bathurstia*† and *Drepanophycus*†, in which one branch of the dichotomy is stem-like with microphylls and the other naked and oriented in a reverse direction, apparently resulting in a rooting structure and a stem from the same axis (Gensel & Berry, 2001). K-branching may thus represent a first step in the evolution of the lycopod root (Fig. 3.1).

Other lycophytes have a root system that is possibly directly comparable to the Zosterophyll K-branching, although it may, alternatively, represent a different system entirely. This is the **stigmarian root system**, for instance, of the arborescent lycophyte, *Lepidocarpon*†. It is produced from a dichotomous division of the embryonic axis (shoot), with one branch becoming root-like (the **rhizomorph**) and one shoot-like (i.e. a leaf-bearing stem). The stigmarian root system therefore seems to be homologous with a stem. This root system bears lateral organs, called **stigmarian rootlets**. It follows that if the root system is a lycophyte stem structure, then these lateral organs are possibly homologous to microphylls. As this root system appears to be a derived feature of rhizomorphic (i.e. lepidodendralean) lycophytes, it possibly represents an independent origin of the root (Raven & Edwards, 2001; Rothwell, 1984; Rothwell & Erwin, 1985).

Is the stigmarian system homologous to any extant root? The closest living relatives of the *Lepidodendrales*† are the Isoetalean lycophytes, *Isoetes* and *Stylites*. Remarkably, the roots of *Isoetes* are similar in structure to stigmarian rootlets, with a single vascular bundle on one side of a large central cavity. They thus may not be homologous to roots of other lycophytes or euphyllophytes (both of which usually have a root cap and root hairs). It may be significant that the leaves of isoetaleans function, in part, as absorptive organs for carbon dioxide (Keeley, Osmond & Raven, 1984), a function performed by leaves in most plants.

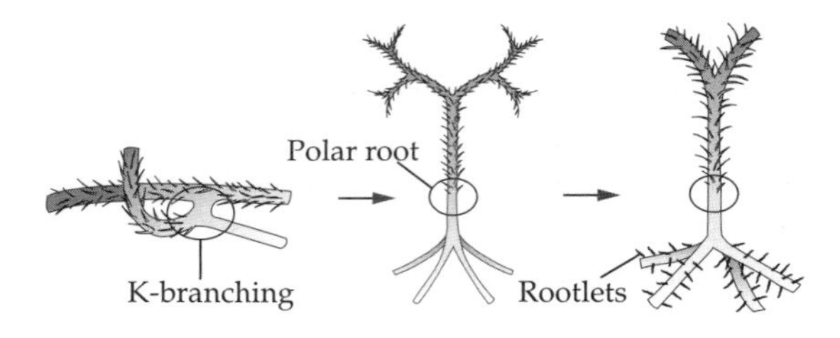

Fig. 3.1. Scheme for the evolution of the lycophyte root from a dimorphic branch system (after Gensel & Berry, 2001).

This curious structure, the **rhizophore** of *Selaginella*, is often mentioned in connection with the origin of roots (Lu & Jernstedt, 1996). However, the rhizophore is a leafless stem that is only superficially root-like. Although it is smooth and leafless like a root, it arises exogenously with a stem-like meristem and has no root cap. The rhizophore meristem is situated at the point where the normal stems bifurcate. After outgrowth, at a later stage in development, it bears true roots. As a synapomorphy for a clade of the genus *Selaginella* (Korall & Kenrick, 2002), the rhizophore is best regarded as a unique structure homologous to a stem. However, in combining features of root and stem, the rhizophore perhaps indicates that these two classes of organ are best viewed, not as discrete entities, but rather as opposite poles of a continuum of characters pertaining to the plant axis.

Are all euphyllophyte roots homologous? The two great crown-clades of the euphyllophytes, the seed plants and the monilophytes, all bear what appear to be similar roots. However, their roots differ in developmental origin (Groff & Kaplan, 1988). The roots of ferns and other monilophytes (and like those of the lycophytes) develop laterally from the stem and are called (in Goebel's terminology) **homorhizic** (i.e. the root is lateral relative to the embryonic axis of the embryo). In contrast most of the root mass of seed plants typically derives from a single, centrally positioned root (i.e. the primary root). Stem-borne roots may be produced later and are termed adventitious roots. Goebel called this system **allorhizic** (shoot and roots are at opposite ends of the embryonic axis). The primary root of seed plants is that root which develops intercalated between the **suspensor** and the **stem**, aligned in direct opposition to the stem axis. If this distinction is taken to be fundamental it may indicate that roots of monilophytes and seed plants are nonhomologous. However, there is very little other evidence for this, and the allorhizic condition may simply be a derived character that has evolved from homorhizy. The formation of the radicle in seed plants is more precocious than the formation of the stem-borne roots of monilophytes and it may be that the central and dominant primary root is a result of **heterochrony** with the root shifting to a more central position as a result of earlier initiation, followed by suppression of subsequent root production. In ferns, in contrast, roots are produced laterally from the stem, each one in association with a leaf (Hou & Hill, 2002, 2004; Fig. 3.2).

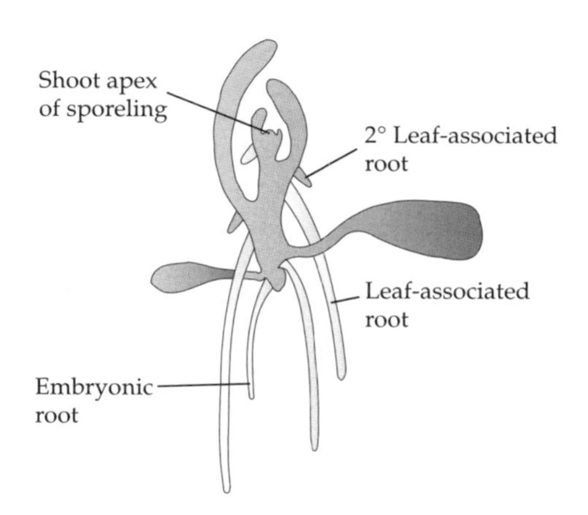

Fig. 3.2. Organization of roots in the fern. Here roots are lateral structures. The first is produced in the embryo, the others associated with leaves (below the leaf) (after Hou & Hill, 2002).

It would be very useful to know the genes responsible for root development in extant lycophytes versus extant euphyllophytes (both monilophytes and seed plants). This would allow assessment of whether there is molecular support for the homology of extant roots. However, it is always possible that orthologous genes have been independently recruited in the evolution of roots in both lycophytes and other vascular plants.

3.4 Differences between root and stem

What are the fundamental differences between root and stem? Many of the obvious differences are only superficial. Roots are generally brown, yellow or white rather than green, but some roots will become green on exposure to light. Roots are generally positively geotropic, in contrast to the negative geotropism of the shoot. However, roots are perfectly capable of growing upward in response to nutrients or oxygen, as in the example of **pneumatophores**. **Root hairs** are only found on roots, as outgrowths of the **root epidermis** or **rhizodermis**. However, these bear many similarities to the **trichomes** of stems. Indeed, the fact that some genes, including *GLABRA2*, affect both trichomes and root hairs suggests a fundamental connection.

The main and most striking difference, then, is in the internal vascular anatomy and in the root apex. The vasculature of roots is usually formed from a single central **procambial strand** which is surrounded by a relatively massive **cortex**, contrasting with young stems (at least in euphyllophytes), which have a ring of procambial strands around the periphery. This very different plumbing has to crossover and realign at the junction of root and stem, called the **cross zone**.

Two rings of cells bound the cortex. These are the **exodermis** on the outside and the **endodermis** on the inside. The endodermis surrounds the vascular system at the center of the root. These cells isolate the cell wall fluid (**apoplast**) of the inner vascular tissue from the outer cortical apoplast. This is achieved by means of a waterproof strip in the cell walls (the **casparian strip**) in all cells of the endodermis. Immediately inside the endodermis lies a tissue called the **pericycle** that surrounds the central vasculature. The pericycle is important as it is the tissue that forms new lateral roots and root buds (adventitious shoots).

A consequence of this centralized vascular system is that **lateral roots** must arise in the center of the root; root branching is said to be **endogenous** (i.e. arising on the inside). New lateral roots thus originate as meristematic divisions in the pericycle deep within the root. The newly formed lateral root then pushes (and possibly digests) its way out through the cortex.

The **root apical meristem** (**RAM**) is somewhat differently organized than the shoot apical meristem (SAM). The most striking difference is that most roots have root cap initials, a set of meristematic cells that are dedicated to the production of a **root cap**. The cap and its initials are not involved in forming the body of the root but (in eudicots) the inner layer of the root cap commonly forms the root epidermis (Clowes, 2000). Other cells of the root cap, the **root border cells**, are shed and they contribute to conditioning

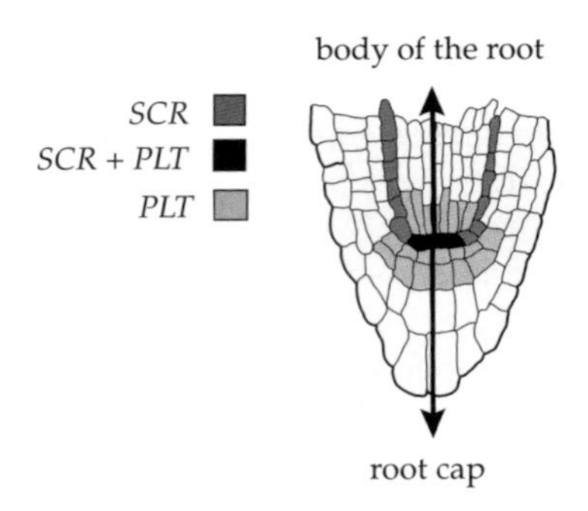

Fig. 3.3. Organization of the root apical meristem (RAM) and the patterns of gene expression associated with the quiescent center. The quiescent centre (QC) occurs at the overlap of cells expressing *SCARECROW* (*SCR*), i.e. the endodermal cell file, and cells expressing *PLETHORA* (*PLT*), i.e. the stem cell population (after Vernoux & Benfey, 2005).

the microenvironment of the root (Hawes et al., 2000). The root cap is also the **gravitropic organ** of the root (Darwin & Darwin, 1880); gravitropy is the ability derived from the sensing of the sedimentation of amyloplasts in the central or **columella** cells of the root cap (Chen, Rosen & Masson, 1999). Finally, it functions to produce **mucilage**, which may assist in lubricating the passage of the root tip through the soil. The root cap and its initials form an important second system of the RAM, one that does not have an obvious counterpart in the SAM.

3.5 The radicle and its origin in the embryo

The radicle is the root portion of the embryo of seed plants. In the seed it usually points toward the micropyle and is the first part of the seedling to emerge on germination. It is also called the primary root, as it forms at an early embryo stage, between the suspensor and the stem, on the same axis as the rest of the embryo. Secondary roots, and all roots of monilophytes, form laterally on stems.

Because of the interesting position of formation and the very early initiation, it is of interest to consider how the radicle meristem is specified in the embryo. In gymnosperms, the early divisions of the zygote result in a mass of nuclei without cell walls being laid down at first. In the angiosperm *Degeneria*, cell walls are laid down but cell divisions are not very orderly and result in a mass of cells. In most flowering plants, however, the first cell divisions of the zygote are very orderly. The first division leads to an apical cell and a basal cell, the subsequent divisions of which follow one of several main patterns. Typically the further divisions of the apical cell give rise to the hypocotyl, cotyledons and SAM while the basal cell gives rise to the suspensor and radicle. However, there is variation. In *Lactuca* the apical cell divides less than the basal cell, and only gives rise to the cotyledons and SAM. In *Sagina*, by contrast, the apical cell gives rise to the whole embryo and most of the suspensor, while the basal cell does not divide at all. This evolutionary plasticity is suggestive that cell lineage is not very important in specifying cell fate within the embryo. Instead, cell position at critical points during development are more likely to be important in determining cell fate.

However, in all seed plants, a **unipolar axis** is soon established within the embryo. This specifies an apex (where the SAM will develop) and a base (the suspensor). Later this changes to a **bipolar axis** when the root apex is established between the shoot apex and the suspensor. In *Arabidopsis* the unipolarity is established as early as the two-celled stage. The cause of this is polar auxin flux from basal to apical cell, associated with the localization of a specific auxin efflux component, PIN7, at the apical end of the basal cell (Friml et al., 2003; Jenik & Barton, 2005). Apical-basal unipolarity is characteristic of all embryos and thus the polar auxin flux that appears to maintain it is likely to be ancient in origin. What is unique to seed plants, however, is the subsequent establishment of a second pole, the root pole, and hence the conversion of unipolar axis to a bipolar one (see also Chapter 5, Section 19; Fig. 5.23).

Remarkably, the second pole appears to be established via the same mechanism as the first pole. A developmentally regulated reversal of PIN7 occurs, which flips the auxin gradient at the base of the embryo, thus establishing the root pole (Friml et al., 2003). In *Arabidopsis* this auxin flip happens at the 32-cell stage of the embryo and the main target of the reversed auxin flux appears to be the **hypophysis** (**hypophyseal cell**), which is the basalmost cell of the embryo at the very top of the suspensor. The hypophyseal cell will later become part of the root meristem (specifically the quiescent centre and the columella root cap). At the same time as PIN7 locates to the basal ends of the suspensor cells, another PIN protein, PIN1, locates to the basal ends of the central provascular cells, helping to drive the basalward auxin flux. As the bipolar embryo is a synapomorphy of the seed plants, we may speculate that the auxin flip is approximately coeval with the origin of this clade.

The end result of the auxin accumulation in the hypophysis is the activation of the *PLETHORA* genes (*PLT1–4* in *Arabidopsis*). The *PLETHORA* genes (AP2-domain transcription factors) not only promote the correct division of the hypophysis to form the RAM, but are also critical for the maintenance of the stem cell population in the RAM, by specifying the **quiescent center** (QC) with its important organizing role (Aida et al., 2004; Aida, Beis & Scheres, 2004; Fig. 3.3). They work additively in a dose-dependent manner (Galinha et al., 2007). *PLETHORA* genes are activated by auxin and in turn promote accumulation of PIN proteins thereby establishing a positive feedback loop to maintain the auxin flux. This is important, as the continued functioning of the RAM is critically dependent on a permanent presence of relatively high auxin levels.

Given their importance in the functioning of the RAM it is interesting that *PLETHORA* genes also act as root identity genes, and, if expressed in the shoot, they lead to the expression of root-like characteristics (Aida et al., 2004). Thus they are key genes in generating the distinct root phenotype. In this, they appear to be the root counterparts of the *WUSCHEL* genes expressed in the SAM (Sablowski, 2004). Like *PLETHORA*, *WUSCHEL (WUS)* is an important identity gene (but for stems) and is also important in maintaining the stem cell population by specifying the **organizing center** (Mayer et al., 1998), which can perhaps be considered the SAM counterpart of the QC. If *WUSCHEL* is expressed in the root, it confers stem-like characteristics (Gallois et al., 2004). However, the two types of genes, although having analogous effects, are not related. *WOX* genes (*WUSCHEL*-like homeobox genes) encode a specific type of homeodomain protein while *PLETHORA* encodes an *APETALA2*-like (*AP2*-like) transcription factor.

3.6 The root apex and root cap

On comparing the longitudinal sections of shoot and root apices, the most obvious differences are, at least in angiosperm roots, the prominent root cap and absence of leaf primordia, together with the apparent great regularity of the root apex with orderly files of cells radiating out from behind the apex. Most roots have a **root cap** of parenchymatous cells protecting the **root apex** proper. The cells of the root cap are continually destroyed as the root pushes through the soil but are replenished by meristematic division at the base of the root cap. In the center of the root cap is a block of cell files called the **columella** that results from these meristematic divisions.

The root apex under the root cap consists of a layer of cells overlying the end of the **stele**. It is in this cortical layer, just underneath the root cap, that the **RAM** resides, and this is the meristem responsible for the bulk of root growth. Typically, new cells are produced in files by rapidly dividing cells around a central **QC** where few divisions take place. As the files of cells lengthen in response to cell division at the apex the root is pushed forward into the soil, accelerated by the fact that the cells grow further by elongation in an **elongation zone** a short distance behind the apex.

Similarities in the basic structure of the root apex can be noted between lineages that diverged between 300 and 400 million years ago, so the root is fundamentally a conservative organ. However, in other respects the structure of the root apex varies considerably, even between quite closely related lineages. In angiosperms the main differences are (1) in the contribution that the **cortex** makes to the root cap and (2) in the contribution the root cap makes to the **rhizodermis** (Clowes, 2000).

In most eudicots the cell files of the root trace to clonally distinct tiers of initials. Such root apices are called **closed meristems** as there is a boundary between the root cap and the rest of the root meristem. However, in other plants (particularly early diverging angiosperms and monocots), **open meristems** may occur. In the open meristem the cell files of the root trace to initials but not to clonally distinct tiers of initials. Instead the central initials form a relatively disorganized mass. Open meristems appear to be primitive and closed meristems advanced (Baum, Dubrovsky & Rost, 2002; Groot et al., 2004; Jiang & Feldman, 2005). However, open meristems do occur in the eudicots, apparently as a derived state. In these it is common for the central initials to expand into the root cap forming a **secondary columella** and blurring the division between the RAM and the root cap (Clowes, 1981).

In most dicots the root cap forms the root epidermis (rhizodermis). However, in monocots the cortex makes the rhizodermis, and many "monoaperturate" dicots have a mixed origin of the rhizodermis. There is a third way of forming the root epidermis, which is rare but found in some waterplants, such as *Lemna*, *Pistia* and *Hydrocharis*. In these plants, the rhizodermis is produced by a separate meristem (Clowes, 1990). Root apices can therefore be thought of as a complex of two or three meristem zones, a root cap meristem and a root apex meristem (and possibly a separate epidermal meristem). However, apart from its possible contribution to the outer layers, the root cap contributes little to the bulk of the root. It is the RAM therefore that is the important creator of the root.

3.7 The RAM and its control

The cortex at the end of the root consists of two cell regions, the meristematic zone where cell divisions occur rapidly and the QC in which the time between cell divisions is some 15 times longer. Laser ablation experiments have shown that the QC is the source of non-cell-autonomous signals that maintain the stem cells in an undifferentiated state (vandenBerg et al., 1997), thus acting as an organizing center for the meristem. In *Arabidopsis*, two GRAS transcription factor genes are vital for the specification and maintenance of the QC and these are *SHORTROOT* (*SHR*) and *SCARECROW* (*SCR*; Nakajima et al., 2001; Sabatini et al., 2003). SHR protein is produced in the stele where it functions as a cell differentiation factor, moving into the endodermis to act as a key regulator in endodermis differentiation, being trapped in the endodermis by protein–protein interaction with SCR (Cui et al., 2007). The SHR–SCR interaction activates more SCR, ensuring that no SHR remains uncaptured (and so the endodermis is tightly limited to a single cell layer). SHR activation of *SCR*, is also important as SCR is required, in combination with *PLETHORA*, to specify and maintain the QC (Benfey, 2005; Birnbaum & Benfey, 2004).

PLETHORA genes are the auxin-regulated root stem cell identity genes (Aida et al., 2004). Where *PLT* is expressed alone, it specifies the active meristematic cell field, but where it is expressed together with SCR, it specifies the QC (Fig. 3.3). Thus auxin, PLT, SCR and SHR act together as a network to specify and maintain the essential structure of the RAM. This network can be extended to include the five PINHEAD-like (*PIN*) genes that act to focus the auxin maximum and therefore PLT expression (Blilou et al., 2005). The *PIN* genes are in turn regulated by PLT creating a highly stable positive feedback loop.

The role of *WUSCHEL*-like (*WOX*) genes in the root is less well known at present. In the SAM, *WUSCHEL* (*WUS*) is a very important regulator. It is expressed in a group of less actively dividing cells that can be regarded as the SAM counterpart of the QC. *WUS* nonautonomously maintains the actively dividing stem population above this region and causes them to express *CLAVATA3* (*CLV3*), which then negatively regulates *WUS* forming a stable control loop for the SAM. In *Arabidopsis* the QC of the embryonic root is known to express a *WUSCHEL*-like gene, *WOX5*, while in rice a *WOX* gene, Os*QUIESCENT-CENTER-SPECIFIC HOMEOBOX* (*QHB*), is expressed in the QC of mature roots (Haecker et al., 2004; Kamiya et al., 2003). Root expressed CLV3 homologues are also known (Casamitjana-Martinez et al., 2003; Hobe et al., 2003). However, the phenotypes of plants mutant for these genes are either mild or not comparable to those of the shoot counterparts. This perhaps indicates that the systems have diverged considerably.

Not only are RAM and SAM governed by different genes, but the functions of these gene sets may also no longer be directly comparable. This implies that the gene systems in the SAM and RAM have been strongly decoupled over evolutionary time. This is perhaps because they diverged at least 300 million years ago, before the complexity characteristic of the angiosperm meristem had evolved. To the extent that they are in the future shown to share gene networks, these will probably be those for ancient and ancestral developmental systems.

3.8 Evolution of the root apex

The three major groups of plants that possess roots (viz. the Lycophytes, the Monilophytes and the seed plants) all have rather different root apices, which raises the question of how the root apex has evolved. In the lycophytes there is a lamentable lack of comparative data on the root apex. A **tetrahedral apical cell** has been reported in some species whereas in others it seems that a prominent apical cell is not evident (Barlow, 1995). In the monilophytes the root apex is always organized around a single apical cell, which is usually tetrahedral, but in the Marattiales and Osmundales the apical cell may be a higher-order polyhedron, generally with five or six faces. A tetrahedral cell cuts off cells in three ranks from its three inwardly facing (**basiscopic**) planes. A more complex polyhedron can cut off cells in more directions and can potentially form a larger and more complex root. All lycophytes and momilophytes thus have a root apex with a single **structural initial** (apical cell).

In the seed plants, the situation is very different (Barlow, 1995). There are almost always multiple structural initials and these structural initials are higher-order polyhedra (usually with eight faces). Furthermore, when the structural initials are very numerous (as in angiosperms and some gymnosperms), these initials are divided into functional initials (rapidly dividing) and quiescent initials (slowly dividing). The functional initials surround the quiescent initials, which form a **QC**. In some gymnosperms, however, particularly the Cycadales and Gingkoales, there are rather few initials and no obvious QC.

3.9 Rhizodermis and root hairs

As we have seen in the previous section, in most plants a specific population of cells in the root apex, either in the root cap or in the cortex, divide to produce the **root epidermis** (**rhizodermis**) as the root grows. **Root hairs** are a characteristic feature of roots. They consist of a single epidermal cell that produces a remarkable tube-like protruberance that may extend through the soil for typically about 1 mm. Root hairs increase the surface area of young roots massively to promote water and nutrient gathering. When seedlings wilt after being transplanted it is probably because the root hairs have been inadvertently destroyed. When the roots have grown sufficiently to produce a new crop of root hairs the plants will recover. Root hairs are not formed at the growing tip of roots as this is being actively pushed through soil by cell division and elongation, a process that would rip off any root hairs. Root hairs therefore form some distance behind the growing tip where the cells are fully elongated.

Root hairs differ widely in their prevalence on plants, they can be very numerous, for instance as dense as 200 mm^2 in pea (*Pisum*). On the other hand they are completely absent in many water plants such as *Azolla* and *Lemna*, and in plants with specialized nutrition such as the parasite, *Cuscuta*, in which adventitious roots form **haustoria**. There is a compensatory relationship between root hairs and mycorrhizae. Most plants are capable of forming both, but in plants that rarely form mycorrhizae, like the Brassicaceae, root hair production is usually particularly well developed. However, there are numerous plants, such as the coconut (*Cocos nucifera*), which have no root hairs at all, and they instead rely on mycorrhizal fungi for nutrient absorption. Root hairs are generally absent, too, from older roots as they, and indeed all the epidermal cells, are fragile. The root epidermis is therefore, after a while, completely destroyed. The outer surface of older roots then consists of the cutinized outer layer of the **cortical parenchyma**: a layer known as the **exodermis**.

In most angiosperms root hairs arise from epidermal cells in no obvious pattern, with some proportion of epidermal cells entering the hair fate pathway. In some plants every epidermal cell forms a root hair, as in *Tropaeolum*, while in others, none do. In such plants the assumption of a hair fate seems to occur relatively late in development and the first sign is a small bulge on the surface of the epidermal cell. However, in some angiosperms, including *Arabidopsis*, root hairs form in very regular patterns, with a regular alternation of root **hair cells** (H) and **nonhair cells** (**nonpiliferous cells**, N; Clowes, 2000). These patterns may result from a vertical asymmetric division into two daughter cells with opposing fates: one daughter (a short cell called the trichoblast) becomes the root hair, and the other daughter (a long cell called the atrichoblast) becomes a nonhair cell. The direction of asymmetry of the division varies. In water lilies and many aquatic monocots, the daughter furthest from the root apex becomes the trichoblast, while in many grasses the daughter nearest to the root apex becomes the trichoblast.

In most plants belonging to the Brassicales and the Caryophyllales, a very different type of patterning occurs. This does not result from an asymmetric division but instead from positional signaling by the cortical cells below the epidermis. Cell files directly on top of cortical cell files become atrichoblasts whereas cell files that straddle two underlying cortical cell files become trichoblasts. This results in a striking pattern of stripes of root hairs on the root (Dolan & Costa, 2001). The best predictor appears to be whether the epidermal cell makes contact with an anticlinal cortical cell wall (i.e. a cell wall on the radius of the root) or just a periclinal (outer) cell wall. If it contacts an anticlinal cell wall, it assumes a hair-cell fate.

In *Arabidopsis* a very beautiful molecular switch has been demonstrated that is able to toggle between the hair-cell (H) and nonhair cell (N) fates (Fig. 3.4). The *GLABRA2* (*GL2*) gene is a transcriptional repressor that specifies N-fate

(Masucci et al., 1996). It does this by repressing genes like *phospholipase D-zeta* that are required for hair development (Ohashi et al., 2003). GL2 production is blocked in H-cells by the action of CAPRICE (CPC) protein (Wada et al., 2002). Remarkably, CPC protein is made in N-cells, but is then exported to the neighboring H-cells to promote H-fate (Kurata et al., 2005). Thus this small-scale cell patterning is achieved by means of an intriguing lateral feedback mechanism involving the intercellular movement of protein (Fig. 3.5).

Numerous genes are involved in the patterning process. For instance, it appears that CPC protein blocks GL2 transcription by inactivating GLABRA3 protein, a key component in the transcriptional complex that activates transcription of GL2. GLABRA3 is also subject to lateral transfer as it is made in the H-cells and moves to the N-cells where, in the absence of CPC, it can form a transcriptional complex with WEREWOLF (WER) and ENHANCER OF GLABRA3 (EGL3) to promote the transcription of GL2 (Bernhardt et al., 2003, 2005). WER appears to be a key controller as it regulates transcription of CPC, thus controlling both H-fate and N-fate (Ryu et al., 2005).

What is not yet known at the time of writing is the nature of the positional signal (presumably originating from underlying cortical cells), which starts the process. However, the gene for its putative cell surface receptor has been found, called *SCRAMBLED* (Kwak, Shen & Schiefelbein, 2005). *Scrambled* mutants have a random distribution of hairs. However, the SCRAMBLED (SCM) receptor is generally distributed so cannot, itself, be the source of the positional information. It is thus likely that the positional information comes instead from a highly localized signal molecule.

Root hairs bear some similarity to the **hairs (trichomes)** of stems and leaves. However, whereas trichomes may be multicellular, the root hair is merely an outgrowth of the epidermal cell itself, so there is never any cell division involved in its formation. Root hair formation and trichome development appear to share a common mechanism, but with intriguing differences (Schiefelbein, 2003). Thus GL2, the transcriptional repressor so important in blocking hairs in the root, is also the same gene that produces hairs in the shoot (Ohashi et al., 2002). In the root of *Arabidopsis*, hairs can perhaps be thought of as the "default state" and have to be repressed to generate hairless cells (when this repression fails, as in *Tropaeolum*, every epidermal cell develops a hair). In the shoot, however, "hairlessness" is the default, and seems to require repression in order to allow hairs to form.

Most of the elements of the trichome pathway and the root hair patterning pathway are identical.

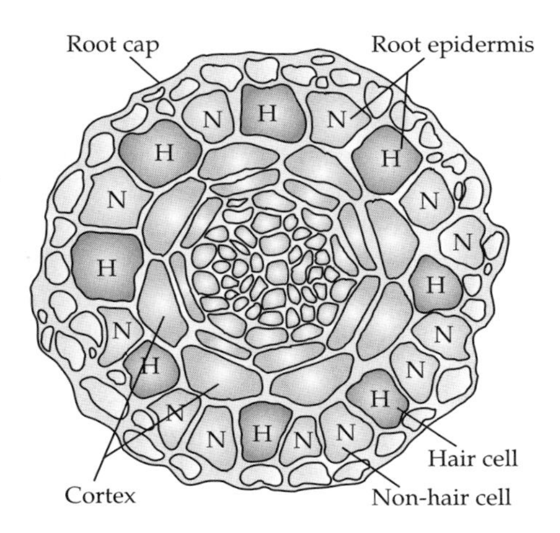

Fig. 3.4. Patterning of the root hairs in alternating rhizodermal cell files. Hair cells are those epidermal cells that contact two cortical cells (after Dolan, 2005).

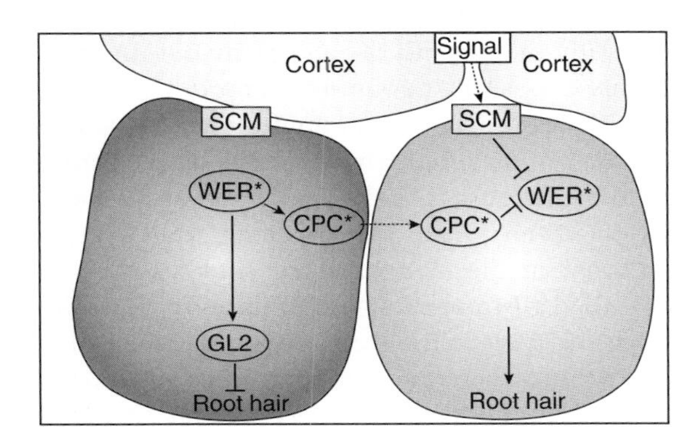

Fig. 3.5. Some of the genes involved in cortical/epidermal signaling in the development of root hairs from H-cells (right) and blocking of development in N-cells (left). See text for explanation (after Schiefelbein, 2003).

However, one gene component has differentiated (although not by much) between roots and shoots. Thus *WEREWOLF* (noted above as a key regulator of root hair formation) is involved in root hair development, but its counterpart in trichome formation is another gene called *GLABRA1* (*GL1*; Larkin et al., 1993). However, these genes are not only closely related (they are both R2R3–MYBs), but they also encode proteins that are functionally identical, as judged by promoter swap experiments (Lee & Schiefelbein, 2001). Only the regulatory regions of these genes are functionally different, causing *WER* to be expressed in roots and GL1 in shoots.

The existence of a common mechanism for epidermal patterning in roots and shoots is truly remarkable given that roots and shoots diverged such a long time ago. As axes ancestral to roots and shoots did not have trichomes or root hairs, roots and shoots have clearly recruited from elsewhere the same epidermal patterning module more recently. It is likely that this patterning module pre-existed for stomatal patterning (as stomates are early evolved features of primitive land plant axes).

3.10 Primary and secondary root systems

The **primary root** is a feature of seed plants only. It is that root that develops from the root meristem present in the embryo. This root meristem forms between the suspensor and the hypocotyl centrally in line with the meristem axis. In some plants, this is long persistent, and produces many orders of branches. In annual plants it is usual for the primary root (and the roots derived from it) to be the only root system. However, in many plants, even when the **primary root system** is considerable, a **secondary root system** (of so-called **fibrous roots**) often develops laterally from the base of the stem. Typically, this forms a mat of surface roots without much secondary thickening. These **cauline roots** are often called **adventitious roots** but this term should be used carefully as it might be taken to imply that there is something unusual about these roots. On the contrary, the development of roots from stem is a very normal feature of development in plants such as monilophytes, which have no primary root; they are the only means of root development. In monilophytes, the roots are formed from the unbranched vertical rhizome (e.g. *Dryopteris*) or the branched horizontal rhizome (e.g. *Equisetum*) and in these cases adventitious root production bears a strict relation to the distribution of leaf production (**phyllotaxis**) on the rhizome.

In angiosperms, it is common for the primary root to abort almost immediately and it is by necessity replaced by cauline roots from the base of the stem. The most common pattern is for a fibrous root system to be formed adventitiously from the very base of the stem at the same time the primary root aborts, usually immediately after germination. Palms (Arecaceae) are a good example of this. They have no primary root and the hemispherical base of the stem is covered with adventitious roots. In the coconut (*Cocos nucifera*) these are said to number up to 6000. *Streptocarpus* (Gesneriaceae), with primary root abortion just after germination, is another example of this. Cauline roots may be continually produced during the life of the plant. A regular relation between root production and the morphology of the stem is found in some rhizomatous monocots such as *Acorus calamus*. However, other geophytes, such as the rhizomatous *Polygonatum,* and bulbs, produce roots liberally over the rhizome or (in the case of bulbs) liberally over the flat stem-disc at the base of the bulb. The relationship between phyllotaxis and root production is sometimes mixed. In the ground orchid, *Orchis,* cauline roots are produced irregularly, but one particular root, the adventitious **root tuber**, is always produced from within a lateral bud of the stem. There may be an adaptive reason for this, producing a direct nutritional link between stem and storage root.

The widespread abortion of the primary root system is a curious and significant phenomenon in plants and, as yet, the physiological or molecular basis of it is not known. It would be interesting therefore to study gene expression in primary roots of related species that differed in the persistence of the primary root. One hypothesis is that genes that promote the formation of adventitious roots are antagonistic to the maintenance of the primary root or vice versa.

A good model system for the study of adventitious and primary root growth is provided by the grasses. They show strong primary root growth, which penetrates the soil typically 30 cm in oats and 60 cm in wheat. However, many grasses have a well-developed, but secondarily produced, root system originating at the root–shoot junction, the buds of which are formed in embryo (typically three buds in oats and five in wheat). Furthermore, grasses show a third type of root, **crown roots**, which arise from nodes at the base of the stem. Very many crown roots can be formed at each node and these penetrate the soil often to form an extensive additional root system. They are particularly noticeable at the base of maize plants where they look like miniature stilt roots. Roots produce large amounts of mucilage and the crown roots of maize are one of the few places where **root mucilage** can be conveniently collected above the soil.

3.11 Patterns of root branching

The branching of roots is an interesting character, both ecologically and morphologically. The **aerial roots** of the Orchidaceae and the roots of duckweeds (*Lemna*) are generally unbranched. However, most other roots are variously branched with the branching under a combination of genetic and environmental control. **Stilt roots** are generally unbranched when aerial, but branch when they reach the ground. When roots branch they do not occur at random on the root surface but may show distinct **rhizotaxy** (the root equivalent of **phyllotaxy**). This is because of the underlying architecture of the root. Lateral roots arise **endogenously** in the **pericycle** but tend to arise more commonly in that part of the pericycle nearest to xylem cells. Roots of dicots commonly have four xylem arms, so roots can frequently be seen to have four equivalent rows of rootlets, each associated with a **xylem bundle**. This is called the **isostichous** arrangement (Guédès, 1979). When the root has only two xylem arms, rootlets may initiate on each side of each xylem bundle. There are still four rows, but as they are in two pairs, this arrangement is called **diplostichous**.

Another pattern of the root system relates to morphological differentiation. If the main and lateral roots do not differ in any major way, the root system is referred to as **homomorphic**. However, if roots are strongly differentiated the root system is referred to as **heteromorphic**. Examples of strongly heteromorphic root systems are found in carrot (*Daucus carota*) and salsify (*Tragopogon porrifolius*), both of which have a swollen **tap-root** (i.e. a dominant vertical primary root) but unswollen lateral roots. In carrot the basal part of the tap-root is swollen but the terminal part remains thin, whereas in salsify the entire tap-root is swollen and only the lateral roots are thin. Pronounced root heteromorphy is found in species with **contractile roots**. Contractile roots are a solution to the problem that seeds of many bulbous or cormous plants germinate near the soil surface but subsequently reposition the plant body at depths of 15 cms or so. Contractile roots are often swollen, so they are effective at making a channel through the soil through which the stem (as bulb or corm) can be pulled (Putz, Huning & Froebe, 1995). However, the main mechanism for movement is a shortening of the root due to a collapse of the cortical cells in bands (Ruzin, 1979; Smith, 1930).

The architecture of the root system as a whole, called **radication**, is influenced not only by the pattern of branching but also how vigorous the branches are when they arise. Both these features are strongly influenced by the soil environment. Roots of most plants respond to a lack of soil phosphate by suppressing growth of the main root and producing numerous lateral branches. Phosphate is not mobile in the soil, and so roots have to "forage" for this ion. More mobile ions, such as nitrate, will tend to diffuse to the root or be washed to the root by soil water. For these ions, a root architectural response is less important (Fitter et al., 2002).

Modification of root architecture by phosphate shortage requires several stages. First the phosphate status must be sensed (Ticconi & Abel, 2004), and then this must lead to the start of cell division in the pericycle through an auxin-dependent upregulation of cell cycle genes. Auxin pathway genes, such as *SOLITARY ROOT/IAA14*, are thus key regulators of the activation of cell division (Fukaki et al., 2002). However, cell proliferation in the pericycle is not in itself enough for lateral root formation, and cell fate of the pericycle cells must also be respecified to a lateral root fate (Vanneste et al., 2005).

Auxin has been known to promote lateral branching in the angiosperm root for some time (Torrey, 1950), and it is tempting to conclude that this is true for nonangiosperms too. However, recent experiments with the fern *Ceratopteris* appear to indicate that fern lateral root formation is auxin independent (Hou, Hill & Blancaflor, 2004). Wider comparative studies would be of considerable interest to determine the role if any of auxin in monilophyte root patterning.

An extreme form of phosphate scavenging by means of root branching is seen in the **cluster roots**. Cluster roots are dense formations of determinate short root branches, which exude carboxylates such as malate and citrate, as organic acids to increase the solubility of soil phosphate and other ions (Skene, 2000, 2001; Fig. 3.6). Phosphate is often unavailable in soil as it exists as insoluble compounds such as calcium phosphate, but it is released by the action of organic acids. Cluster roots were first observed in the family Proteaceae, and for this reason are often called **proteoid roots**. With the exception of the Proteaceae (an early diverging rosid family), plants forming cluster roots tend to be in the group of angiosperms in which bacterial nodulation occurs (the eurosid 1 clade). This association with nodulation (even though cluster roots are not themselves nodulated) may indicate that common gene pathways are involved. Cluster roots are known in the Betulaceae, Casuarinaceae, Eleagnaceae, Leguminosae, Moraceae, Myricaceae and Proteaceae. Recently *Lupinus albus* (white lupin) has emerged as a model organism for studying cluster roots. *Lupinus albus* cluster roots actively secrete citrate. The putative organic acid transporter gene has recently been identified (Uhde-Stone et al., 2005). This gene, Lupinus albus *MULTIDRUG AND TOXIN EFFLUX* (La*MATE*), is, as expected, strongly upregulated in response to phosphate deficiency (Uhde-Stone et al., 2003).

3.12 Secondary thickening of roots

The filamentous roots of grasses and those of many annual eudicots show minimal or no **secondary xylem**. However, in perennial herbaceous plants secondary thickening of the primary root by the **root cambium** can lead to large

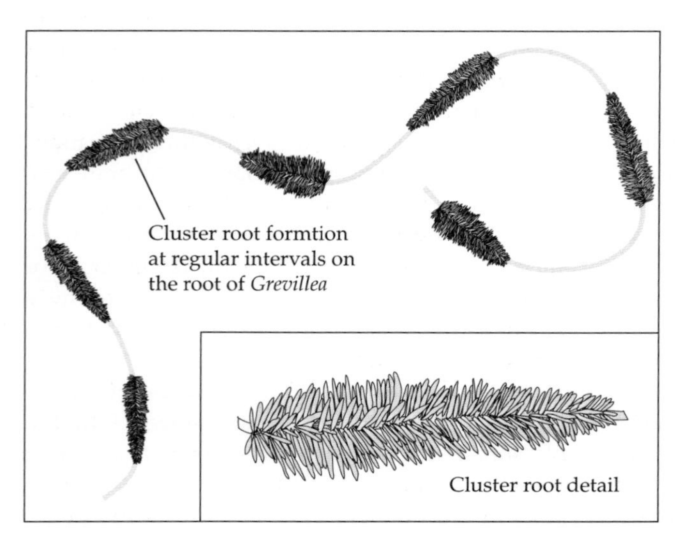

Fig. 3.6. Pattern of cluster root formation on the root system of *Grevillea* (after Skene, 2000).

tap-roots, either **fusiform** (spindle-shaped) as in carrot, or **napiform** as in the extremely swollen example of the turnip. Alternatively, lateral or adventitious roots can thicken to form specialized storage organs called **root tubers** by a proliferation of parenchymatous growth. Some root tubers have stem shoot buds on them, which use up the reserves when they develop. Other root tubers, such as *Dahlia*, have no stem buds of their own. Instead it is the buds situated on the same stem that produced the root tuber that use their reserves. Many important tropical crops are root tubers, including the sweet potato (*Ipomoea batatas*) and cassava (*Manihot utilissima*).

The roots of monocots, as well as some dicots with specialized roots such as **Daviesia** (Pate et al., 1989), have multiple root cambia and these species are said to have **anomalous secondary thickening**.

Tree roots may form a **root trunk** indistinguishable from stem trunk. An example of this is found in trees that germinate on tree stumps and the root grows down to the soil. When the tree stump rots away a normal looking tree is left and it is difficult, if not impossible, to tell that the base of the tree trunk is of root origin. This is due to the fact that the root cambium in light tends to produce stem-like wood. In contrast, roots underground tend to have wider vessel elements and so underground root wood is usually clearly distinguishable from trunk wood. Most of the secondary thickening of trees occurs near to the main trunk, and so root thickness tapers rapidly away from the trunk. Secondary thickening in this **zone of root taper** (**ZRT**) may be due in part to stimulation by mechanical stresses imposed by the canopy above, such wood being called **reaction wood**. An extreme form of root secondary thickening, that appears to have mechanical significance, is that of **root buttresses** resulting from extreme vertical secondary thickening of surface roots near the trunk.

Fine roots of trees are the source of much of the nutrient absorption within forest ecosystems as well as a sink for much of the forest's primary productivity. When first produced they have no secondary tissue, and as turnover of fine roots is high, many die before secondary growth commences, leading to a marked dichotomy (**heterorhizy**) of root diameters. It is those fine roots that are formed with a greater amount of **protoxylem** (the first formed xylem of newly formed organs) that are more likely to survive and secondarily thicken (Hishi & Takeda, 2005). However, when secondary thickening commences, a **cork cambium** (**phellogen**) forms a corky layer (**periderm**) on the root, which ends its function as an absorptive organ. Furthermore, the **epidermis** and **cortex** of the root are lost (and with them the ability to form mycorrhizal associations). In large, old roots, as in mature trunks, the root bark may be formed from expended **phloem** tissue rather than by the action of a cork cambium.

3.13 Evolutionary loss of roots

Mosses, liverworts and hornworts are entirely without roots and this appears to represent a primary condition rather than a secondary loss. In contrast, roots are almost universal in the vascular plants. Roots have, however, been secondarily lost sporadically across the vascular plants. In some ferns, leaves have usurped the function of roots. Thus roots have been lost in two epiphytic groups of filmy ferns (Hymenophyllaceae; Schneider, 2000). Filmy ferns have delicate leaves that are capable of absorbing water and nutrients. Furthermore, some rootless species have evolved novel structures such as root-like shoots and adhesive hairs. Roots have also been lost in ferns in consequence of an aquatic habit. In *Salvinia* the function of roots has been taken over by specialized submerged leaves. At each "node" of the shoot two floating leaves are produced along with an axillary bud and a submerged leaf (Lemon & Posluszny, 1997). The floating leaves are flat structures produced by a marginal meristem whereas the submerged leaves are highly dissected, functioning as both absorptive and stabilizing organs. *Psilotum* appears to be an example of a terrestrial rootless monilophyte. However, it has an underground organ that is difficult to classify as root or stem (Takiguchi, Imaichi & Kato, 1997), which may represent an anomalous, developmentally mixed structure.

In angiosperms there has also been secondary loss associated with the epiphytic and aquatic habits. Spanish moss (*Tillandsia usneoides* in the Bromeliaceae) is an example of an epiphyte that has lost roots. The root function is performed by the shoot, which absorbs water and nutrients from the air. *Tillandsia* is not completely rootless as there is a rudimentary radicle in the seed. However, this does not develop (Cecchi et al., 1996). *Ceratophyllum* is a genus of two species of water-plant, both of which have lost roots. The submerged shoot fulfils the absorptive function while modified branches in the substrate provide anchorage. Other rootless water plants include *Wolffia arrhiza* (Lemnaceae), *Utricularia* (Lentibulariaceae) and *Aldrovandra* (Droseraceae). *Aldrovandra* and *Utricularia* are both carnivorous plants and so have alternative means of mineral nutrition.

Loss of roots in angiosperms is caused by the failure of the embryonic radicle to develop, coupled with the absence of adventitious roots. Many monocots and some dicots display an early cessation of primary root growth and rootless plants may be seen as an exaggeration of this feature. However, in plants with early primary root abortion, secondary roots usually appear from the radicle/hypocotyl junction. In rootless plants this does not occur. There are numerous mutants known which suppress lateral root growth, including those of the auxin pathway discussed previously. It would therefore be interesting to functionally examine the genes of rootless plants that are homologous to those for lateral root outgrowth.

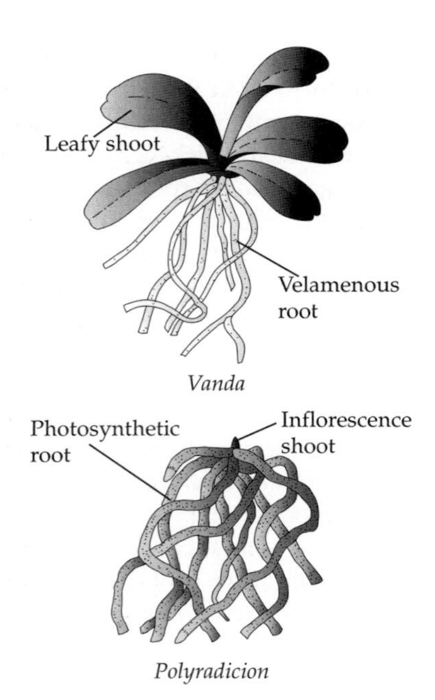

Fig. 3.7. Velamenous and photosynthetic roots of *Vanda* and *Polyradicion*, respectively (after Benzing et al., 1983).

3.14 Roots in light and air

Roots have frequently evolved into specialized structures that are often nutrition or water related. The **aerial roots** produced by orchids and aroids provide an example of roots with the function of atmospheric water collection. In this they are assisted by a **velamen**, a thick external layer of dead parenchymatous cells (developed from the **rhizodermis**). These cells form a microhabitat that is not only highly absorbent, but becomes a whole ecosystem, complete with fungi, bacteria, cyanobacteria and protists. Directly below the velamen is the **exodermis**. In most roots with an exodermis, the exodermal cells are suberized and serve to protect the root. However, in roots with a conspicuous velamen, some cells of the exodermis are thin-walled passage cells important for absorption. In addition to their absorptive function, some velaminous orchid roots are green and serve as important photosynthetic organs (as in *Taeniophyllum*, *Polyrhiza* and *Angraecum*), in which case they have adaptations for gas exchange. Gas exchange is effected through **pneumathodes**, which are groups of air-filled, as opposed to water-attracting, cells in the velamen (Benzing et al., 1983; Fig. 3.7).[2]

Such *green roots* may be strongly flattened and ribbon like, as in the *thallus-root* of the extraordinary aquatic plant family Podostemaceae. Green roots indicate that roots retain a capacity to become photosynthetic, and that absence of

[2] *Polyrhiza* (*Polyradicion*) *lindenii* is the famous ghost orchid of the Florida everglades. It has no leaves, subsisting entirely by means of its photosynthetic roots.

assimilatory ability is not a fundamental difference between roots and shoots. This is confirmed by the fact that many normal roots, such as those of the bog-bean (Menyanthes) will become green when exposed to light. It is curious that many roots (and perhaps all roots) possess an ability to respond to light by greening, as this can rarely be adaptively significant. Blue light is very effective in greening Arabidopsis roots (Usami et al., 2004). It would be interesting to survey systematically plant groups for the ability of roots to green in light but as far as I am aware no one has done this. It does, however, indicate that the root shares basic physiology with the stem, an organ that predates it in evolution. If the root evolved by modification of stems, as seems likely, it is not surprising to find vestigial stem-like behavior in the root. However, we would not expect to find any potentially root-like behavior in the stem.

The most remarkable of the green roots are the thallus-roots of the Podostemaceae. These crustaceous green thalli are formed from root tissue that assumes an assimilatory and anchoring role in the absence of strongly developed vegetative shoot. The evolution of the thallus involves dorsiventral flattening of the root and the concomitant loss of the root cap (Rutishauser, 1997). As the Podostemoids occur typically on rocks in very fast flowing rivers and waterfalls, the thallus functions as a holdfast, and is often disc-like, with adhesive "rhizoids." The roots may have a marginal meristem flanking the RAM, thus allowing for lateral growth (Koi & Kato, 2003). Remarkably, when vegetative and flowering stems form, they form endogenously on this thallus.

Roots may also be important for gas exchange. In waterlogged soil conditions, roots growing vertically into the air allow oxygen to diffuse into the root. These are the **pneumatophores**, found best developed in mangroves, such as *Avicennia* and *Sonneratia*. Pneumatophores bear a spot called **pneumathodes**, composed of spongy tissue that allows air passage into the root. The **knee roots** of *Taxodium*, although different in detail, provide a gymnosperm example. The aquatic plant Ludwigia has an interesting adaptation for gas exchange. It has dimorphic roots consisting of first, negatively geotropic roots anchoring the plant in the substrate and second, positively geotropic roots that are gas filled in consequence of their being composed of aerenchyma tissue (parenchyma with a massive amount of intercellular space; Ellmore, 1981a). These roots may reach the water surface and act as a conduit for atmospheric oxygen to the nodes in the mud below (Ellmore, 1981b).

Another function of roots, although very rare, is protection from large herbivores. In such cases, adventitious roots are metamorphosed into thorns. These **root-thorns** are found in *Acanthorhiza* (Arecaceae) and *Myrmecodia* (Rubiaceae). In *Acanthorhiza* the thorns are branched and are produced at the base of the stem between leaf scars. A more common adaptation is the transformation of roots into organs of support, as discussed below.

3.15 Roots specialized for mechanical support

The root system in the soil provides an important mechanical function distributing the stresses and strains produced by the aboveground canopy as a result of its weight and movement in response to wind. However, certain roots are specialized to provide additional mechanical support. A good example of roots promoting mechanical stability is the **stilt root**. Stilt roots are adventitious roots arising often high on the plant body. They may initially function as aerial roots, and are generally unbranched initially, but on reaching the soil they branch (Greig & Mauseth, 1991). In tropical figs (*Ficus*) they often descend from branches vertically and on reaching the ground they form new trunks, supporting the limbs above them. By this means an individual tropical fig can cover large areas. In *Pandanus, Piper* and *Rhizophora* mangroves they exit the main stem obliquely forming **prop roots**. Prop roots are particularly important in arborescent monocots with limited secondary thickening of the primary stem. Thus in palms like *Socratea* and *Iriartea* they permit vertical growth earlier in development than would otherwise be possible (Schatz et al., 1985; Fig. 3.8).

Some stilt roots, as in certain Ficus species, grow intimately associated with the trunk of the tree thus forming root buttresses rather than prop roots. When mature they are virtually indistinguishable from stem buttresses. Normal root buttresses are formed by asymmetrical secondary thickening near the trunk by surface roots. These are extremely important for the mechanical support of large tropical trees. Their functional effectiveness is increased by the occurrence of sinker roots put down vertically at the ends of the buttress roots, so increasing anchorage strength (Crook, Ennos & Banks, 1997; Fig. 3.9).

Climbing plants frequently use roots for attachment to a support. Certain tropical plants climb by means of root-tendrils. Roots tendrils may act by twining along or around supports, or may act as root hooks, which catch onto other plants. Root tendrils are found in orchids such as *Vanilla*, Melastomataceae (*Medinilla, Dissochaeta*, etc.) and Loranthaceae (e.g. *Struthanthus* and *Phthirusa*). Root-tendrils have been around for a long time, apparently being the mechanism of climbing in the pteridosperm *Callistophyton*† (Krings, 2003; Krings et al., 2003).

Ivy (*Hedera*) produces **clinging roots** on its stems that adhere tightly to surfaces. The adhesion mechanism is also found in the specialized clinging roots of twig epiphytes such as the bulbophylline orchids. Similarly, the climbing fig, *Ficus pumila*, produces adventitious roots in pairs on young stems. The tips of these bear root hairs that stick together to form an adhesive pad. These root hair pads secrete a sticky mucilaginous substance that adheres to the substrate (Groot, Sweeney & Rost, 2003; Fig. 3.10).

Adventitious roots may also be modified into flotation devices. Floating roots are produced from the nodes of certain water plants such as *Jussiaea*. They support the plant in water by virtue of the mass of aerenchyma tissue developed in the roots, the air spaces of which are gas filled.

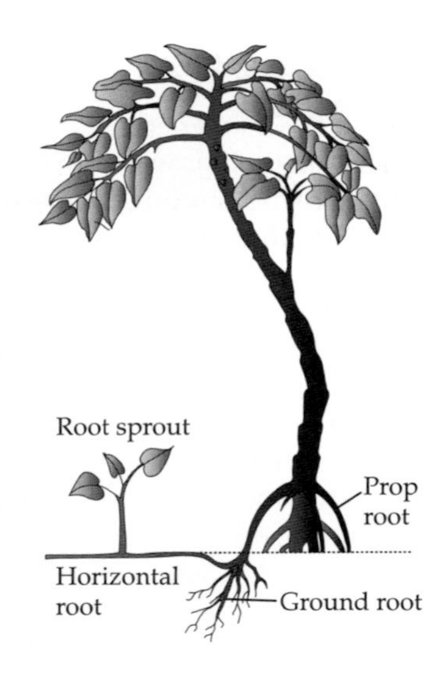

Fig. 3.8. Three types of root in *Piper* (after Greig & Mauseth, 1991).

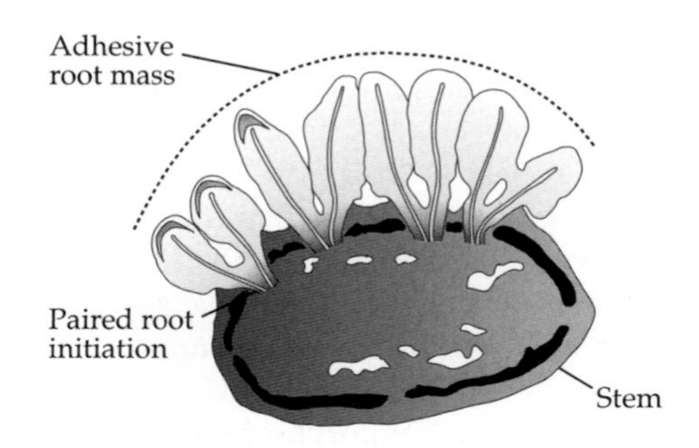

Fig. 3.10. Adventitious adhesive roots in *Ficus pumila* (after Groot et al., 2003).

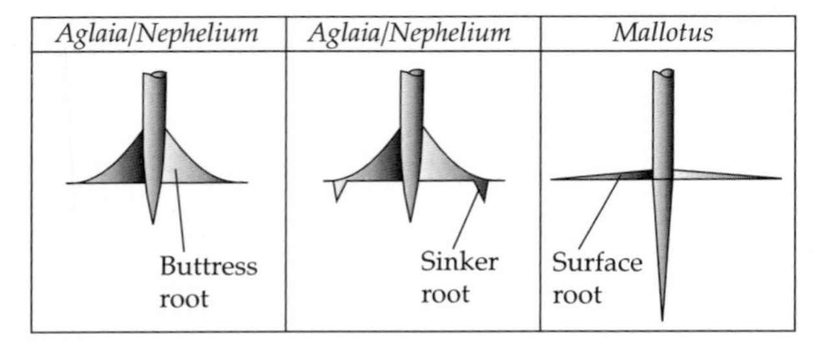

Fig. 3.9. Root buttresses and sinker roots in tropical trees (after Crook et al., 1997).

3.16 Root morphology and disease

Many root diseases are known to affect root morphology. Root-inhabiting nematodes such as the genus *Meloidogyne* produce **root knots**. The fungus *Plasmodiophora brassicae* produces root hypertrophy known as **club root**, while the virus Beet Necrotic Yellow Vein Virus (BNYVV) produces **rhizomania** (root proliferation) in sugar beet. However, the root disease that is best studied is the bacterial disease, **hairy root** caused by *Agrobacterium rhizogenes*. This is of particular developmental interest because *A. rhizogenes* acts as a natural genetic engineer to induce roots.

A. rhizogenes is a soil-dwelling bacterium which has a piece of extrachromosomal DNA called a root-inducing (Ri) plasmid. Upon infection, a portion of the Ri plasmid DNA is incorporated into the host cell genome (Chilton et al., 1982). This portion, the transferred DNA (T-DNA), carries as many as 18 open reading frames (ORFs), the expression of which promotes the formation of adventitious roots. These roots have the characteristic hairy root morphology. They are fast growing and have a reduced geotropic response and increased branching. Four of these ORFs, named *root locus* A-D (*rol*A-D) have the greatest effect on root morphology. In addition, an extra ORF, *orf13*, appears to promote meristem formation by inducing the expression of *KNOX* genes (Stieger et al., 2004). Hairy roots can be maintained indefinitely in culture and are thus a promising experimental system for the study of root function and development (Uhde-Stone et al., 2005).

3.17 Roots for specialized nutrition

3.17.1 Haustoria

Roots penetrate soil and are also the organ of choice for penetrating other plant tissue. **Root haustoria** are roots specialized to provide **parasitic** or **hemiparasitic** connections with other species. They may be relatively superficial organs or they may grow and branch within the host to form a significant **endophyte**. For instance, dodder (*Cuscuta*) is an entirely leafless parasite that produces haustorial lateral organs. These **intrusive organs** are considered to be homologous with **adventitious roots**. From the tip of the intrusive organ, structures reminiscent of root hairs develop, known as **search hyphae**. These may grow between the cells of the host, or even penetrate cells as intracellular structures (Kuijt, 1977).

Haustoria are very variable in structure. The remarkable tree mistletoe of Western Australia, *Nuytsia* (Loranthaceae) has a notable haustorial development. Two opposing folds of the parasite root cortex grow out. These grow round each side of the host root, completely encircling it like a pincer, and eventually forming a continuous collar around the host root. Within the collar a sickle-shaped sclerenchymatous cutting tissue develops. As the collar thickens it constricts the host root and drives this **haustorial sickle** through the stele of the host (Calladine & Pate, 2000; Calladine, Pate & Dixon, 2000). This severing allows close connection between host and parasite vascular systems. The sickle is so sharp that when removed from the root it can cut tissue paper (Calladine & Pate, 2000; Fig. 3.11).

The Orobanchaceae are a family of haustorial parasites that includes both obligate achlorophyllous parasites and facultative hemiparasites

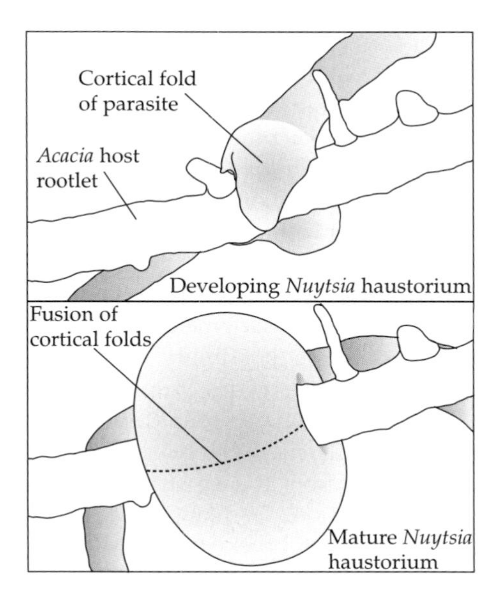

Fig. 3.11. Haustorium of *Nuytsia* (Santalaceae) (after Calladine & Pate, 2000).

that are capable of free-living existence. In obligate parasites like the witchweed, Striga, germination is stimulated by root hydroquinones and the tip of the radicle becomes a primary haustorium, ceasing growth until root tissues are penetrated. In contrast to the primary haustorium of obligate parasites, the haustoria of hemiparasites are all formed laterally and the root itself continues to grow.

The stages of haustorium development in hemiparasites such as owl clover (*Tryphysaria*) are as follows. First there is a cessation of root tip growth. Second, a dense ring of epidermal hairs (root hairs) forms just below the root tip (when mature these secrete an adhesive substance which sticks the haustorium to the host root). Third, cortical cells under the ring of hairs expand to form a bulge in the root. Next, meristematic divisions commence to form the intrusive organ and, finally, the root apex recommences growth. In *Triphysaria* this whole process takes about 12 h (Tomilov et al., 2005). The intrusive part of the haustorium penetrates to the vascular tissue of the host and forms a xylem bridge to the host vasculature (Heide-jorgensen & Kuijt, 1995).

It is small organic molecules from the host that trigger haustorium formation. These **haustorial induction factors** (**HIFs**) leach out of plant roots into the **rhizosphere**. In *Triphysaria*, the most potent HIFs are a quinone and an anthocyanidin (2,6-dimethoxy-*p*-benzoquinone and peonidin; Albrecht, Yoder & Phillips, 1999). This process is somewhat reminiscent of the triggering of nodulation genes in rhizobial bacteria by flavonoids from the legume host.

Parallels with micro-organisms in the manner of haustorial development and induction have suggested the possibility that haustoria evolved by capturing genes from fungi or bacteria as a result of lateral gene transfer (Atsatt, 1973). However, such genes have not yet been found and at present it seems more likely that regulatory mutations in plant genes have led to the alteration of preexisting developmental pathways to form novel structures in parasites. For instance, auxin is a key regulator of lateral root formation and it also appears to be involved in haustorium formation. In experiments with *Triphysaria*, exogenous application of auxin has been shown to promote haustorium formation and the expression of a reporter gene driven by an auxin-responsive promoter has been shown to increase during haustorial induction (Tomilov et al., 2005).

3.17.2 The mycorrhizal root

The plant root is fundamentally a symbiotic organ, termed a **mycorrhiza** or "fungus root." The fungus root can be considered a distinctive organ with unique morphological and functional characteristics. The intimate association between fungi and roots is likely to have been present since the evolutionary origin of the root. Early fossils of rootless organisms show indications of mycorrhizal associations in their rhizomes. This association is facultative, as roots are capable of functioning without mycorrhizae. Indeed some groups of plants, such as the Brassicaceae, appear to have lost the ability to form mycorrhizal associations entirely. Nevertheless, in most cases the root functions much more efficiently when fungal hyphae, with their vast extent and huge surface area, are co-opted to forage for mineral nutrients.

It is estimated that more than 80% of land plants have mycorrhizal roots (Newman & Reddell, 1987). It is fungi, therefore, that capture most of the phosphate required by green plants. Green plants need phosphate to make the enzyme, **ribulose bisphosphate carboxylase** (**rubisco**), the enzyme responsible for drawing down carbon dioxide from the atmosphere. In return for the flow of phosphate to the green plant, there is a counter-flow of fixed carbon from autotroph to mycorrhizal fungi, amounting annually to some 5 billion tons globally (Bago, Pfeffer & Shachar-Hill, 2000). However, this is only a small part of the approximately 120 billion tons of carbon fixed annually by the autotrophs.

There are two main types of mycorrhizae. The first type comprises those made with the *Glomeromycota*, known as arbuscular or VA mycorrhizae (AM or VAM). The second type comprises those made mainly with the *Basidiomycota*, called ectomycorrhizae (ECM). However, these are merely extremes and there is a big diversity of minor but distinctive types such as those found in the Ericaceae. Bizarrely, the Australian monocot *Thysanotus* (Anthericaceae) has a type all to itself, in which the fungal hyphae form a layer under the epidermal cells (McGee, 1988). In some conifers the mycorrhizae take the form of nodules (Furman, 1970), similar to the actinorhizal nodules of eurosid angiosperms. Interestingly, one eurosid angiosperm (*Gymnosperma* in the Casuarinaceae) has mycorrhizal nodules in addition to actinorhizal nodules (Duhoux et al., 2001; Fig. 3.12).

Successful mycorrhizal infection alters the morphology of the host root in interesting ways. In VAM, this is variable, but generally they become fatter, shorter and devoid of root hairs. In ECM the pattern is more striking, leading to a pronounced heteromorphic root system of nonmycorrhizal long roots and mycorrhizal short lateral roots, which may be dichotomously branched (as in the genus Pinus; Fig. 3.13). The short roots appear to grow very slowly, which may assist the formation of the mycorrhiza. Conversely, the faster-growing long roots usually form a periderm as they age, so preventing the formation of mycorrhizae.

ECMs develop as a result of hyphae recognizing and sticking to the surface of young roots. A hyphal sheath eventually forms around the root called the **mantle**. From the mantle hyphae penetrate between the cells of the epidermis and cortex where they ramify to form a finely branched and anastomosing structure called a **hartig net**. In contrast, VAM do not have a mantle but there is direct penetration of hyphae through slits that open between epidermal cells. Details vary depending on the fungal genus, but typically the fungus follows the following infection pattern. Once in the cortex, the hypha forms an **appressorium** on a cortical cell allowing it to enter the cell. Once in the cell the hypha is accommodated within a **perifungal membrane** that is continuous with the plasma membrane. By this means the fungus proceeds intracellularly to the inner cortex where it emerges into the intercellular space and may spread freely about the root. At intervals these hyphae then penetrate cells of the inner cortex and form branched intracellular structures called **arbuscules**. The hyphae may also form **vesicles** that function as storage sacs.

The establishment of mycorrhizal interactions involves complex signaling between the partners. The presence of the plant root increases branching in hyphae nearby by means of the diffusion of a **branching factor** from the root (Giovannetti et al., 1993). In turn, the presence of hyphae induces expression of early nodulation genes such as *EARLY NODULATION11* (*ENOD11*), evidence for a fungal signaling factor (Kosuta et al., 2003). Curiously, when the fungus is not in contact with the root, *ENOD11* expression is widespread in the root. However, when hyphae contact the root, *ENOD11* expression

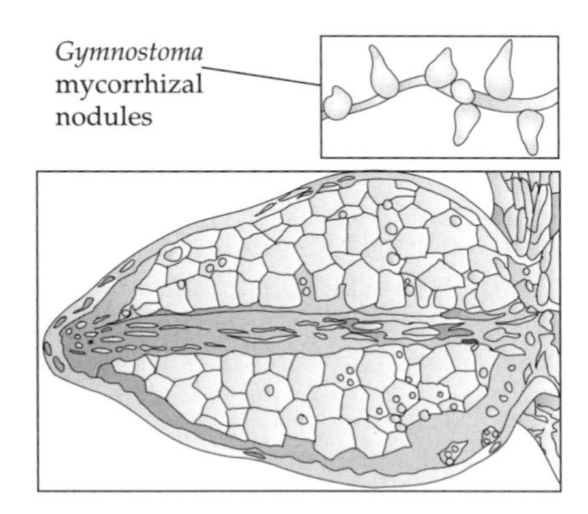

Fig. 3.12. Mycorrhizal nodules of *Gymnostoma* (Casuarinaceae) (after Duhoux et al., 2001).

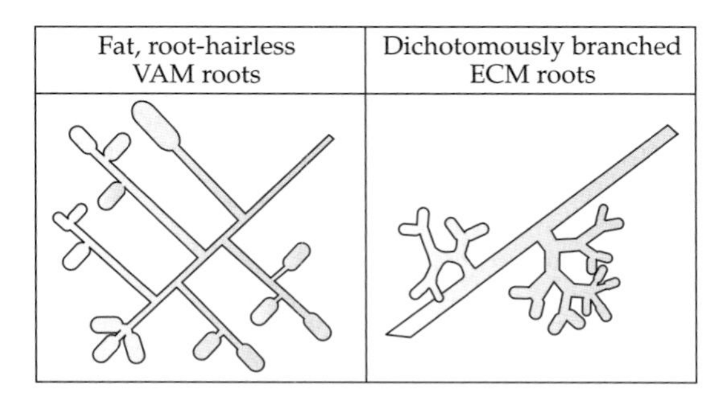

Fig. 3.13. Morphology of mycorrhizal roots.

becomes restricted to the infection positions only (Chabaud et al., 2002; Genre et al., 2005), implying that a negative regulatory mechanism is activated on contact to target gene expression more precisely.

Remarkable progress has been made recently in understanding how the fungal signal may be transduced through a signaling cascade within the plant. The progress results from work on nodulation in legumes as many of the nodulation defective mutants that have been detected are also VAM defectives (Parniske, 2004). This indicates that the mycorrhizal signaling pathway shares a common core with the nodulation pathway. As mycorrhizal infection is much older than nodulation, we must assume that when nodulation evolved (in the rosid one clade) it co-opted a large part of a preexistent molecular system for mycorrhizae. At the core of this common system is a proposed signal transduction cascade in which three key plant genes have been identified (see Box 3.1). These are the plant receptor-like kinase, *SYMBIOSIS RECEPTOR KINASE* (*SYMRK*; Endre et al., 2002; Stracke et al., 2002), the predicted ion-channel *DOES NOT MAKE INFECTIONS1* (*DMI1*; Ane et al., 2004), and a calcium and calmodulin dependent protein kinase *DOES NOT MAKE INFECTIONS3* (*DMI3*; Levy et al., 2004; Mitra et al., 2004). A likely hypothesis is that the fungal signal is sensed directly or indirectly by *SYMRK*, which activates the ion-channel *DMI1* causing a calcium spike, which is transduced by *DMI3*.

The genetic evidence from mutant analysis suggests that a large number of downstream control genes are activated at various points in the infection process. By this means, the ingress of the fungus into the root is tightly controlled by the plant, which marshals the growth of hyphae through the tissues with precision.

3.17.3 The mycoheterotrophy and the coralloid root

The standard mycorrhizal association is a mutualistic one. Some plants have taken this use of fungi to a logical extreme by becoming entirely **mycoheterotrophic** (Bidartondo, 2005). Mycoheterotrophic plants are typically without chlorophyll and derive all their carbon from fungi. In such cases, plants can be thought of as managing fungi in a somewhat analogous way as that by which attine ants farm fungi. Mycoheterotrophy is phylogenetically extremely widespread, including the liverwort *Cryptothallus* and the moss, *Buxbaumia*. Many monilophyte gametophytes are mycoheterotrophic including *Psilotum*, *Ophioglossum* and *Schizaea*.

In angiosperms mycoheterotrophy is also widespread, occurring notably in the families Burmanniaceae, Ericaceae, Gentianaceae, Orchidaceae and Triuridaceae. Mycoheterotrophy has

Box 3.1 Major genes of the early signaling cascade of legume mycorrhizal infection (Lj = *Lotus japonicus*; Mt = *Medicago truncatula*; Ps = *Pisum sativum*).

1. Receptor kinase. Probable location: plasma membrane. Genes: Lj*NOD-FACTOR RECEPTOR1* (*NFR1*), Lj*NOD-FACTOR RECEPTOR5* (*NFR5*); Mt*NOD-FACTOR PERCEPTION* (*NFP*); Ps*SYM10* (Amor et al., 2003; Madsen et al., 2003; Radutoiu et al., 2003)
2. Leucine-rich receptor kinase. Probable location: plasma membrane. Genes: Lj*SYMBIOSIS RECEPTOR-LIKE KINASE* (*SYMRK*); Mt*DOES NOT MAKE INFECTIONS2* (*DMI2*); Ps*SYM19* (Capoen et al., 2005; Endre et al., 2002; Stracke et al., 2002)
3. Cation channel. Probable location: plastid membrane. Genes: Lj*POLLUX*, Lj*CASTOR*; Mt*DOES NOT MAKE INFECTIONS1* (*DMI1*; Ane et al., 2004; Imaizumi-Anraku et al., 2005)
4. Calcium-calmodulin dependent kinase. Probable location: nuclear membrane. Gene: Mt*DOES NOT MAKE INFECTIONS3* (*DMI3*; Levy et al., 2004)
5. GRAS transcription factor. Probable location: nucleus. Gene: Mt*NODULATION SIGNALING PATHWAY1* (*NSP1*; Smit et al., 2005)

a major impact on the morphology of the root. Such plants often have **coralloid roots** with many short stubby branches and so resembling a lump of coral. An example is the mycoheterotrophic orchid *Corallorhiza*, which is named for its characteristic root form. Another good example of the coralline root form is found in the snow plant (*Sarcodes sanguinea*).

Coralloid roots provide an intriguing link between fungal and bacterial symbioses, as some actinorhizal roots (for instance *Eleagnus* roots with **actinorhizal** nodulation with *Frankia*) also show a strongly coralline form. The nitrogen-fixing cyanobacterial symbiosis of cycads also involves coralloid roots. Cycads produce coralloid roots in the absence of cyanobacteria, but colonization is required to complete development (Vessey, Pawlowski & Bergman, 2004). Coralloid roots of cycads have a highly distinctive morphology (Fig. 3.14). They are covered with a **papillose sheath** and grow upward (**apogeotropic growth**). Cyanobacteria (usually species of *Nostoc*) enter the root from the soil, probably through the prominent **lenticels**. Inside the root they form a **cyanobacterial zone**, which is a specialized layer of the outer cortex. Curiously, even though the coralloid roots are produced underground, the cyanobacterial zone is still green, as cyanobacteria synthesize chlorophyll even in the absence of light.

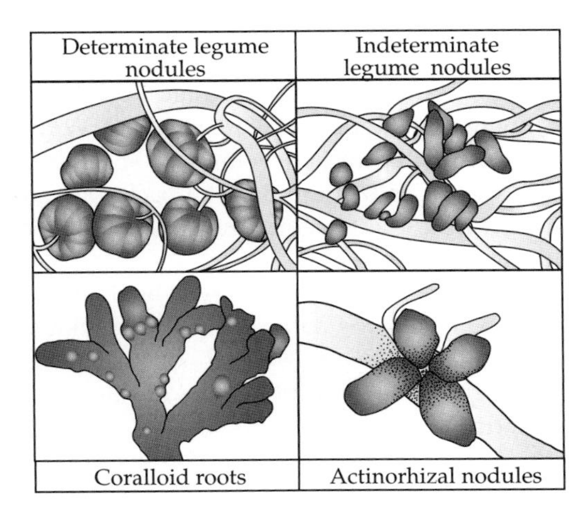

Fig. 3.14. Morphology of prokaryotic root associations (after Vessey et al., 2004).

3.17.4 Actinorhizae and bacterial nodules

Root nodules are stubby root outgrowths stimulated by symbiotic infection by *rhizobia* (rhizobial bacteria) or *actinobacteria* (hypha-forming bacteria) from the soil (Vessey, Pawlowski & Bergman, 2004; Fig. 3.14). Well-known examples of root nodules are those of *Alnus* and papilionoid legumes. When nodules are root-like, as in the case of many of the actinobacterial associations, they are known as **actinorhizae** (literally "actinobacterial roots"). All actinobacterial symbionts are members of the genus *Frankia* whereas rhizobial symbionts are members of the rhizobial complex (*Allorhizobium, Azorhizobium, Bradyrhizobium, Mesorhizobium, Rhizobium* s.s. and *Sinorhizobium*). All the symbionts are nitrogen-fixing (diazotrophic) bacteria with molybdenum-iron nitrogenases encoded by *NIF* genes that function to convert atmospheric nitrogen to ammonium that can be used by plants. In return, the plant supplies the bacteria with nutrients and habitat.

All nodulating plants come from families in the eurosid 1 clade. The trait is usually considered to have been independently derived many times within this clade. If correct, the multiple origin of nodulation in a single clade is evidence for some molecular predisposition in this clade for the evolution of root nodules. This phenomenon has been called a **latent homology**, but the molecular basis of this particular latent homology has not yet been discovered. However, it should be noted that although the multiple gain reconstruction involves the fewest evolutionary changes, it is impossible, on phylogenetic grounds alone, to rule out a single origin followed by a great many losses. The metabolic cost of maintaining the nitrogen-fixing symbiosis is a plausible reason for the evolutionary loss of nodulation under conditions of adequate soil nitrogen (Layzell, Gaito & Hunt, 1988; Vessey, Pawlowski & Bergman, 2004).

The rhizobial nodulating families are Leguminosae (many genera) and Ulmaceae (*Aphananthe, Celtis* and *Parasponia*; Trinick, 1973). Actinorhizal nodulators are found in the families Rosaceae, Eleagnaceae, Rhamnaceae, Betulaceae, Casuarinaceae, Myricaceae,

Coriariaceae and Datiscaceae. Nodulation is usually scattered in these families rather than universal. For instance in Rhamnaceae the majority of genera do not nodulate, but nodulation has been reported in at least some species of *Adolphia, Ceanothus, Colletia, Discaria, Kentrothamnus, Retanilla, Talguenea* and *Trevoa*. As noted above, the nodulation symbiosis has co-opted parts of the older mycorrhizal molecular pathway. Many diazotrophic bacteria are free-living, associated with the vicinity of roots (**rhizosphere**) or the root surface (**rhizoplane**). It is therefore possible that the nodulation symbiosis started when a molecule from a soil bacterium incidentally tripped the mycorrhizal developmental pathway, producing improved habitat for the bacterium or even bacterial ingress into the plant. The symbiosis is now exquisitely specific and precisely engineered through coevolution, and a particular species of bacterium generally has a relatively small range of host plants it can successfully infect.

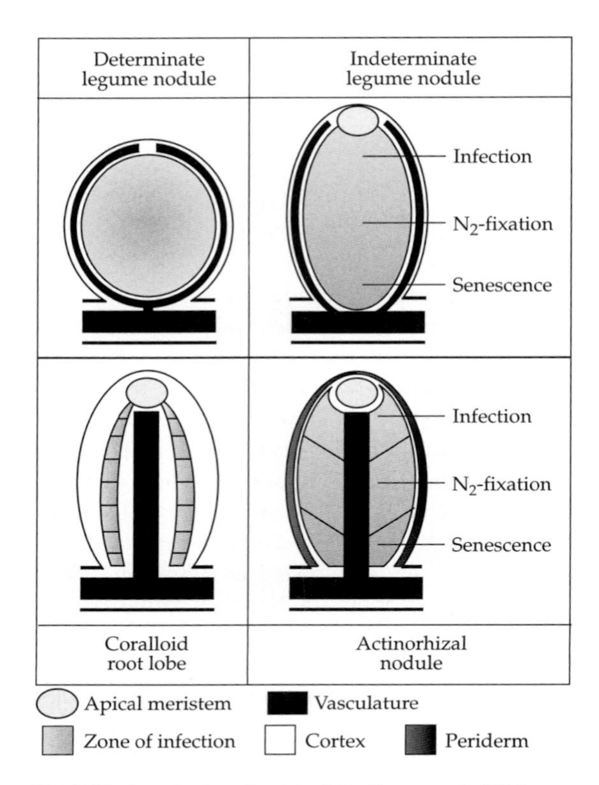

Fig. 3.15. Organization of nodules (after Vessey et al., 2004).

The presence of bacteria within the root causes the formation of **nodule primordia**, usually opposite the **protoxylem poles**. Actinobacterial nodules form from primordia in the **pericycle**, as do lateral root primordia. In contrast, rhizobial nodule primordia arise in the cortex. In initiation and subsequent growth, there is an important difference between **indeterminate nodules** (found in *Medicago, Pisum* and *Trifolium*) and **determinate nodules** (*Lotus, Glycine* and *Phaseolus*). Indeterminate nodules form deep within the cortex and initiation may involve some pericycle cells as well as inner cortical cells (Fig. 3.15). Furthermore, indeterminate nodules have a persistent apical meristem and continue growing while the nodule is functional, resulting in a somewhat elongated shape. Determinate nodules initiate in the outer cortex, do not have persistent growth and are therefore spherical in shape.

In both actinorhizal and rhizobial symbioses, bacteria may gain entrance to the root (depending on the species) either **intercellularly**, through cracks in the surface, or **intracellularly**, through an invagination process. The most remarkable of the latter is the **shepherd's crook** a root hair invagination, in which the tip of a root hair senses the presence of bacteria on the surface and bends over to sandwich the bacterium. An **infection thread** then forms (Fig. 3.16). This is an intercellular tube formed of cell wall material that conducts the bacteria down into the outer cortex (in the case of determinate nodules) or the inner cortex (indeterminate nodules). Within the nodule the bacteria develop into specialized forms known as **bacteroids**. The bacteroids are enclosed in a plant membrane called the **peribacteroid membrane**).

All actinorhizal nodules, as well as the rhizobial nodules of the Ulmaceae, are obviously modified lateral roots. Even though they lack a root cap, they are root-like in development, originating as cell divisions in the pericycle, as do lateral roots. They are also root-like in internal anatomy, having central vascularization. Furthermore, they may developmentally switch into structures that are typical of roots

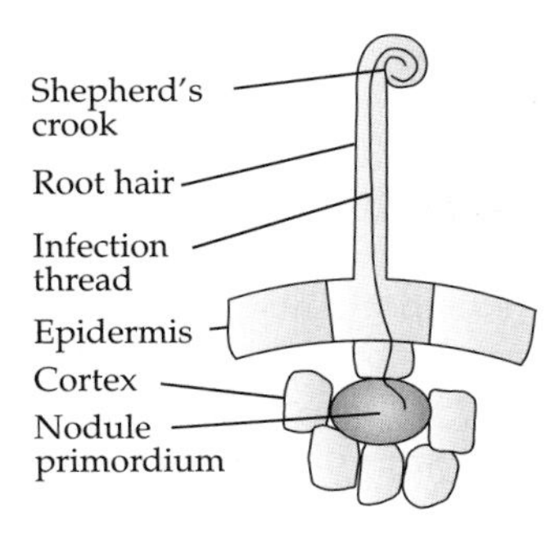

Fig. 3.16. Infection process in nodule formation (after Vessey et al., 2004).

in all features except apogeotropic growth. This morphological phenomenon is found in several species including Myrica and Casuarina; and the clusters of nodules crowned with a tangle of upward-growing roots are very striking.

The homology of legume nodules is more problematic. They have peripheral vascularization, like stems, and originate from cortical cell divisions. These characters have led to them being homologized as root-borne shoots. However, the probable common origin (and general similarity) with actinorhizal nodules should make us wary of such an interpretation. An alternative view is that they are homologous to lateral roots but in a highly derived state. The peripheral vascularization may be a result of **heterotopy**, of clear functional importance as it permits the central part of the nodule to be reserved for the symbionts. The cortical origin of the legume nodule is not very clear-cut, as in many species with indeterminate nodules the zone of stimulated cell division extends down from the cortex to the pericycle of the parent root. Furthermore, actinorhizal nodules develop by first forming a **prenodule**. The prenodule results from cortical cell divisions and only later does the nodule properly arise from pericycle divisions. Thus the cortical origin of the legume nodule could be explained by **heterochrony**, with such a nodule displaying juvenile features relative to actinorhizal nodules (persistent juvenility is a phenomenon known as **pedomorphosis**).

A third possibility is that legume nodules are homologous to neither lateral roots or to root-borne stems, but are instead merely **hypertrophic** proliferations of the cortex of the parent root. Running counter to this is the possession of a persistent meristem at the tip of indeterminate legume nodules. This implies a degree of organization beyond that of mere hypertrophy. On present evidence, it seems most likely that the legume nodule is indeed homologous to a lateral root, albeit in a highly derived condition. However, the test of this hypothesis will come from determining whether genes characteristic of lateral root initiation are also activated when nodules form.

The development of the *Lotus* root nodule has been the object of intense study at the molecular level as there are large quantities of genomic information on both *Lotus* and on the rhizobial bacterium responsible for inducing nodules (nodulating bacteria). A complex set of cell–cell signaling occurs between host and bacterium that causes cellular proliferation of the root (Geurts, Fedorova & Bisseling, 2005; Stacey et al., 2006).

The first step is the secretion of chemicals, typically flavonoids, from the host root. These act as bacterial attractants and activate the bacterial *NOD* genes. As a result of *NOD* gene expression the bacterium secretes lipochitooligosaccharide nod-factors. The nod-factors trip the host's nod-signal receptor (a plasma membrane receptor kinase) and thereby cause a cascade of molecular changes within the plant. Many elements of this cascade are now known from analysis of nod-defective mutants (Stacey et al., 2006; see Box 3.1). It is thought that (1) the **receptor kinase** activates (2) a **leucine-rich receptor kinase** that in turn activates (3) a **cation channel** and so causes a **calcium spike** that (4) activates a **calcium-calmodulin dependent kinase** that (5) activates a **GRAS transcription factor** that is a key regulator of the further, as yet poorly characterized, downstream gene expression required for nodule formation. This cascade is remarkably fast with the calcium spike occurring within minutes of the application of the nod-factor.

The downstream genes activated by this cascade are responsible for nodule development and for the synthesis of the late **nodule-specific proteins** (late **nodulins**) that include the **hemoglobins**. The hemoglobins function in **oxygen control**, which is one of the most interesting aspects of nodule biology. Nodules have an oxygen paradox. The energy for nitrogen fixation must come from oxidative phosphorylation, which requires oxygen. Nitrogen fixation itself, however, requires highly reducing conditions and the exclusion of oxygen. Most nodules have a **periderm** that is probably impervious to oxygen and in such cases oxygen diffusion into the nodule is thus entirely through **lenticels**, in a spatially controlled manner. Within the nodule the microdistribution of free oxygen appears to be controlled by hemoglobin. The very high oxygen affinity of hemoglobin is therefore the mechanism that eliminates oxygen from regions of the nodule where it is not physiologically required. The presence of relatively large amounts of hemoglobin is the reason that many nodules turn strikingly pink on being cut open.

CHAPTER 4

The organography of the leaf

4.1 The nature of the leaf

The word "leaf," as used generally, is a vague term comprising many independently derived but functionally and morphologically similar organs in different groups. Leaves form laterally on stems and their primary function (at least in leaves which have not become evolutionarily specialized for other functions) is photosynthesis.

As photosynthetic organs, leaves are generally more efficient than stems. The relatively thin plate of light-harvesting cells that is characteristic of many leaves results in an increase of photosynthetic area for comparatively little investment in biomass. They are analogous in this respect to the ears of African elephants (*Loxodonta africana*) which may comprise 20% of the surface area of the elephant but only 2% of the body mass. Elephants' ears are for cooling (Phillips & Heath, 1992), rather than photosynthesis, but the principle is the same.

Leaf-like organs grouped under this term include the **phyllidia** of *bryophytes* (mosses and liverworts), the **microphylls** of the *lycophytes*, and the **euphylls (megaphylls)** of the *monilophytes* (ferns and horsetails) and the *spermatophytes* (seed plants). Monilophytes and spermatophytes together comprise the *euphyllophyte clade*. In this book, the word "leaf," when used without qualification, refers to the euphyllophyte leaf.

Euphyllophyte leaves are the lateral units of the shoot. They are evolutionarily derived from stems but with many distinctive features that differentiate them from stems. This is in contrast to lateral roots, which are morphologically similar to primary roots. The nearest roots come to producing distinctive lateral organs is in the production of specialized organs such as nodules. Leaves are produced by the growing apex of stem, and therefore form in strict **acropetal succession**, a feature not found in roots. When newly formed, leaves grow faster than the stem, and so young leaves usually hide the **apical meristem**. By contrast, the primary root apical meristem is never concealed entirely by lateral roots. Furthermore, leaves are produced in a regular arrangement (**phyllotaxy**) whereas lateral roots are typically produced in a more irregular fashion, in response to local water and nutrient conditions in the soil. Leaves share with stems the characteristic of being uniformly **exogenous** in origin (i.e. formed from outer tissue layers) whereas roots are **endogenous**.

One very remarkable feature of the euphyllophyte leaf, but only in seed plants, is the constant association of leaves with lateral stem meristems in the form of **axillary buds**. Leaves themselves, however, are without other organs except by lateral displacement, or when adventitious meristems are formed. They tend to be **dorsiventral** structures, appearing as flattened **chlorophyllous** plates, not strongly lignified, often unbranched and never with **nodes** and **internodes**. Because they are determinate organs with limited life spans they are, of necessity, **caducous** (falling off) organs, often with an **absciss layer**.

Because of the requirement for photosynthetic gas exchange the euphyllophyte leaf has many holes or **stomates**, typically concentrated under the leaf (**hypophyllous**). Such stoma arrangements lead to **hypostomatous** leaves. However, **epistomatous** leaves occur (with stomates on the upper surface) as well as **amphistomatous** (stomates more or less equally on both surfaces). In bearing numerous stomates euphyllophyte leaves contrast with the phyllidia of mosses, which never bear stomata (gas exchange is not a problem as phyllidia are typically one cell-layer thick). For this reason, stomata in bryophytes (or at least their analogous gas-exchange pores) are features of either **sporophyte** stems, as in mosses, or of thick **gametophyte thallus** as in *Marchantia*.

Leaf ground tissue is generally **parenchymatous** with abundant air spaces and the vascular tissue mostly **schizostelous** (i.e. composed of **meristeles**, vascular units with few **vascular bundles** and no **medulla** (**pith**)). Secondary thickening is not extensive but there may be some thickening of the meristeles. In contrast to stems, leaves are generally determinate in growth and there is therefore very limited apical growth. Instead of having organized tip growth, leaf meristematic activity tends to quickly become diffused, with **laminal**, **marginal** and **basal meristem** zones replacing **apical meristematic activity**. An exception to this is found in the fern *Lygodium* in which the leaves are extremely stem-like in behavior forming indeterminate branching and climbing units that scramble up into the forest canopy (Fig. 4.21). In *Lygodium japonicum*, an Asian species invasive in Florida, these leaves may be up to 20 m long.

4.2 Homology of leaf-like organs

4.2.1 Phyllidia and microphylls

Our understanding of the homology of plant leaves comes from mapping leaf-like structures onto the phylogeny of land plants. The inclusion of leafless fossil types is particularly important to this exercise. As new fossils become available and as ideas on the phylogenetic position of fossils change, so will our views on the homology of leaf-like organs. Hornworts (Anthocerotophyta) have no leaf-like organs but other groups of the bryophyte grade do have leaf-like organs, called **phyllidia**. The phyllidia of mosses (i.e. Bryophyta in the strict sense) and leafy liverworts (order Jungermanniales of the Hepaticophyta) share many similarities of both structure and development and so are plausibly homologous. However, the phylogenetic reconstruction of the earliest diverging land plants that is supported by current evidence suggests that the phyllidia-bearing mosses and jungermannialean liverworts are well separated by leafless marchantialean and metzgerialean liverworts and hornworts.[1] This suggests independent origins for moss phyllidia and liverwort phyllidia. The earliest unequivocal liverwort megafossils are thalloid (metzgerialean) as are the earliest liverwort-like fossils (Bateman et al., 1998; Edwards, Duckett & Richardson, 1995).[2]

The **microphyll**, characteristic of the lycophytes, is dissimilar developmentally and morphologically from the phyllidium and thus there is no suggestion that these are homologous structures. Furthermore, many of the earliest macrofossils of vascular plants, such as *Cooksonia*† from the Silurian period, are completely leafless. This is further evidence that the vascular plant leaf is a *de novo* structure, unconnected with anything in bryophyte clades. The microphyll, in contrast to more elaborate vascular plant leaves, has only a simple vascularization, typically consisting of a single unbranched vascular strand. The junction of this vascular strand with the main vascular system of the stem is not marked by any perturbation of stem vasculature. This is in contrast with euphyllophytes in which the vascularization of the leaf is associated with changes in the stem vascular system, giving rise to a distinct nodal anatomy.

An elegant theory for the origin of the microphyll is the **sterilized sporangium theory** of Kenrick and Crane (Kenrick & Crane, 1997). This suggests that lateral **sporangia** progressively lost their sporangial nature and became leaf-like. Support for this theory comes from the fact that the related leafless zosterophylls

[1] This takes an entirely phylogenetic view of homology. However, using Van Valen's view of homology as the transfer of information from ancestor to descendant (see Chapter 1, Section 1.5), it is possible that there is "homology" between phyllidia and microphylls if they share common elements of their developmental pathways.

[2] One possibility, first suggested by Hagemann, is that some characteristics of sporophyte stems (ability to branch) and leaves (dorsoventrality) first evolved in the flat thalli of early land plant gametophytes and were co-opted into the sporophyte.

(Zosterophyllales†) bear many stalkless lateral sporangia on their reproductive axes, a condition putatively plesiomorphic for the lycopsids. It is therefore possible to imagine how an axis covered with sporangia could become an axis covered by microphylls. The main difficulty with this theory, however, is that microphylls in lycophytes are either sterile or bear a sporangium on their adaxial surface. If the microphyll results from sterilization of a sporangium, where do the sporangia of microphylls come from?

Kenrick and Crane solved this problem with two ingenious scenarios. One proposes that the fertile microphyll results from the expression, in one organ, of two developmental pathways: for both a sterile and a fertile sporangium (an ontogenetic fusion). The other suggests that developmental pathway for the fertile microphyll might derive from the partial sterilization of paired sporangia. In this scenario the microphyll is homologous to sporangium number one and the sporangium homologous to sporangium number two.

An alternative theory for the evolution of the microphyll is that it represents the end point of an extreme **reduction of a fertile lateral branch**. Such a theory would predict the existence of intermediates with complex branched microphylls on the lycopsid stem lineage. There are lycopsids, such as *Leclerqia*†, which have branched microphylls, but this appears to be nested well within the core lycopsid clade and therefore the branched microphylls probably represent an elaboration of simple microphylls rather than a reduction from a more complex structure. More promising candidates are the fossils referred to the genus *Estinnophyton*†, which have branched leaves bearing multiple sporangia and have traditionally been considered lycopsids. However, these fossils are quite possibly primitive sphenopsids, in which case the leaves would be reduced **megaphylls** rather than complex microphylls (Hao, Wang & Wang, 2004).

A third theory for microphyll origin is that proposed by the plant morphologist F. O. Bower (1855–1945), called the **enation theory** (Bower, 1935). This suggests that microphylls derive from the vascularization and planation of *de novo* outgrowths of the stem called **enations**. Such outgrowths in the form of bumps, spines and other processes are widespread in primitive fossil vascular plants, but it is not clear whether these bumps represent an organogenetic dead-end or lead to extant organs. The main evidence for the enation theory is that the fossil genera *Sawdonia*†, *Asteroxylon*† and *Drepanophycus*† can be arranged in a plausible series of progressive leaf vascularization, implying that vascularization of microphylls is a derived state. The enations of *Sawdonia* are unvascularized whereas *Asteroxylon*† has a vascular strand that extends to the base of the leaf but not into it. *Drepanophycus*, on the other hand, has well-vascularized spine-like leaves. Under the sporangial sterilization hypothesis, the leaves of *Sawdonia* have either secondarily lost vascularization or represent nonhomologous structures (enations).

4.2.2 The megaphyll

The megaphyll is the form of leaf possessed by the seed plants and the monilophytes (ferns and horsetails), together forming the euphyllophyte clade. The structural complexity attained by the megaphyll (especially at the level of vascularization) implies that it is not homologous to the microphyll. Taking this view allows the microphyll to be considered a synapomorphy of the lycopsids. Bower suggested (Bower, 1908) that the megaphyll evolved via increasing elaboration of the microphyll. However, the discovery of numerous leafless fossils of otherwise high complexity prompted Bower to change his view and postulate separate origins for the megaphyll and the microphyll (Bower, 1935). The megaphyll is thought to have arisen from lateral branch systems by processes of the sort enumerated by the telome theory. The evolution of leaves requires that lateral branch systems become determinate, dorsiventral and laterally expanded. These changes correspond to the telome theory postulates of "reduction," "planation" and "webbing". At the level of lateral branch systems all megaphylls are probably homologous, as all the euphyllophytes are derived from an ancestor that was probably *Psilophyton*-like with lateral branches. However, as there are fossil stem-group seed plants, stem-group equisetaleans and stem-group ferns with lateral branches that are not fully leaf-like, it seems that the derived leaf-like features of all three groups evolved independently.

The fossils that are of vital importance to this interpretation are (1) for the seed plants, leafless progymnosperms† such as *Aneurophyton*† (Aneurophytales); (2) for the ferns, leafless stem fern fossils such as *Pseudosporochnus*† and *Rhacophyton*† and (3) for equisetaleans, leafless stem equisetalean fossils such as *Ibyka*† and *Archaeocalamites*†. Although these fossils are called "leafless" they do have determinate or subdeterminate lateral branches from which conventional megaphylls could be plausibly derived. However, the lateral branches do not show evidence of planation or webbing. As the leaf-like features of the three groups therefore appear to have been independently obtained, a case could be made for distinguishing these features as spermatophylls, filicophylls and sphenophylls.

All spermatophylls appear to be strictly homologous as current evidence points to spermatophytes being monophyletic and derived from within the leaf-possessing pteridosperm clade. However, the situation in the monilophytes is far from clear. Recent phylogenetic work has suggested a sister group relationship between the adder's-tongue/whisk fern clade and the other monilophytes including ferns and horsetails. This would strongly suggest that the leaf in the adder's-tongues (Ophioglossales) and whisk ferns (Psilotales) represents another independent derivation of the leaf. Similarly the leaf in the marattialean clade (possibly a sister group to the sphenopsids) may also have independently derived its advanced leaf-like features.

In the most extreme view, the lamina-bearing megaphyll may be considered independently derived in spermatophytes, ferns, marattialeans, ophioglossoids and sphenopsids (the multiple-megaphyll hypothesis). If this is accepted, it raises the question of how such multiple evolutionary convergence came about at the genetic level. One possibility is that the genetic apparatus for the developmental processes of determinacy, dorsiventrality and laminal outgrowth are preexistent to the evolution of megaphylls, having evolved for other functions within the plant. It is then possible to hypothesize that these pathways were independently recruited into megaphyll evolution in several lineages.

An alternative to the multiple-megaphyll hypothesis is the early megaphyll hypothesis, namely, that the megaphyll evolved early in plant evolution and the leafless fossils represent reversals. This is supported by the existence of megaphyllous plants very early in the fossil record of vascular plants, such as *Eophyllophyton*† of the early Devonian (Hao, Beck & Wang, 2003) and by the existence of the complex-leaved progymnosperm *Archaeopteris*† in the fossil record along with leafless Aneurophytalean progymnosperms†. Under the early megaphyll hypothesis these plants represent the stem group of the euphyllophytes and leafless fossils represent reversals. However, under the multiple-megaphyll hypothesis these plants represent additional early derivations of the megaphyll on separate lineages. It also looks increasingly likely that *Eophyllophyton*† represents a derived lycophyte with elaborate microphylls, thus removing the problem posed by this taxon.

4.3 Molecular basis of relationship with the shoot apical meristem

In the section above, the possibility was noted that the molecular developmental pathways required for leaf evolution might predate the leaf. The first of these mechanisms to consider is the mechanism that distinguishes a leaf from a stem as an organ of separate identity (even though it is derived from a stem). A major facet of that separate identity lies in the determinate rather than indeterminate growth of leaves versus stems. Indeterminacy of stems is facilitated by the expression of *KNOX* type 1 genes (knotted-like homeobox genes of the type associated with meristems), particularly those *KNOX* type 1 genes that are expressed in the central zone (CZ) of the meristem. In eudicots with simple, highly determinate leaves, *KNOX* gene expression is excluded from the leaf primordium, a process that appears to be very important in promoting leaf over stem identity and determinate organ fate. *KNOX* genes are excluded from the primordium by the action of *ARP* genes (MYB transcription factors of the type represented in *Arabidopsis* by ASYMMETRIC LEAVES1, in *Zea* by ROUGH SHEATH2 and in *Antirrhinum* by PHANTASTICA). The KNOX-inhibitory role of *ARP* genes involves interaction with a histone chaperone suggesting an epigenetic component of this effect (Phelps-Durr et al., 2005).

Reciprocally, CZ-type *KNOX* genes are known to repress *ARP* genes in the meristem (Byrne et al., 2000; Kim et al., 2003b). The *ARP–KNOX* regulatory module therefore works as a reciprocal duet that is probably functionally highly conserved in plant evolution.

Some *KNOX* gene expression is, however, found in leaves of some species, particularly compound leaves (Bharathan et al., 2002), and overexpression on *KNOX* genes leads to leaf lobing (Lincoln et al., 1994; Fig. 4.1). It appears that interactions between *KNOX* and *ARP* genes are required to govern the correct development of complex leaf morphologies (Kim et al., 2003a). Crudely speaking *KNOX* and *ARP* genes act as "accelerator" and "brake" to govern spatial variation in leaf development. Two related species, namely, the compound-leaved *Cardamine hirsuta* and the simple-leaved *Arabidopsis thaliana* provide a useful pair in which to examine evolving *ARP–KNOX* interactions. The differences in leaf architecture between these two species are due to changes in *KNOX cis*-regualtory elements, which allow *KNOX* gene expression to reappear in the *Cardamine* leaf and not in the *Arabidopsis* leaf (Hay & Tsiantis, 2006).

It is known that type 1 *KNOX* genes are widespread (probably ubiquitous) in land plants, including mosses (Champagne & Ashton, 2001)

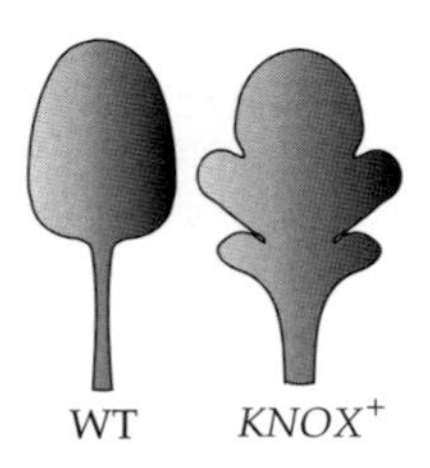

Fig. 4.1. Effect of *KNOX* gene overexpression on leaf morphology (after Tsukaya, 2005).

but the ubiquity of the *KNOX–ARP* interaction has been less certain. In the fern *Anogramma chaerophylla*, as in compound-leaved angiosperms, there is no *KNOX* downregulation at leaf inception (Bharathan et al., 2002). In the fern *Osmunda regalis* it has been shown that *KNOX* genes and *ARP* genes are coexpressed in the meristem and in the leaf primordium, contrasting with the mutually exclusive pattern characteristic of angiosperms (Harrison et al., 2005a). It may be that ferns and angiosperms have a similar ARP–KNOX regulatory pattern but that in ferns the onset of *KNOX* repression is later, consistent with the idea that highly divided fern leaves are fundamentally less determinate than the simple leaves of angiosperm model organisms. Leaves of *Lygodium*, which may be up to 20 m long (Fig. 4.21), offer a graphic example of this lack of determinacy in fern leaves.

The question is then raised of whether *ARP–KNOX* interactions extend to microphylls, widely supposed (for the reasons given above) to be nonhomologous with megaphylls. The lycophyte *Selaginella kraussiana* has two type 1 *KNOX* genes (Harrison et al., 2005a). One is expressed in the stem and the other is expressed in the apex (in the cells around the apical cell) and there is one *ARP* gene. As in angiosperms, but not ferns, *ARP* genes but not *KNOX* genes are expressed in the leaf and primordia, consistent with possible interactions causing the strict determinacy of *Selaginella* microphylls (Harrison et al., 2005a). However, *ARP* and *KNOX* genes are coexpressed in the apex (as in ferns), implying either that lycophyte *ARP* and *KNOX* genes lack an angiosperm-type interaction, or that in plants with a dichotomizing axis, such as *Selaginella* stems and fern rhizomes, *ARP* expression is required in the apex to control meristem behavior during the dichotomization process (Harrison et al., 2005a).

Although expression studies do not prove *ARP–KNOX* interaction they are highly suggestive, as is the remarkable result that *Selaginella ARP* gene can function in *Arabidopsis* to complement *ARP* knockout mutations. As microphylls and megaphylls have independent origins, in would appear that *ARP–KNOX* interactions have been independently recruited as mechanisms of leaf development. This in turn implies that *ARP–KNOX* interactions were preexistent to the evolution of leaves and so present for recruitment in multiple lineages. One possibility is that *ARP–KNOX* interactions controlled dichotomous branching in leafless Rhyniophytes† and so were able to control anisotomous branching and, eventually, leaf formation.

4.4 Axes of polarity in leaf development

The word polarity was first applied to plants in 1888 (Voechting, 1878–1884) referring to experiments with the stems of willow (*Salix*) that "remembered their original growth direction and produced roots and shoots appropriately when hanging upside down in a damp chamber." **Polarity** is therefore equivalent to **asymmetry**, in this case developmental asymmetry: one part of a structure is not the same as another and therefore the structure is asymmetrical. The establishment of polarity therefore is equivalent to the breaking of symmetry and is accomplished, in a developmental context, by asymmetrical gene expression. There are three axes of polarity in a leaf. The **proximodistal axis** runs from the leaf base (proximal to the stem) to the leaf apex (distal to the stem). The proximodistal axis is marked by transitions from leaf base and petiole to leaf lamina and leaf apex. In the previous section it was noted that the differences in expression between *KNOX* genes characteristic of the stem and *ARP* genes characteristic of the leaf could be responsible in part for the establishment of such a polarity.

Another major axis of polarity is the **mediolateral**, across the breadth of the leaf. In many leaves the **mediolateral axis** involves a clear transition from the rachis or midrib through the lamina to the leaf margin. However, it is common for leaves to be **bilaterally symmetrical** about the midrib and have equivalent but mirror-reversed axes on each side of the midrib. However, this is not always the case, and in leaves of *Begonia*, *Ulmus* and many *Eucalyptus* species the leaves show strong left–right asymmetry about the midrib in the mediolateral axis (Fig. 4.2).

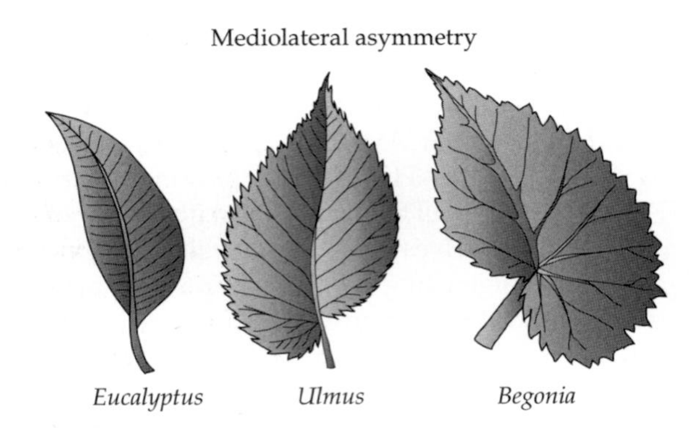

Fig. 4.2. Examples of mediolateral asymmetry in leaves.

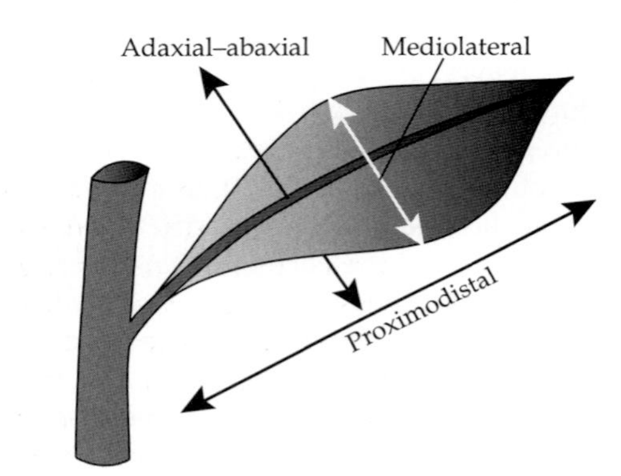

Fig. 4.3. Axes of polarity in the leaf.

The third axis of polarity in the leaf is the adaxial/abaxial or dorsiventral axis (Fig. 4.3). The botanical terms adaxial (toward the axis) and abaxial (away from the axis) are to be preferred as less ambiguous than the zoological terms dorsal and ventral, when used in a botanical context. Leaves have no "back" (dorsus) or "stomach" (venter). However, to the extent that leaves do have a "back" it is the lower side of the leaf. Thus in the older literature the ventral surface of the leaf always refers to the upper (adaxial) surface. However, in recent usage the reverse is sometimes erroneously intended. It is evident from turning over leaves that very commonly there is a big difference between adaxial and abaxial leaf surfaces. In general terms this is because the adaxial surface functions as a light-receiving surface and to shed water, whereas the abaxial surface functions in gas exchange. There are exceptions of course, but in most seed-plant leaves this difference in function holds and translates into a striking difference in surface morphology.

Before the evolution of the leaf, stem systems had a single continuous surface. This raises the question of which of the two surfaces of the bifacial leaf is homologous to the stem surface and which is the new surface. It is easy from a cursory examination of a selection of plants to see that the abaxial surface is the "old" surface and the adaxial surface has evolved *de novo*. The stem surface usually shows a smooth transition with the petiole and lower leaf surface. By contrast, the leaf adaxial surface is usually abruptly distinguished from that of the stem. The evolution of the adaxial leaf surface as a new plant structure is therefore one of the key transitions in terrestrial evolution. Developmentally, the distinction between abaxial and adaxial leaf identity is maintained by the interplay of abaxial genes of the *YABBY* (*YAB*) and *KANADI* (*KAN*) types and adaxial identity genes of the HD-ZipIII family (Fig. 4.4).

YAB and *KAN* genes determine the "outer" identity of plant organs (Eshed et al., 2004). *HD-ZipIII* genes (e.g. *PHABULOSA* and *PHAVOLUTA* in *Arabidopsis*) negatively regulate *YAB* and *KAN* genes in the adaxial regions of the leaf allowing the differentiation of adaxial cell types and maintenance of adaxial identity (McConnell et al., 2001). In turn these genes are prevented from affecting the abaxial regions of the leaf by cleavage of their messenger RNA directed by a 21-bp micro-RNA and mediated by ARGONAUTE proteins (Kidner & Martienssen, 2004).

It is interesting to note that another *HD-ZipIII* gene, *REVOLUTA* (*REV*), is necessary for the correct development of internal polarity in both the meristem and in developing vasculature. Therefore, this antagonistic system, originally for the maintenance of correct radial polarity in the stem, was likely to have been preexistent to leaves. It would therefore have been available for recruitment as a "ready-made" system to regulate bifacial identities in leaves. In this context, it is interesting that the micro-RNA sequence that negatively regulates *HD-ZipIII* genes is highly conserved in all land plants (Floyd & Bowman, 2004). *REV* also takes part in an elegant negative feedback loop to stabilize its activity levels. *REV* promotes transcription of the *LITTLE ZIPPER* gene *ZPR3* which heterodimerizes with *REV* dampening its activity (Wenkel et al., 2007). Given the importance of *HD-ZipIII* genes in the morphological patterning of land plants it is highly likely that this feedback loop will be found to be deeply conserved.

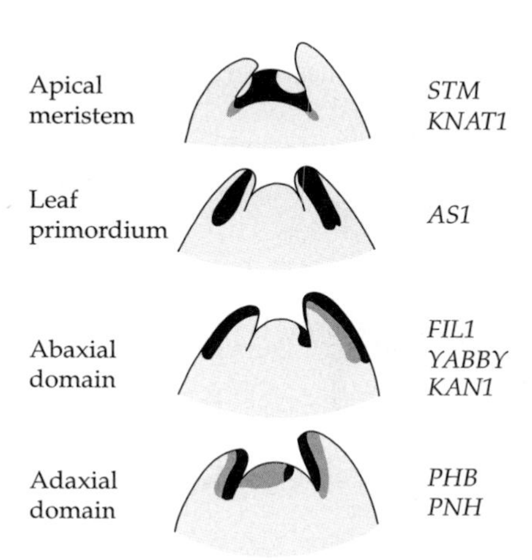

Fig. 4.4. Expression domains of major leaf-related developmental genes. Note the mutually exclusive pattern of the *YABBY* genes (e.g. *FIL1*) versus the *HD-ZIPIII* genes such as *PHB* (after Tsukaya, 2002). *STM = SHOOTMERISTEMLESS, KNAT1 = Antidropsis KN1-like, AS1 = ASYMMETRIC LEAVES1, FIL1 = FILAMENTOUS FLOWER1, KANl = KANADI1, PHB = PHABULOSA, PNH = PINHEAD.*

4.5 Laminal extension and the bifacial theory

There appears to be a connection between **bifaciality (adaxial–abaxial polarity)** and the production of a lamina (Fig. 4.5). This is evident from those plants, often monocots, which naturally produce **unifacial** leaves, such as onions (*Allium* spp.) and rushes (*Juncus* spp.). These plants tend to have radial leaves without a flat lamina. The association is also evident from unifacial leaf mutants that are abaxialized or adaxialized due to expression changes in the abaxial or adaxial identity genes (Golz et al.,

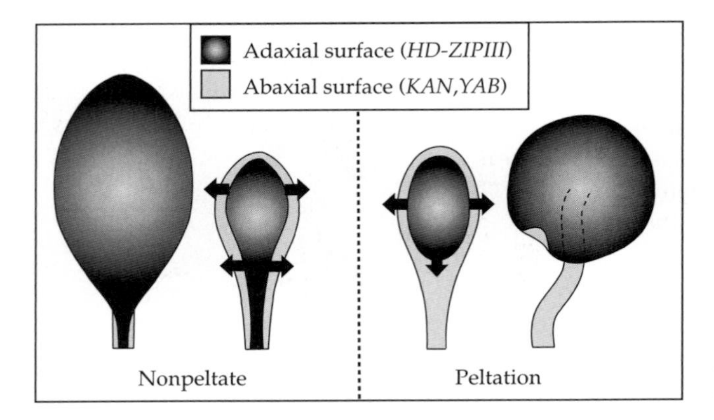

Fig. 4.5. Diagram showing the dependency of lamina extension on bifaciality and how the different distribution of bifaciality gives rise to different leaf morphologies.

2004; McConnell et al., 2001). These tend to have radial, needle-like leaves. It is also evident from surgical manipulations that abaxialize leaf **primordia** by isolating them from the leaf apex at an early stage (Sussex, 1954, 1955).

In developing *Drosophila* wings, there is an analogous specification of dorsal and ventral identity. At the boundary between dorsal and ventral cell types, there is an interaction that is required for the formation of a developmental organizing center at the dorsiventral boundary of the wing (Brook, DiazBenjumea & Cohen, 1996). This raises the question of whether there might be a similar interaction at the boundaries of adaxial–abaxial zones that is required to promote lamina outgrowth, an outgrowth that requires the localized effect of *KNOX* genes to maintain meristematic activity at the margin of the developing lamina. In a ground-breaking paper Waites and Hudson (1995) proposed that such a juxtaposition of cells with adaxial and abaxial identity are required for lamina outgrowth. They based this on their study of abaxialized and radialized *phantastica* mutants in *Antirrhinum*, with defects in the ARP gene *PHANTASTICA*.

The role of *ARP* genes in abaxial–adaxial patterning is still unclear. In *Arabidopsis* and some other species, in contrast to *Antirrhinum*, *ARP* genes appear only to have minor effects on adaxial–abaxial polarity and may function by interacting with other genes rather than by direct action. Furthermore, the abaxialization of primordia that have been surgically separated from the apex indicates that signaling between the apex and the leaf is required at early stages to promote adaxial identity. This is no surprise as the adaxial part of the leaf is defined by its proximity to the apex. In addition, the flattening of the leaf primordium commences very early, at a stage when the influence of the apex is to be expected.

The **bifacial theory of lamina outgrowth** is supported by other abaxial–adaxial patterning mutants, such as *leafbladeless* in maize, which develops ectopic patches of abaxial identity on the adaxial surface of the leaf. At the edge of these patches (i.e. at the boundary of the two identity types) ectopic lamina forms (Timmermans et al., 1998). Similarly, the begonia mutant, *Begonia hispida* var. *cucullata*, also develops ectopic patches of abaxial cell identity on the adaxial surface. Around these patches lamina develops as little cups on the leaf surface, with the abaxial surface innermost, contiguous with the abaxial patches.

What is not yet clear is exactly how adaxial–abaxial signaling acts to regulate *KNOX* effects in the leaf. One candidate for part of this pathway is the gene *SERRATE*, a zinc-finger protein that not only works in the micro-RNA pathway regulating adaxial identity genes, but which also acts to limit the competence of cells to respond to *KNOX* gene expression (Grigg et al., 2005). Thus, *SERRATE* provides a link between leaf adaxial–abaxial polarity and the developmental effects of *KNOX* genes, of a sort required by the bifacial theory of lamina outgrowth.

4.6 The parts of the leaf

4.6.1 The leaf base or sheath

In a "complete" angiosperm leaf there are three parts: the base, the petiole and the lamina (Fig. 4.6). The variation of angiosperm leaves is such that one or more of these parts may be undeveloped or missing. When the base of the leaf is clearly differentiated from the petiole it is usually sheathing to some degree. Its close proximity to the stem makes this inevitable. When a **leaf sheath** is well developed, as in grasses, it can be a very distinctive and important organ. Often, as in grasses, the junction of the leaf sheath and the rest of the leaf has the action of a **pulvinus** to regulate the angle of the lamina. In many dicots, the entire leaf base is pulvinus-like (e.g. *Aesculus*).

A leaf sheath is formed when the base of the **leaf primordium** occupies a large part of, or even all, the circumference of the apical meristem. As the stem matures, this leaf tissue inevitably develops to be wrapped around the stem at maturity. If the leaf primordium completely encircles the apical dome a **tubular sheath** or **vagina** can be formed. The subsequent leaves and stem then have to grow out through this tube. If the internodes are short this results in a nesting of tubes one inside the other. Leaf sheaths may thus provide protection for the apical meristem, which may be completely enclosed in a succession of sheaths. Lateral buds too are enclosed in the sheath and may thus be protected from damage by desiccation or herbivores. When these sheaths are robust they can even form a **false stem** or **pseudocaul**, as in the banana (*Musa*). In banana, the apical bud is near groundlevel and above it towers a column (pseudocaul) of overlapping leaf sheaths. In the pseudocaul the sheath has an obvious function of mechanical support. However, it has also been shown that in grasses also the sheaths contribute considerably to the rigidity of the stems (Niklas, 1998).

Many dicots do not have a leaf sheath as their primordia are small points on a relatively large apical meristem, and these leaves therefore consist only of petiole and blade. Although well-developed sheaths are found in a few dicots they

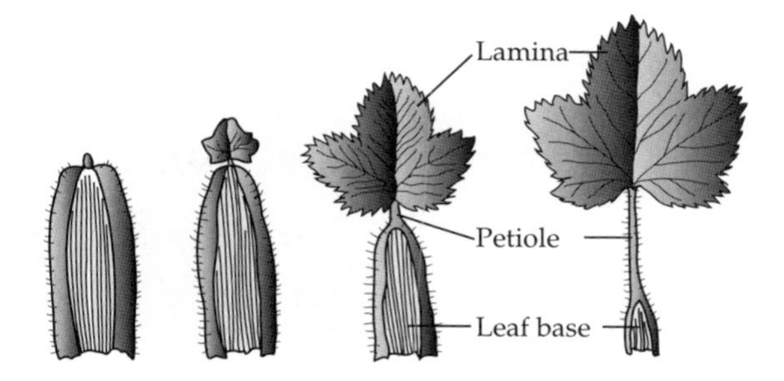

Fig. 4.6. Diagram showing the equivalence of scale leaves in gooseberry (*Ribes*) with the leaf base of foliage leaves.

are particularly common in the monocot clade, as this clade often has broad-based primordia. Palms (Arecaceae), gingers (Zingiberaceae) and grasses (Poaceae) are all examples of groups of plants having a pronounced sheath. In the grasses, the leaf sheath is particularly well demarcated, being separated from the blade by an **angular deflexion** and the presence of a **ligule** and **auricle** (lateral outgrowths at the base of the blade). The vast majority of grasses have a ligule although it is occasionally reduced to a ring of hairs. When present, this small flap of tissue is always situated at adaxial side of the sheath–blade junction.

The sheath represents a major part of the patterning of the leaf along the proximodistal axis. Like the petiole, a case can be made for the sheath having a greater degree of stem identity than the lamina, possibly due to homeobox-gene-like function. This is supported by the change from blade to sheath identity that occurs when a *KNOX1* homeobox gene that normally does not occur in leaves, *ZmLIGULELESS3* (*LG3*) is leaf expressed (Muehlbauer et al., 1999). The conversion to sheath only occurs if *LG3* is expressed early in development. Later expression produces ligule, auricle or altered lamina, in that order, implying a progressive loss of competency during development to change identity along the proximodistal axis, with sheath identity being set first (Muehlbauer, Fowler & Freeling, 1997).

The very precise boundary between blade and sheath in grasses, marked by a precisely positioned ligule and auricle, represents a very

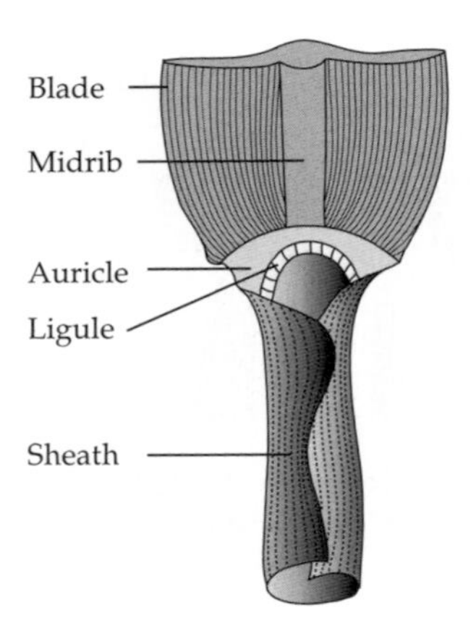

Fig. 4.7. Highly diagrammatic view of the parts of a grass leaf, focusing particularly on the morphological complexity at the sheath–blade boundary.

beautiful system in plant development which is not yet fully understood (Foster et al., 2004) (Fig. 4.7). The boundary is set by a complex network of genes including *ZmLIGULELESS1* (*LG1*), *ZmLIGULELESS2* (*LG2*) and *ZmEXTENDED AURICLE1* (*ETA1*; Harper & Freeling, 1996; Osmont, Jesaitis & Freeling, 2003). Curiously, while *eta1* mutant plants resemble *knox* overexpression mutants, no *KNOX* expression can be detected, implying that *ETA1* may be providing a downstream link with the *KNOX* pathway.

4.6.2 Petiole

The **petiole**, or **leaf stalk**, is the organ between stem and **leaf blade** (**lamina**). It may be directly connected to the stem, or in those plants with sheathing leaves there may be a leaf sheath or base between the petiole and the stem. It supports the lamina as well as positioning and orienting it. It is therefore a highly adaptive organ for optimizing light capture. Unsurprising, given their role in support, petioles are rather woody and stem-like. The woodiness comes from the high proportion of **meristeles**, compared to other tissue, that run through the petiole carrying water to irrigate the lamina, which often has a high water demand. Like stems they are often unifacial being composed entirely or mainly of abaxial surface (i.e. the surface topologically equivalent to that possessed in the stem). Petioles occasionally have a swollen base called a **pulvinus** or **leaf-cushion**.[3] Pulvini are characteristic of the Leguminosae where they are implicated in changes to the angular deflexion of the leaf on a diurnal cycle (the so-called sleep movements). Pulvini also occur on petiolules, the petiole-like leaflet-stalks in compound leaves.

Most petioles are rounded in cross section, another stem-like feature, with a clear distinction between leaf blade and petiole. However, some petioles are **winged** (**alate**) as in *Citrus* and many other Rutaceae. These flattened petioles are much more blade-like but in the case of *Citrus* there is a clear demarcation between the petiole and blade by means of a constriction. In *Citrus* this demarcation is an indication of the **unifoliolate** nature of this leaf.

In other plants the distinction between petiole and lamina is blurred by the possession of a **decurrent lamina**, where the leaf blade narrows gradually into petiole. Thus it is sometimes difficult to say where the petiole ends and the blade starts. Does a rachis with a very narrow lamina constitute a winged petiole or a narrow blade? The difference here is that the developmental programs responsible for blade and petiole are either sharply demarcated in time and space, as in *Citrus*, or alternatively, responsive to a gradient so that petiole identity declines as lamina identity increases. The development of a lamina requires a highly active lateral meristematic activity at the edge of the leaf primordium whereas the petiole has little or no lateral meristematic activity. It would be interesting to know the genes responsible for this and how they are spatially controlled.

There is another characteristic form that the petiole may take and that is the **phyllode** (Fig. 4.9). Phyllodes are leaf-like developments of the petiole usually associated with loss of the lamina. However, "leaf blades" that have developed evolutionarily from petioles are sufficiently different from the real thing to be easily

[3] It is often difficult to determine whether the pulvinus develops from what is morphologically leaf base, petiole or the junction between the two, especially when these parts of the leaf are not well developed or differentiated.

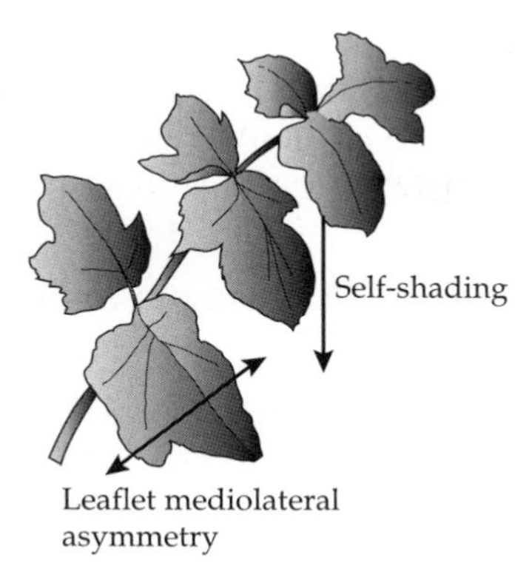

Fig. 4.8. Functional analysis of leaflet mediolateral asymmetry in *Heracleum* (hogweed).

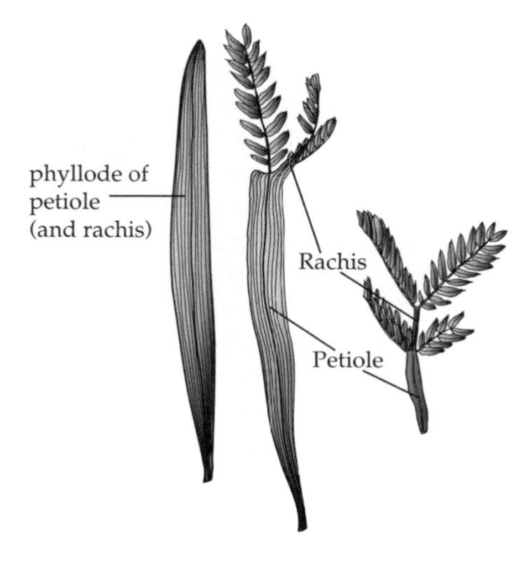

Fig. 4.9. Leaf series from *Acacia* showing the derivation of the leaf-like phyllode from expanded petiole and rachis.

distinguished. For one thing they are sometimes unifacial, like many petioles. The evolution of phyllodes, like cladodes, has the effect of making a radical departure from the developmental canalization that may limit the evolutionary potential of the lamina. The ecological and physiological consequences are significant as many phyllodes are drought tolerant or spiny, and the character is found in Mediterranean or desert environments, as in Australian *Acacia*. Phyllodes are also found, but more rarely in mesic environments as in *Lathyrus aphaca* of Europe.

4.6.3 Leaf blade

The **leaf blade,** or **lamina**, is usually the main part of the leaf and is generally flat. Unlike the petiole and the sheath, which are never branched, the lamina may either be unbranched, as in a **simple leaf**, or be extensively branched to form a **compound leaf** (Fig. 4.8).

The leaf lamina has an enormous impact on the carbon cycle of the planet being the place in which most fixation of carbon dioxide takes place via the enzyme **ribulose bisphosphate carboxylase-oxygenase (rubisco)**. Leaves require so much rubisco as to make this the most abundant protein on the planet. The carbon dioxide concentration in the atmosphere is measured at frequent intervals at Mauna Loa, Hawaii. The resultant "Keeling curve" (after its discoverer, Charles Keeling) shows the well-known year-by-year rise of carbon dioxide associated with human activities. However, within this rising curve a marked annual cycle is visible as CO_2 levels drop in spring and summer as temperate forests leaf out, and rise in autumn and winter as biomass decays. This annual cycle, which has been described as "planet breathing" is due almost entirely to the activities of leaf laminas.

The surface morphology of the leaf lamina appears to be strongly tied to the physiological functions of light interception, gas exchange and photosynthesis. In this, leaf **bifaciality** is extremely important, providing two distinctive physiological surfaces. **Stomates** tend to be more abundant on the lower (**abaxial**) surface of the leaf (the **hypostomatous** arrangement), where water loss can be controlled by epidermal features, for instance, by hairs and scales which, on the lower surface, do not affect light interception. However, some leaves are **epistomatous** with stomates on the upper (**adaxial**) surface and others **amphistomatous** (both surfaces).

4.7 Leaf insertion

Leaf insertion is a term that refers to how the organs of the leaf relate to the stem. Usually the junction between the leaf and the stem is at the sheath or the petiole. However, where these are lacking and the lamina sits directly on the stem, the leaf is termed **sessile** as opposed to **petiolate**. Members of the family Boraginaceae (e.g. *Myosotis* or forget-me-not) have mostly sessile leaves, with the lamina meeting the stem.

The lamina of sessile leaves may take on special forms where it meets the stem. A **decurrent** leaf is one in which the lamina extends down on either side of the leaf insertion to form **wings** on the stem. This is due to the primordium developing laterally on the shoot apex so that the primordia are contiguous. When the internodes lengthen, a small piece of lamina therefore remains present all along the internode as in *Verbascum thapsiforme* and *Lathyrus tuberosus*.

Where leaves are inserted at the same level as in **opposite** or **whorled** leaves, the leaf bases of the different leaves can grow together by postgenital or congenital fusion to become **connate**. Examples of connate leaves are to be found in *Lonicera* (Fig. 4.10), *Dipsacus*, *Succisa* and *Montia perfoliata*. Alternatively, individual leaves can entirely encircle the stem, as leaf sheaths do. Such leaves are called **perfoliate**. This is caused by blastozones at the basal part of the leaf primordium causing expansion and fusion. The difference between perfoliate and sheathing leaves is that the encircling portion in a perfoliate leaf is clearly lamina and is flat or cup-shaped rather than sheathing. The genus *Bupleurum* (Apiaceae) contains many examples of perfoliate plants (e.g. *Bupleurum rotundifolium*). If the leaf does not completely encircle the stem but nevertheless has basal growth of the lamina that clasps the stem, it is known as **amplexicaul**, as in *Sonchus oleraceus*.

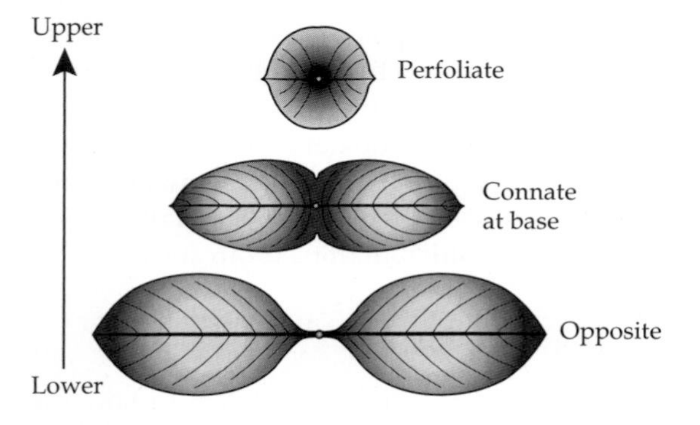

Fig. 4.10. Progressive leaf connation in *Lonicera* (honeysuckle). The perfoliate leaves are found subtending inflorescences.

4.8 The shape of the simple lamina

The shapes of simple leaves are important for taxonomy and clearly represent a major axis of evolutionary variation. The shapes of leaves are the result of the differential activities of lateral and distal meristem growth in different places around the leaf outline at different times. The spectrum of different shapes can be modeled as a series of transformations from the basic ellipse. Probably the commonest leaf shape is the **elliptical** or **oval** shape, widest in the middle and narrowed at the tip and where the blade meets the petiole. Lateral meristematic activity in elliptical leaves is greater in the middle of the lamina and this activity reduces smoothly toward either end. The length of elliptical leaves is typically 1.5 to 3.0 times the width, due to a greater proximodistal meristem activity than lateral meristem activity (e.g. *Cornus sanguinea*, *Fagus sylvatica*).

If proximodistal growth is reduced, an **orbicular** leaf may be formed in which the length/breadth ratio is 1:1 (e.g. *Rhus cotinus* and *Lysimachia nummularia*). On the contrary, if proximodistal growth is considerable, an **oblong-elliptic** or **lanceolate** leaf is formed. These leaf types may be arbitrarily defined as either length 3 to 4 times the width (oblong-elliptic) or length more that 4 times the width (lanceolate). *Aucuba japonica* is an example of an oblong-elliptic leaf and *Salix alba* is an example of a lanceolate one.

In all the cases above, the lateral meristematic activity declines symmetrically toward the lamina tip and base so the widest part of the leaf is in the middle. It is common, however, for the lateral meristematic growth to be greater nearer the base causing the widest point of the leaf to be located toward the base. In this case, an **ovate** (**egg-shaped**) leaf is the result, as in *Syringa vulgaris*. More rarely, but still frequent, the activity may be greatest nearer the tip and this is called **obovate** (i.e. like an inverted egg) as in hazel, *Corylus avellana*. A special and extreme case of obovate is **spathulate** (**spatula-shaped**), exemplified by the common lawn-daisy *Bellis perennis*. This is an "obovate" leaf with the top section orbicular and the bottom section long extended.

From the paragraphs above, it is clear that two variables, the position of the widest point and the length/breadth ratio, can together be used to define the majority of the variation between the common leaf shapes of different species. All these leaf shapes are based on modifications of the basic ellipse and have parabolically curved margins. However, straight-sided leaves also occur and thus curvature (or lack of it) is another important type of variation in describing leaf shape. For instance, a variant of the ellipse with straight sides is the rhomboid or diamond-shaped leaf, found quite commonly in plants (Box 4.1).

All the terms discussed so far assume that the leaf is a two-dimensional object. However, in many cases there is no pronounced lamina production and so three-dimensional shape terms are required. An example is the **ensiform** (**pennant shaped**) leaves of *Iris*, *Xyris* (Sajo & Rudall, 1999) and other monocots (including some orchids). These are unifacial leaves that do not develop a lamina in the conventional sense (consistent with the bifacial theory of lamina development). However, meristematic activity on the adaxial and abaxial side of the leaf produces a false lamina oriented in the adaxial–abaxial plane rather than the mediolateral plane. Unifacial leaves that develop no lamina of any kind and bifacial leaves in which lamina development is minimal may be termed **subulate** (**awl-shaped**) as in *Juniperus communis* and *Ulex europaeus*, or **acerose** (**needle-shaped**) as in *Pinus* and *Picea*.

Another leaf oddity that should be mentioned here is the occurrence of **asymmetrical leaves**. The vast majority of leaves are symmetrical along their midrib. There is in other words no left–right polarity in the mediolateral axis. However, a few plants have blades that are prominently unequal-sided. Examples are found in *Ulmus*, *Begonia* and *Eucalyptus* (Fig. 4.2). This asymmetry takes two major forms (with intermediates between). The petiole may be straight

Box 4.1 The curvature of the leaf margin.

In elliptical leaves the outline describes a **parabolic curve** that reduces from the widest point to the narrow ends. Depending on the length/width ratio this type of leaf may vary from **orbicular** to **narrowly elliptical**. In some species, parts of the ellipse are represented by nearly **straight lines** and the leaf is therefore **diamond-shaped** or **rhomboid**, as in some species of *Betula*. The two axes, length to breadth ratio and position of the widest point are important in both curved and straight-line leaves. Long narrow leaves are **narrowly rhomboid** and short ones **broadly rhomboid**. If the widest point of the leaf is toward the bottom, the leaf is **trowel-shaped** or **trullate**. This is the straight-margin equivalent of the **ovate leaf**. If the leaf has the widest point at the very bottom it will be **triangular** (specifically taken to be an isosceles triangle—*Atriplex hortensis*) or **deltoid** (an equilateral triangle—*Populus deltoides*). If it is widest near the top it is **obtrullate** (the straight-margin equivalent of **obovate**). If this is extreme, an **obtriangular** leaf results. Obtrullate and obtriangular leaves are shaped like an inverted trowel or inverted triangle respectively. Inverted triangular leaves are sometimes also called **cuneate** or **wedge-shaped**, as in *Ginkgo biloba*. These leaves above all have margins divergent or convergent over most of their length. However, it is very common for leaves of monocots, as well as some other groups, to be more or less parallel sided over their entire length. In such leaves, there is no single place that can be identified as the widest point. Such leaves are described as **linear** if they are relatively short as in many conifers such as *Taxus* and some grasses (Poaceae). If they are very long, such leaves are usually termed **ligulate** (**strap-shaped**) as in *Typha* and *Butomus*.

and there may be unequal outgrowth of lamina on each side. Alternatively, the whole leaf may be curved and hence asymmetrical, even though the differential growth of lamina on either side of the midrib may be small. This is a common pattern in many *Eucalyptus* species.

4.9 Meristems, blastozones and growth of the simple leaf

We have seen in the section above how the spatial distribution of lateral and proximodistal growth leads to different leaf shapes. To understand leaf growth it is therefore necessary to understand meristematic activity in the leaf. Stems grow at the tips as a result of apical meristem activity. Many leaves, especially those of ferns and plants with dichotomous venation, also grow at the tips and margins, as a result of the activities of **tip meristems** (increasing cell numbers in lengthwise files) and **marginal meristems** (increasing cell numbers in lateral files). In the angiosperms, however, the situation is less clear.

Early studies of eudicot leaf development, for instance on tobacco (*Nicotiana*; Avery, 1933), emphasized the activities of the tip and marginal meristems (especially at the beginning of leaf development). Later studies, however, such as those on *Xanthium* (Maksymowych & Wochok, 1969) that used the incorporation of radioactively labeled thymidine to localize DNA synthesis and therefore cell division, suggested a more complex picture with lamina growth being driven by the cell division by what was called a **submarginal meristem** as well as more dispersed cell divisions in a so-called **plate meristem**. Likewise, studies suggested that in some species lengthwise growth is driven by the activities of **intercalary meristems**. The history of leaf developmental studies has been one of dispute about the relative importance and even existence of these meristems, particularly the marginal meristem, as cell divisions are never completely restricted to one location. The current view is that the margin does not constitute a discrete meristem, but it is a very important source of brassinosteroid signaling that affect lamina growth and shape (Reinhardt et al., 2007).

In angiosperms marginal meristematic activity is short lived and later histogenetic cell division is diffuse. *Arabidopsis* has been recently studied using transformation with a cyclin::beta-glucuronidase reporter construct to determine spatial and temporal patterns of cell division in the developing leaf (Donnelly et al., 1999). Using this technique it is apparent that in *Arabidopsis* laminal expansion is correlated with marginal cell division, while later histogenetic cell division is more diffuse (perhaps equivalent to the **plate meristem** concept). In eudicots such as *Arabidopsis* the pattern of cell maturation is notably basiplastic (i.e. maturation from tip to base; Donnelly et al., 1999).[4]

On the other hand, some fern and gymnosperm leaves do grow by prolonged marginal meristematic activity and a marker of this phenomenon is **dichotomous venation** that runs to the margin (as in *Ginkgo*) indicating the formation of new vein material at the margin of the leaf where growth is occurring, driven by marginal auxin production. In these cases, the pattern of cell maturation is likely to be **acroplastic** (base to tip). In contrast, the reticulate venation of angiosperms requires the formation of new vein material in midlamina and indicates more diffuse growth.

In angiosperm compound leaves, leaflets arise by the formation of specialized meristems called marginal **blastozones** (i.e. morphogenetic meristems as opposed to histogenetic meristems; Hagemann & Gleissberg, 1996). These meristems are different from regions of histogenetic cell division that occur in the developing leaf as blastozones have specific organogenetic potential. They are also distinguished by specific molecular markers. *UNIFOLIATA* (*UNI*) is thus a marker of blastozone formation in pea (*Pisum*).

In most dicots the tip meristem (an apical blastozone with apical and subapical initials) is only active, for a short time, in the very early stages

[4] Although euphyllophyte leaves are thought to have evolved from branch systems, in deep evolutionary history, the basiplastic development of most leaves, in contrast to the universally acroplastic development of axes, is a strong indication that they are no longer developmentally comparable.

of **primordium development**, providing length to what is, at this stage, a primordium roughly cylindrical at the distal end. Soon afterwards, lateral growth is established followed by growth distributed throughout the lamina. This last type of cell division is the last to cease, and it ceases as a result of controlled arrest. Arrest usually starts at the tip, consistent with the basiplastic pattern of leaf maturation, and propagates to the base in what is known as the **arrest front**. Maintenance of cell division in organs may involve an as yet unknown signal, produced by the *Arabidopsis* cytochrome P450 *KLUH* gene, which acts as a stimulator of plant organ growth. The cessation of organ growth may therefore be regulated, at least in part, by the expression dynamics of *KLUH* (Anastasiou et al., 2007).

To form petioles, and in the formation of many monocot leaves with parallel venation, there is strong activity of intercalary meristems with cell divisions oriented along the proximodistal axis. Intercalary meristems are often situated at the base of the leaf, in which case they are called **basal meristems**.

The final part of leaf growth is cell expansion, which adds the greatest part to the dimensions of the leaf. Leaf expansion tends to be greater in the proximodistal axis and this is particularly marked in grasses in which longitudinal expansion of the cells adds greatly to leaf length and little to leaf width (Woodward, 1979).

Apical and marginal leaf blastozones are somewhat reminiscent of shoot apical meristems, SAMs (apical blastozones). These blastozones are more persistent in ferns and some gymnosperms. The situation is somewhat different in dicots with complex compound leaves or persistent tip growth like *Chisocheton* and *Guarea* in the Meliaceae (Fisher, 2002; Steingraeber & Fisher, 1986). In *Chisocheton* the leaf may resume tip growth to add an extra pair of leaflets every year for several years. It therefore shows indeterminate growth and a persistent apical meristematic organization. However, in the context of angiosperms, the indeterminacy of *Chisocheton* is a derived feature.

Compound-leaf growth by means of blastozones may be the primitive means of leaf growth in megaphylls. However, in angiosperms with simple and entire leaves, such as *Antirrhinum*, *Arabidopsis* and *Nicotiana*, blastozone formation does not occur and leaf growth is only by much more distributed meristematic growth, controlled by arrest fronts. The molecular initiation, suppression and coordination of the disparate blastozones and other regions of cell division to generate precise patterns of leaf size and shape is one of the great challenges of plant molecular organography.

4.10 Molecular control of simple leaf size and shape

Many mutations affect leaf size and shape as any defect in cell cycling or cell expansion will have an effect. Auxin is a key regulator of leaf growth so many of these effects will be auxin related. What is not yet clear is which of these genetic pathways are responsible for the evolution of differences in leaf size and shape between related species.

Candidates for such effects fall into four classes: (1) genes which affect the overall activity of specific meristems by their effects on the frequency of cell cycling, (2) genes that affect the polarity of cell divisions, (3) genes that affect the polarity and extent of cell expansion and (4) genes that control the spatial and temporal progression of the distal to proximal growth arrest front. The analysis of mutants in *Arabidopsis* has identified genes in all four classes. What is now required is careful developmental and genetic work to locate adaptive evolution relative to these candidate pathways.

One gene that has an important role in extending the duration of cell cycling downstream of auxin is *AINTEGUMENTA* (*ANT*), while *ANT* expression in turn is prolonged by the expression of the auxin inducible gene *ARGOS* (Hu, Xie & Chua, 2003). The *ANGUSTIFOLIA3* (*AN3*) gene has been shown to be required for the normal functioning of the plate meristem. Mutants defective in this gene tend to have narrower leaves as a result (Horiguchi, Kim & Tsukaya, 2005). Other genes appear to affect lamina meristematic activity in specific ways. Overexpression of *LEAFY PETIOLE* (*LEP*) causes lamina to form on parts of the petiole where it is normally lacking. It is thus a candidate for promoting proximal meristem activity (van der Graaff et al., 2000). Conversely *JAGGED* (*JAG*) overexpression increases lamina growth in distal regions (Dinneny et al., 2004).

ROTUNDIFOLIA4 (*ROT4*) and other *ROTFOUR-LIKE* (*RTFL*) genes appear to promote cell division in a mediolateral direction and so overexpression mutants have relatively broad (round) leaves (Narita et al., 2004). On the other hand, *ROTUNDIFOLIA3* (*ROT3*) is an unrelated gene that affects cell expansion along the proximodistal axis (Tsuge, Tsukaya & Uchimiya, 1996), while *ANGUSTIFOLIA* (*AN*) affects expansion in the mediolateral direction (Kim et al., 2002).

The control of the arrest front is a promising candidate for evolutionary fine-tuning of leaf shape. The gene *CINCINNATA* (*CIN*) has been identified from *Antirrhinum* as a gene whose expression precedes the arrest front and it appears to sensitize cells to an arrest signal (Nath et al., 2003). Loss of *CIN* function causes leaf margins to delay arrest and the greater marginal growth causes an imbalance that leads to curved or crinkly leaves (Fig. 4.11). *CIN* is negatively regulated by a micro-RNA (*miR-JAW*). Overexpression of *miR-JAW* therefore causes the same crinkly phenotype. Interestingly, *CIN* appears to have different action in petals where it influences cell differentiation and promotes growth (Crawford et al., 2004), rather than inhibiting growth as in leaves. It therefore appears that rather than having a fundamental role, *CIN* is recruited into diverse underlying pathways as a key developmental regulator. Its function would therefore be regulator with extremely precise spatial and temporal expression, of a sort that is required for the development of very precise organ shapes.

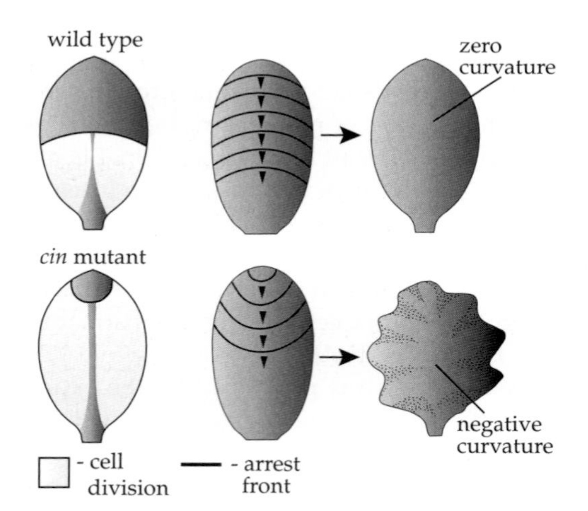

Fig. 4.11. Control of leaf development by the *CIN* mutant, which is responsible for the orderly arrest of cell division, from apex to base, in the developing leaf. This is an essential part of leaf development necessary for the production of a perfectly flat leaf (after Nath et al., 2003).

4.11 Heterophylly, heteroblasty and hormones

Heterophylly refers to the production of two or more forms of leaves on one plant. Commonly, this is in response to environmental feedback. Thus, many water plants have narrow or even highly dissected underwater leaves and broader, laminate floating leaves. The underwater leaves function to minimize resistance to water current and to promote diffusion of oxygen and carbon dioxide. A good example of this is provided by the aquatic *Ranunculus* species in the subgenus *Batrachium* (Cook, 1969) and also by *Potamogeton*, *Callitriche*, *Nuphar* and *Sparganium* to mention only a few.

Integration of environmental signals into a leaf morphogenetic response appears to be accomplished by hormones or light. Two important signaling molecules in aquatic plant heterophylly are **ethylene gas** and **abscissic acid** (**ABA**). Ethylene promotes the submerged leaf form. It is poorly soluble in water and does not readily diffuse out of plant tissues underwater. However, it readily diffuses out of plant tissues in air. In *Ludwigia arcuata*, underwater leaves have been produced by artificial application of ethylene (Kuwabara et al., 2003a; Kuwabara & Nagata, 2005). When underwater leaves are exposed to air not only does ethylene diffuse away but also the leaves are droughted. ABA is a drought hormone and accordingly ABA treatments induce the formation of terrestrial leaves (Kuwabara et al., 2003a,b).

This is potentially an excellent system with which to study the genes controlling leaf morphogenesis. Unfortunately, however, as ABA and ethylene have multiple roles within the plant, the genes upregulated in response to these hormones extend far beyond those concerned with leaf morphogenesis. The laboratory of Bai-Ling Lin (Hsu et al., 2001) has identified numerous putative ABA-responsive heterophylly genes from the water-fern *Marsilea quadrifolia*. Many of these are transcription factors that could potentially be involved in leaf morphogenetic switching, but this will remain a speculation until functional studies are carried out.

Heteroblasty is a particular kind of heterophylly that is plant-age related rather than environmentally related. In other words, leaf morphogenesis is responding to internal physiological signals relating to plant size, architecture and condition, rather than to external environmental cues. Leaf morphogenesis may, for instance, respond to plant phase change. Phase changes are conditioned changes in the SAM. As leaves develop on the flanks of SAMs we would expect links between the condition of the SAM and leaf morphogenesis. An example of a phase change is the change of a SAM from vegetative to flowering, a process that is usually associated with a change in leaf morphology.

One of the most dramatic examples of heteroblasty is in the New Zealand lancewood, *Pseudopanax crassifolius* (Gould, 1993; Fig. 4.12). So dramatic in fact, that when the plant was discovered, the juvenile and adult phases, with long thin leaves and shorter wider leaves, respectively, were placed in different genera.[5] It would be very exciting to discover the differences in gene regulation in the SAM and in the leaf responsible for these dramatic differences. Without genomic tools for *Pseudopanax*, the study of heteroblasty

[5] As seedlings, the leaves start out ovate and pass through a palmately three-lobed stage and then a simple and linear stage. In the juvenile phase (marked by a tall unbranched stem), the leaves are very long (~60 cm), linear and strongly deflexed. The adult phase is marked by a branching tree habit and the production of short (~15 cm) oblanceolate leaves (Gould, 1993).

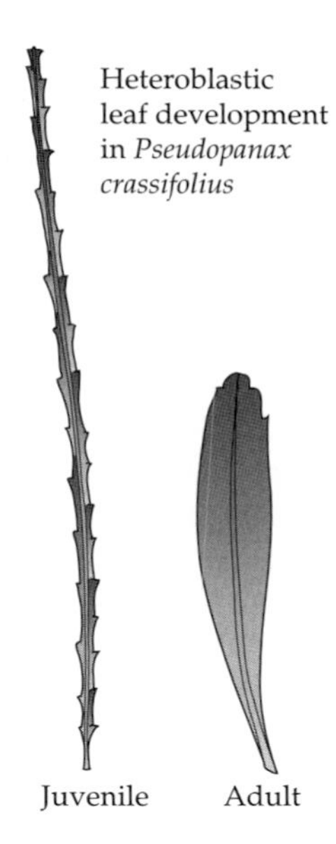

Fig. 4.12. Heteroblasty in *Pseudopanax* (lancewood) (after Gould, 1993).

may for the present be confined to model organisms such as *Arabidopsis*, which has slight, but nevertheless quantifiable, heteroblasty (Tsukaya et al., 2000). However, any candidate genes discovered in *Arabidopsis* could at a later stage be investigated in plants, like *Pseudopanax*, which exhibit marked heteroblasty.

The **embryo** may be considered the first "phase" of plant growth with cotyledons the leaves of this phase. The *Arabidopsis* gene *LEAFY COTYLEDON1* (*LEC1*) promotes the embryonic phase of the life cycle and is required for normal embryo development. It encodes a subunit of the CCAAT-binding transcription factor (Kwong et al., 2003). The *LEC1* mutants have **trichomes** on their **cotyledons** and are more like typical leaves than normal cotyledons (Meinke, 1992; Meinke et al., 1994). Thus, cotyledons may be considered part of a heteroblastic succession of leaf development, characteristic of the embryonic phase. One interpretation of the *LEC1* mutants is that when the embryonic phase is disrupted, the cotyledons are freed from their phase constraint to adopt a more general leaf developmental program marked by the presence of trichomes (Tsukaya et al., 2000).

4.12 Morphology of the leaf tip and base

The **leaf tip** or **leaf apex** is an interesting morphological entity. The variety of forms is important for plant systematics and may have adaptive significance. This variety results from a corresponding variety of meristematic behavior at the distal end of the leaf, the genetic control of which is poorly understood.

If there is little lamina production at the distal end of the leaf, the apex is **acute** (**sharp**). This is the characteristic apex of lanceolate, ovate, rhomboid or triangular leaves. However, if lamina production extends to the end of the leaf, the leaf apex will be **obtuse** (**blunt**), characteristic of orbicular, obovate or spathulate blades.

An extreme form is the **acuminate** (**tapered** or **drawn out**) apex. In this the apex extends further than would be expected from examining the curvature in the immediate neighborhood of the tip. Examples of this are found in many tree genera including *Tilia* and *Ulmus*. If the acuminate tip is downcurved it may function to shed water from the leaf as a so-called **drip tip**. A long acuminate tip is usually taken as an indication that the apical growth of the leaf primordium, which usually ceases at an early stage, is persistent. In monocots, where the main part of the lamina is **bifacial** the tip may be **unifacial**, represented by a small tip circular in cross section. This is the **forerunner tip**, often taken as a vestigial indication that the bifacial leaf in monocots has evolved from a unifacial ancestral form.[6]

An extreme form of apex is the **emarginated** (**notched**) leaf exemplified by *Liriodendron tulipifera* (Fig. 4.13). This is the opposite of the acuminate form, as the tip ceases growth abnormally early, rather than late, and is overtopped by lateral marginal growth. The formation of the emarginated leaf of *Liriodendron* is due to the complete arrest of the growth of the apical tip of the leaf primordium. By contrast, strong growth occurs in the distal leaf lobes, which dwarf and overtop the minute mucro-like leaf apex which soon becomes scarcely visible in the gap between the enlarged lateral lobes (Pray, 1955).

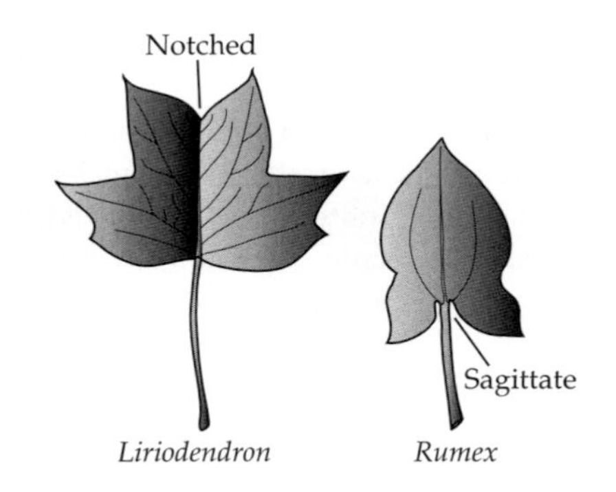

Fig. 4.13. Notching at apex and base of leaves.

The lamina base is another very interesting part of the leaf. This is because, just as at the tip, the proximal part of the leaf is subject to special meristematic processes that create structures quite different from the main part of the lamina. For instance, although many leaves have a sharp transition from lamina to petiole, other species have lamina that is **decurrent on the petiole**. In this the lamina tapers out very gradually as a winged petiole. We have seen in a previous section how there are lamina meristem genes that are active particularly at the proximal end of the leaf such as the *LEAFY PETIOLE* (*LEP*) gene of *Arabidopsis* (van der Graaff et al., 2000) and these are candidate loci for determining the difference between species with decurrent and nondecurrent lamina.

Just as at the leaf tip, a fundamental distinction at the lamina base is between **acute** (**sharp**), characteristic of rhomboid, lanceolate or cuneate leaves, and **obtuse** (**blunt**) bases typical of ovate or orbicular leaves. Acute and obtuse lamina bases have the lamina meeting the petiole at less than or greater than 45°, respectively. When the

[6] Eichler (1861) first suggested that the monocot leaf blade is formed from a different part of the primordium compared to nonmonocots. Arber in a series of papers argued that monocot leaves are phyllodinous (Arber, 1918, 1922, 1925, 1950). Indeed eudicot leaves derived by expansion of the leaf base (such as *Plantago*) are often strikingly monocot-like. However, the phyllodinous theory is currently disfavored. Rather, that monocots appear to form leaves out of the "lower leaf zone" (unterblatt) whereas eudicots form leaves out of the upper primordium zone (Kaplan, 1973).

lamina meets the petiole at a right angle the leaf base is described as **truncate** (**cut off**), which is typical of triangular or deltoid leaves.

Hypertrophy of the lamina at the lamina base is very common and the molecular basis of this phenomenon is unknown. However, it appears that lamina meristem activity frequently evolves a positive response to markers of proximal polarity. Where there is basal hypertrophy, the lamina base is described as **notched**, and notched leaves are due to the activity of basal meristems on either side of the petiole. Owing to various types of basal hypertrophy, the lamina base may take many forms (see Box 4.2).

4.13 Venation

A leaf has to be porous to a certain extent for gas exchange during photosynthesis. This porosity entails water loss so the leaf has to be efficiently irrigated by a network of veins that may be under evolutionary pressure to develop in a way that offers least resistance to the passage of water. In plants with thin leaves, the main veins are often thicker than the lamina and project out of the lower surface of the leaf. A substantial vein of this sort is called a **rib** or **costa**. These have a secondary function of providing support (analogous to a skeleton) for the relatively flimsy lamina. Well-developed costae are therefore needed to support the large leaves of plants such as bananas (*Musa*) and aroids (Araceae).

A leaf generally has a **main vein** centrally placed on the lamina. This is called the **midrib**. It extends from base to apex and generally divides the lamina into two equal halves (except in unequal leaves). From the midrib extend the **primary veins**, which, if they form

Box 4.2 The form of the lamina tip and base.

Tip. The tip is usually either **acute** or **obtuse** depending on whether it makes an angle of less than or greater than 90°. Alternatively, it can be long drawn out into an **acuminate** tip. Where the leaf tip is abruptly tapered from broad to a narrow point this is referred to as **mucronate**, the point being a **mucro**. If there is an abrupt tapering from broad to narrow without a mucro, the leaf apex is referred to as **truncate** (**cut off**). If the apex is depressed the tip is **emarginate** and extreme forms are called **notched**, as in *Liriodendron* (Fig. 4.13).

Base. In addition to the common angles of junction between lamina and petiole at the base of the blade (**acute, obtuse**, and **truncate**), hypertrophy may occur at the lamina base resulting in leaves that are **notched** at the base, that is, the basal lamina lobes extend below the insertion of the lamina onto the petiole. Different types of notched leaves may be described as cordate/reniform, sagittate, hastate or auriculate. **Cordate** (heart-shaped) and **reniform** (kidney-shaped) are leaf descriptive terms that have the same basal morphology but differ in their tip form. They both have blunt basal lobes. However, a cordate leaf has an acute tip whereas a reniform leaf has a rounded tip. Examples of cordate leaves include *Aristolochia sipho* and *Petasites officinalis*. Cordate leaves are very common as this leaf form is an elaboration of the even more common **ovate** form, by increased meristematic activity at the base. In the same way, the reniform leaf represents a basal elaboration of the orbicular leaf. Examples of reniform leaves include *Glechoma hederacea* and *Nuphar luteum*.

Remarkable basal lamina shapes include **sagittate (arrowhead-shaped)** and **hastate (halberd-shaped)**. Both of these have acutely pointed narrow segments (laciniae) at the base. In the sagittate type these are downwardly directed (e.g. *Sagittaria sagittifolia*) and in the hastate type they are horizontally directed (e.g. *Rumex acetosella*). Finally, leaves may have small basal lobes resembling little ears (**auricles**). The auricles of **auriculate** leaves often clasping the stem (e.g. *Salvia officinalis* and *Solanum dulcamara*).

well-developed "ribs," may be termed **lateral** costae (from the Latin *costa* for rib). Veins that are not thickened into ribs are sometimes called **nerves** and thus **nervation** is a synonym of **venation** in denoting the general layout of veins in a leaf. As well as the midrib and primary veins there may be further orders of vein branching, finally ending in **veinlets**, which may form loops or may be blind ending.

A typical angiosperm leaf has primary veins coming off the midrib along its length like branches of a feather from the main shaft and are thus termed **feather-veined** or **penninerved** (e.g. *Quercus*). However, venation has to follow leaf shape and in palmate leaves the primary veins tend to branch off the midrib at the base, as in an *Acer saccharum* leaf. Leaves of this sort are said to be **palmately veined** or palminerved (Fig. 4.14).

Parallel venation is typical of monocots although there are many monocots that do not conform to this. In parallel venation, the veins run the length of the leaf and meet up at the top. In parallel-veined plants with linear leaves, the leaves are straight-veined (**rectinerved**) as in grasses. However, parallel-veined plants with broad leaves are **curved-veined** (**curvinerved**) as in *Hosta* and *Convallaria*. This difference is a consequence of the activity of a plate meristem providing extra lamina between the veins in curvinerved species.[7]

Auxin has an important role in venation patterning. It is produced by the margins of primordial, and pathways of polar auxin transport set the routes of vein development (Scarpella et al., 2006; Scheres & Xu, 2006). A striking local occurrence of auxin precedes vein differentiation in developing *Arabidopsis* leaves (Mattsson, Ckurshumova & Berleth, 2003). Further evidence that the auxin transport pathway is central to vein patterning comes from mutants. For

[7] In response to the need for exquisite precision in describing fossil leaves, in which venation may be one of the few characters available, paleontologists have elaborated a fine-scale classification on venation patterns (Hickey, 1973, 1974; Hickey and Wolfe, 1975). It should be noted that venation patterns tend to follow patterns of meristematic activity (and auxin distribution).

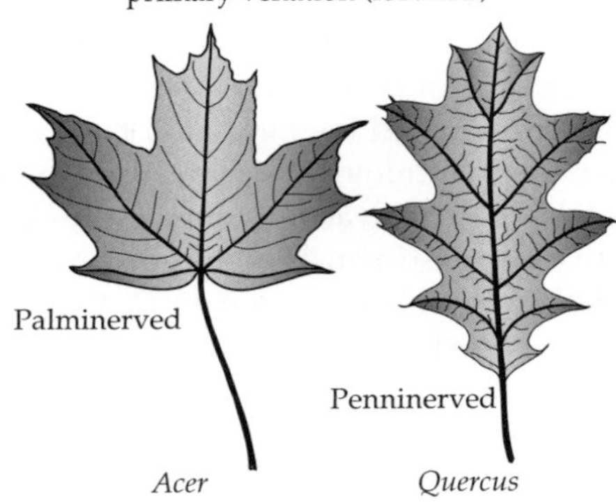

Fig. 4.14. Palminerved and penninerved leaf lobation in *Acer* (maple) and *Quercus* (oak).

instance, *SCARFACE* (*SCF*) is a regulator of vesicle trafficking required for normal auxin efflux that is also required for normal vein patterning (Deyholos et al., 2000; Sieburth et al., 2006). Given this mode of patterning, we can seek to explain two major patterns of venation. The two types are (1) parallel-dichotomous venation that extends straight to the margin, as in many ferns, many extinct seed-ferns and *Ginkgo* and (2) reticulate venation in which not all veins are oriented to the margin and more complex patterns of veinal growth occur. In the first type, it is likely that leaf growth is strictly from marginal meristematic growth and auxin transport is marginal to basal. In the reticulate pattern, nonmarginal patterns of leaf growth and more complex patterns of auxin concentration and transport are likely to occur.

4.14 Lobes: leaf dissection related to primary venation

When considering the **dissection**[8] of a simple leaf (or leaflet) the most important distinction is whether the dissection of the lamina is

[8] The term "dissection" is a misleading one because leaf dissection usually results from unequal growth and not from cutting or division. However, in palms and certain aroids active division of the leaf does occur, by an abscission-like process (palms) or by apoptosis (Araceae), see Section 4.19.

independent of the **primary veins** or alternatively whether it follows the primary venation (**vein-related dissection**). Primary veins are the main veins that branch directly from the **midrib** (center vein). If it is independent it represents superficial heterogeneity in marginal growth, as in the formation of teeth (see next section). However, if it reflects the primary venation it makes the simple leaf akin to a compound leaf and there seems to be a continuum between the dissected leaf and the completely divided compound leaf. An example is the 5-palmatilobed leaf of some *Acer* species (Sapindaceae). Many other members of the Sapindaceae (including some species of *Acer*) have compound leaves and this state is probably an ancestral condition in the family. Pinnate leaves also occur in other members of the genus *Acer*. It is thus plausible that the palmatilobed *Acer* leaf has evolved from a compound ancestor.

As this sort of dissection is vein related, it is of interest to examine the connection between primary venation and lobing. It might be thought that it simply resides in the auxin production of lobe primordia, which then guides major vascular differentiation in the lobe. However, simple leaves also have primary veins, so it appears that there is a morphogenetic template for primary venation in the absence of lobation. Alternatively, therefore, one can view the phenomenon as suppression of lamina growth in the **sinuses** (gaps between lobes). The spatial pattern of these sinuses may be responding to the same morphogenetic template as is governing the formation of primary veins, but in a reciprocal way. Where lobed leaves have evolved from simple-leaved progenitors it may be useful to see lobed leaves from a sinus-driven perspective: a lobe simply being the absence of a sinus and a lobed-leaf akin to an incompletely developed simple leaf.

There are two main axes of variation used to describe vein-related dissection (see Box 4.3). The depth of dissection varies, and this has given rise to a variety of terms to describe it. Second there is a difference of whether the lobes originate from a well-developed midrib (pinnate), or from the lamina base in cases where the midrib is not well developed (palmate).

The degree of dissection is of presumed adaptive significance. A highly dissected leaf is more efficient at dissipating heat by convective cooling under conditions of high insolation. In *Geranium sanguineum*, forms exist with either palmatilobed or palmatisect leaves. The pinnatisect plants are commoner in continental parts of Europe subject to high summer temperatures and the palmatilobed types are more common on the cooler Atlantic coast.

Box 4.3 Lobing of leaves.

The most significant axis of variation of primary vein-related dissection is the depth of division of the lamina. This is arbitrarily described as **lobed** (**lobate**) if the divisions do not extend halfway to the midrib. If the divisions extend about halfway down the primary veins toward the midrib it is described as **split** (**fissate**). If the lamina is deeply divided, significantly more than half way to the lamina it is known as **cut** (**sectate**). Deeply cut leaves, which have very narrow lobes (i.e. **laciniae**) presumably due to low marginal meristem activity, are known as **laciniate**. The term laciniate is usually used for **teratomorphs** in which there is often also abnormal dissection along the secondary veins as well as the primary ones. Such teratomorphs are curious: although lamina formation has been deeply disturbed the pattern of primary venation has not, implying that the two are developmentally separable.

To complete the description of a dissected leaf the distribution of the primary veins, whether **palmate** or **pinnate**, can be added to the term. Thus we may have leaves that are described as **palmatilobed** (e.g. *Hedera helix*), **pinnatilobed** (*Quercus robur*), **palmatifid** (*Acer pseudoplatanus*), **pinnatifid** (*Quercus rubra*), **palmatisect** (*Ranunculus acris*) or **pinnatisect** (as in many *Taraxacum* leaves).

The degree of dissection results from the relative balance between tip and marginal growth in the lobe primordium. This appears to be evolutionarily quite labile, and under relatively simple genetic control, as cut-leaved (**sectate**) or slashed-leaved (**laciniate**) variants of cultivated plants are commonly grown.

Furthermore, the transition between compound leaves and dissected simple leaves occurs very frequently. Also common are transitions between pinnately (or palmately) compound leaves and pinnatisect (or palmatisect) leaves. However, what is not yet clear is whether this transition should be regarded as **heterochronic**, with lobed leaves being the result of incomplete development of compound leaves at maturity (**paedomorphosis**). Alternatively, an active process of suppression of lamina growth in sinuses between lobe primordia may be involved. There appears to be a relationship between the degree of lamina outgrowth in the sinuses and the degree of midrib development. Strong midrib development is correlated with less sinus lamina tissue, but the causes of this relationship are underexplored.

4.15 Teeth: leaf dissection independent of the primary venation

Leaves with a **toothed** margin are extremely common in temperate regions whereas **untoothed** (**entire**) margined plants predominate in the tropics. So marked is this pattern that it has even been used as an indicator of paleoclimate in paleobotanical assemblages. **Teeth** have an **angle** (the point of the tooth) and the **sinus** (the hollow between the teeth) (Box 4.4).

The formation of teeth may be more due to suppression of growth in the sinus than enhancement of growth at the tooth itself. The suppressor gene *CUP-SHAPED COTYLEDON2* (*CUC2*), also known to be involved in the separation of shoot lateral organs, suppresses growth in the sinuses, thus causing serration in *Arabidopsis*. *CUC2* expression is regulated by *miRNA164a* and it is probable that the balance between regulator and target (miR164a and *CUC2*) determines the extent of toothing (Nikovics et al., 2006).

Teeth are considered to be of adaptive significance to increase gas exchange for photosynthesis and to promote convective cooling of the leaf.

Box 4.4 Description of toothing.

The morphological classification of teeth depends on the form of the angle and the sinus, specifically whether the angle they make is acute or obtuse (Fig. 4.15). A leaf with an obtuse angle and an obtuse sinus is called **repand** (e.g. *Alliaria petiolata*). Where the angle and sinus are both acute the leaf is called **serrate** (**saw-toothed**, as in *Urtica dioica*). In between are leaves with an acute angle and an obtuse sinus (**dentate**) or an obtuse angle and an acute sinus (**crenate**). Examples of dentate plants include *Castanea sativa* and **Ilex aquifolium**. Crenate plants include *Caltha palustris* and *Althaea rosea*. A characteristic of serrate leaves is that the teeth are often themselves toothed. Such leaves, with toothed teeth, are described as **biserrate**, to indicate the two orders of serration.

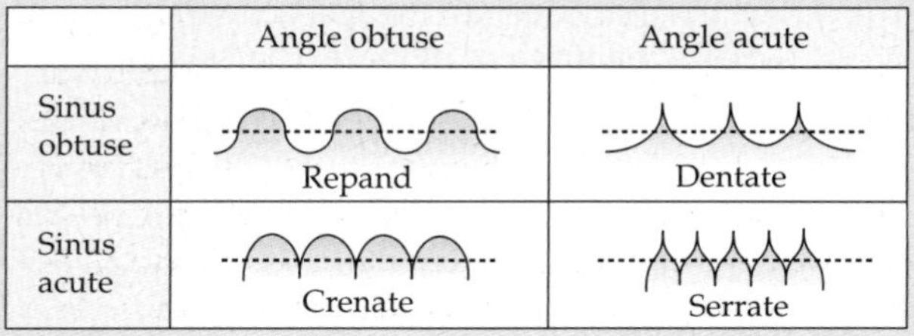

	Angle obtuse	Angle acute
Sinus obtuse	Repand	Dentate
Sinus acute	Crenate	Serrate

Fig. 4.15. Common classes of leaf serration according to the sharpness of the sinuses and angles.

They do this by promoting turbulent airflow across the leaf therefore breaking the boundary layer, even at very low air speeds.

Teeth are rare in the *paleodicots* and *monocots*, and so developmental mechanisms that elaborate teeth appear to be a feature primarily of the *eudicots*. The feature is caused by topical variation in marginal meristem activity at the leaf margin. Neither the developmental mechanisms nor the underlying genes are well known, but it is a character that appears to change readily in plant evolution with losses appearing to outnumber gains. Thus toothing is often characteristic of families although there may be many toothless species scattered within otherwise toothed families. There are also numerous examples of toothless leaves occurring in genera that are otherwise toothed. Such species are often denoted by the epithet *integrifolius*.

Teeth may be veinless or be supplied with a veinlet, which may be shy of the margin or may run up in the angle of the tooth. The vein may even end in a thickening (gland) that projects out of the tooth. An example of this is the **gland-tipped tooth**, which, because it is characteristic of the salicoid clade of the Salicaceae (*Salix*, *Populus*, *Idesia*, *Carrierea*, etc.), is sometimes referred to as the **salicoid tooth**.

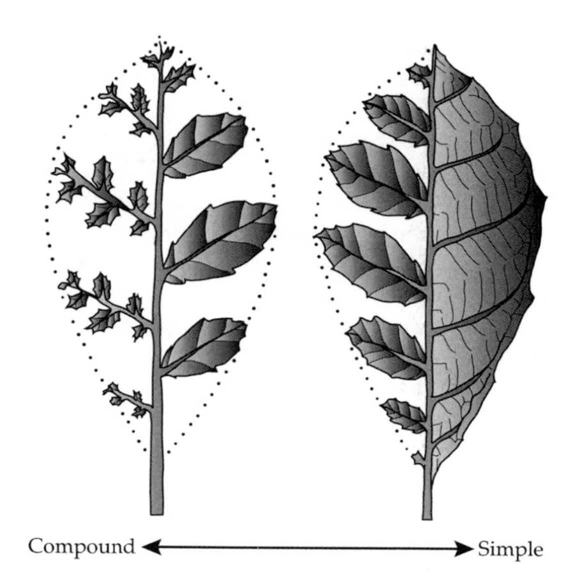

Fig. 4.16. Progressive levels of leaf dissection and compounding caused by the relative preponderance of blastozone formation versus lamina expansion (after Corner, 1959).

4.16 Compound leaves

The euphyllophyte leaf has evolved from a branch system and therefore may retain some features of stemminess in its developmental patterns, as in the strong indeterminacy found in the leaves of some ferns such as *Lygodium* (Fig. 4.21). This thinking has sometimes been extended, but with little justification, to the **compound leaf** of angiosperms. However, the compound leaf is probably best thought of in terms of the behavior of different leaf-specific meristems. Compound leaves form marginal blastozones (organogenetic meristems) in addition to undergoing meristematic lamina expansion. Growth by blastozones allows the formation of repeating modules (**leaflets**) each reminiscent of a simple leaf and each borne on an individual **leaflet stalk** called a **petiolule**. A petiolule is thus defined as the stalk subtending an individual leaflet. All other stalks within the laminate portion of the compound leaf are termed **rachis**. The petiolule may be very short in cases where the leaflets are **sessile**. The term petiole is thus reserved for the stalk lying between the stem and the laminate portion of the compound leaf that is homologous to the petiole of a simple leaf.

Petiole and rachis may be anatomically indistinguishable but morphologically it is important to separate the two, as the rachis is equivalent to the midrib of a simple leaf and not to the petiole. Thus leaflets are not outgrowths of the petioles (if so they would be called **petiolar stipules**). Instead they are better thought of as equivalent to dissections of the lamina in lobed leaves, caused by the formation of specific marginal blastozones in the developing leaf (Fig. 4.16).

If compound leaves have a second or third order of branching they are termed **bipinnate** or **tripinnate** respectively[9] (or if palmate, **bipalmate**

[9] Examples of bipinnate leaves are *Dryopteris filix-mas* and *Levisticum officinale* (lovage). Tripinnate leaves are those of *Conium maculatum* (hemlock) and *Anthriscus sylvestris* (cow parsley)

and **tripalmate;** Box 4.5). The additional rachises so created are equivalent to the **primary veins** of a simple leaf (in the bipinnate case) or to the **primary and secondary veins** of a simple leaf (in the tripinnate case).

The developmental processes responsible for the difference between simple and compound leaves are relatively simple. They involve the dissection of the simple lamina into a lamina of rachises and leaflets. The simple lamina is formed from tip growth accompanied by marginal meristematic growth to give a continuous sheet of tissue. The compound lamina has a greater amount of tip-meristem activity (a characteristic of stems) and restricted and highly localized blastozone activity to give individual leaflets (Box 4.6). In compound leaves new lateral tip meristems form on the primordium at the points where primary veins would be inserted in a simple leaf. This provides the branches and contrasts with the more generalised meristematic activity promoting a continuous margin in the simple leaf.

Determinacy of the compound leaf may be provided by abortion of the tip or by the conversion of the tip into a leaflet. The latter is by far the most common. In both palmate and pinnate compound leaves, the number of leaflets is mostly odd as there is a terminal leaflet. Compound leaves with a terminal leaflet are termed **imparipinnate** or **imparipalmate** (e.g. *Rosa, Robinia*). It is thus this terminal leaflet that bestows clear determinacy on the stem-like compound leaf. However, in some species, such as *Chisocheton* in the Meliaceae, a tip meristem is persistent and a terminal leaflet does not form. The leaf is thus effectively indeterminate. Alternatively, in some compound leaves the determinacy is provided by an abortion of the tip meristem, or the conversion of the tip into a tendril or spine, rather than the conversion of the tip meristem into a terminal leaflet. These plants are known are **paripinnate** (e.g. *Pisum sativum*).

As leaflets often mirror the shape of simple leaves in the same family it is likely that similar developmental programs are used in leaflets as in simple leaves. It also means that the same set of morphological-descriptive terms that are appropriate for simple leaves are also appropriate for leaflets.

In some species **interrupting leaflets** are found. These are small leaflets that occur on the rachis between the ordinary leaflet pairs as in *Solanum tuberosum, Potentilla anserina* and *Filipendula ulmaria*. Such leaves are referred to as **interruptedly pinnate**. These are leaflet-like rather than stipule-like and should not be confused with **stipellas** which are stipules of leaflets in a compound leaf. These occur very rarely but may be seen in *Robinia pseudacacia* and

Box 4.5 Types of compound leaf.

Compound leaves are either **palmate** or **pinnate**. In pinnate leaves there is a strongly developed main rachis and the leaflets are disposed on either side of this. If pairs of leaflets are inserted into the **rachis** at the same point, the leaves are referred to as **jugate** and pairs of leaflets as **juga** (literally "yokes"). **Bijugate** leaves (a common type) thus have two pairs of leaves.

In palmate leaves the leaflets are all inserted by their petiolules onto the top of the petiole. The palmate leaf is most easily explained as being equivalent to a pinnate leaf in which the main rachis has not developed or at least is very short. Examples of palmate leaves include the horse chestnut, *Aesculus hippocastanum* (with seven leaflets) *Cannabis sativa* (with nine leaflets) and *Lupinus polyphyllus* (with many leaflets).

Second-order branching also occurs in both pinnate and palmate types, leading to **bipalmate** and **bipinnate** leaves in which the same morphology is successively iterated. There are, however, interesting cases where the morphology changes. An example is the **palmato-pinnate** legume, *Mimosa pudica*, in which a number of pinnate rachises are palmately arranged at the top of the petiole.

Box 4.6 Leaflet number in compound leaves.

In leaves with two leaflets (**bifoliolate**) or three leaflets (**ternate** or **trifoliolate**), it is scarcely meaningful to distinguish between palmate and pinnate classes. With such a small number of leaflets, the rachis cannot be developed so the leaf is, by default, palmate. The broad bean (*Vicia faba*) is an example of a bifoliolate leaf and *Trifolium* and *Menyanthes trifoliata* example of ternate leaves. Some compound leaves have been reduced in evolution to a single leaflet and are called **unifoliolate** (e.g. *Citrus*, some *Lupinus* spp.). Unifoliolate leaves can be distinguished from simple leaves by the presence of a joint (articulation) representing the rachis (specifically the point of insertion of missing lateral leaflets).

Other special terms for particular leaflet numbers are **quadrifoliolate** and **quinate**. Quadrifoliolate refers to a palmate leaf with four leaflets (e.g. the water-fern *Marsilea* or the supposedly lucky "four-leaved" clover). Quinate refers to a palmate leaf with five leaflets as in *Akebia quinata*. *Aegopodium podagrarium* has a bipalmate leaf based on a ternate construction in which each leaflet is ternate again. This is termed **biternate**.

Thalictrum species. They are found at the base of the petiolules or at the base of the leaflets.

4.17 Homology of the compound leaf

The compound leaf of euphyllophytes is thought to have evolved from lateral branch systems. These lateral branch systems were in turn derived from dichotomously branching ones, which became **monopodial** by unequal branching. At the ends of the branches increasing dorsiventrality evolved, finally leading to lamina outgrowth which evolved to give distal leaflets (pinnae and pinnules). A **heterochronic** shift at the start of lamina outgrowth to an early developmental stage would result in a continuous lamina. This scenario is similar to that envisaged in Zimmerman's "telome theory," except that his concept of "ontogenetic webbing" to give continuous lamina does not seem to have any realistic empirical basis.

The question can be asked whether any of these stem features of early leaf evolution can be seen in the modern leaf, particularly the compound leaf. The ubiquity of simple leaves in so many early-diverging angiosperm clades leads to the conclusion that the most recent common ancestor of extant angiosperms was simple leaved. However, many angiosperms, particularly those in the eudicot clade, have evolved very complicated leaves. A good example is chinaberry, *Melia azedarach,* in the Meliaceae. This has leaves up to 60 cm long that are bipinnate. Thus the rachis is branched and these rachis branches branched again into leaflets.

What is going on here? The simple leaf (which represents in evolution a condensed branch system) has changed the activity and timing of its lamina-producing growth and the activity of marginal organogenetic blastozones so that the simple leaf is expanded, by a reiteration of developmental programs. In angiosperms this appears to be an advanced feature. An angiosperm compound leaf is therefore no nearer to the photosynthetic branch systems of the progymnosperms than is a simple leaf.

The simple leaf is homologous to the entire compound leaf. The simple leaf is not homologous (at the organ level) to a single leaflet of the compound leaf, despite the, often uncanny, similarity between a leaflet and a simple leaf. This similarity between a leaflet and a simple leaf is to be expected, as the developmental processes are the same. The heterochrony that expands a simple leaf to a compound one causes the iteration of the lamina development processes later in development to form leaflets, rather than early in development to form a simple leaf. The processes themselves are, however, the same.

The interconversion between simple and compound leaves can run in both directions, and frequently has during eudicot evolution. Furthermore, as the interconversion is a heterochronic process all intermediates between extremes are possible. As the genetic control of leaf development has changed during evolution, so too might have the mechanism by which heterochronic evolution happens during evolution. Also, compound leaves can mimic simple leaves by being **unifoliolate**, as in *Citrus*. This is a completely different type of transition.

It is also evident that novel mechanisms have evolved to give pseudocompound leaves (see Section 4.19) that mimic those uncondensed by heterochrony. If the term "compound" is reserved for the latter, the more general term "dissected leaf" may be used to cover all complex leaf types while being neutral about mechanism. From these considerations we would expect the molecular control of leaf dissection to differ in different clades. Nevertheless, the work of Neelima Sinha and her colleagues has shown some remarkable commonalities of pattern, the most striking of which involve the expression of *KNOX* genes (Bharathan et al., 2002).

4.18 Molecular control of compound-leaf development

The expression of *type-I KNOX* genes is characteristic of the SAMs of branch systems. The expression of *KNOX* genes in leaf blastozones (Hareven et al., 1996) could be taken as some indication of the evolution of leaves from branch systems but is more probably simply a requirement for meristem function. It should be noted that even in compound-leaved angiosperms there is a transient downregulation of *KNOX* genes at leaf primordium inception. In simple-leaved plants, *KNOX* genes are never reexpressed.

Very roughly, it appears that there is a correlation between the complexity of the leaf primordium and the delay in final cessation of *KNOX* expression (Bharathan et al., 2002). The cessation of *KNOX* expression therefore indicates the adult phase of leaf development with the establishment of determinacy and lamina production. In simple-leaved species, such as *Arabidopsis*, *KNOX* is never expressed in the developing leaf. This can be viewed as an extreme example of peramorphosis, or the attaining of adult features (absence of *KNOX* expression) at a juvenile developmental stage. Conversely, compound leaves may be considered examples of pedomorphosis: the retention of juvenile characteristics (e.g. *KNOX* expression) late into development. (See Fig. 4.17.)

KNOX gene expression is part of a larger regulatory network that is not yet fully elucidated. However, it is known that *KNOX* acts, in part, within a *KNOX*/gibberellin antagonistic regulatory module. In tomato, *KNOX* gene expression downregulates gibberellin (GA) and increases leaf complexity. Conversely, the application of GA results in a decrease of leaf dissection, by antagonizing the *KNOX* gene mediated increase in leaf complexity (Hay et al., 2002). It would be interesting to known whether other compound-leaved species show a similar reaction to GA application and hence the generality of the *KNOX*/GA regulatory module in leaf development.

There are two important twists in the *KNOX* story. One involves secondarily simple leaves and the other involves legumes. Some simple-leaved plants such as anise, *Pimpinella anisum*,

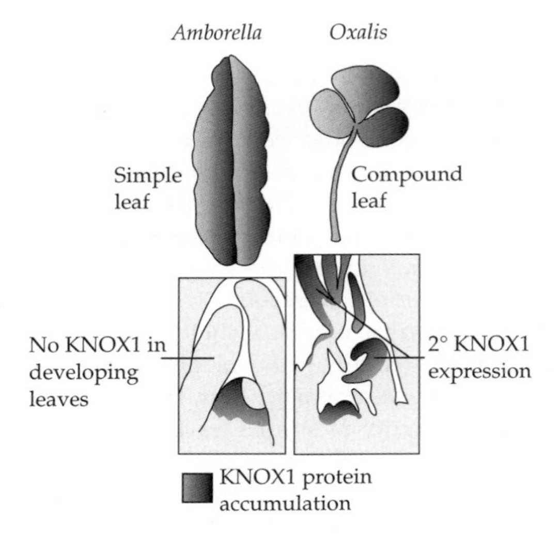

Fig. 4.17. Secondary *KNOX* expression in compound leaves, as opposed to simple leaves (after Champagne & Sinha, 2004).

have complex primordia with *KNOX* expression (Bharathan et al., 2002). This condition would normally be associated with dissected leaves. However, processes late in development gave rise to simple leaves despite the complex start. Other *Pimpinella* species, and other members of the family Apiaceae, have highly dissected leaves and it is reasonable to suggest that simple leaves in anise may be secondarily derived, in which case the complex primordium is a vestige of their compound-leaved ancestry.

The other exception is found in legumes. The compound-leaf primordia of pea do not express *KNOX* genes. Instead this role appears to be taken by *UNIFOLIATA* (*UNI*), the pea homologue of Ath*LEAFY* (*LFY*). This is very curious, as *LFY* is better known as a gene affecting the vegetative-floral transition and its loss-of-function mutant does not appear to have any leaf phenotype in *Arabidopsis*. Nevertheless, there are some intriguing hints that *LFY* may have some subtle role in leaf development in nonlegumes. First, there is a slight leaf phenotype (altered compoundness) in tomato associated with loss-of-function of the *LFY* homologue in tomato (Molinero-Rosales et al., 1999). Second, the gene *UNUSUAL FLORAL ORGANS* (*UFO*), which is linked with *LFY* in function, has a leaf phenotype in *Arabidopsis*. Furthermore, this leaf phenotype is not expressed in plants that have no *LFY* expression (Lee et al., 1997). These hints suggest that *LFY* has some slight role in leaf development. In turn this may explain why it has been co-opted in pea, and at least some other legumes, to regulate indeterminacy (blastozone formation) in place of *KNOX* genes.

There is still some confusion over the role of *ARP* genes in the development of compound leaves. *ARP* genes are responsible for downregulating *KNOX* genes. They are thus important candidates for a role in the development of leaf complexity, by restricting *KNOX* expression and promoting simple rather than complex leaves. However, the reverse appears to be the case in tomato (*Lycopersicum esculentum*). In the absence of the tomato *ARP* gene Le*PHAN*, there is no leaflet development, as evidenced by the *wiry* mutant of tomato (Kim et al., 2003b). This implies that both *ARP* and *KNOX* genes are required for leaflet formation and that, at least in tomato, they act constructively rather than antagonistically. However, an adaxial surface does not form in wiry mutants of tomato, as Le*PHAN* is required for correct development of abaxial–adaxial polarity. Bifaciality is required for lamina formation so it may be that leaflets cannot form in tomato as an incidental consequence of *ARP* effects on another aspect of leaf development, adaxial–abaxial polarity. Indeed variation in the pattern of bifaciality in diverse species is highly correlated with leaf form (Kim et al., 2003a).

A similar mutant to *wiry* of tomato is found in pea (Tattersall et al., 2005). *CRISPA* is the pea *ARP* homologue, and loss-of-function mutants have compromised adaxial identity, like tomato wiry mutants. However, leaflets form in the pea mutants while they do not in the tomato mutants. There are two possible considerations to be borne in mind in contemplating this difference. First, in pea the adaxial identity may not be sufficiently compromised to prevent leaflet outgrowth. Second, the basipetal (i.e. tip to base) process of leaflet formation in tomato may occur at a developmentally later stage than the acropetal process in pea (where the first leaflets form basally and early). If this is the case then leaflet development in tomato may be more lamina dependent than leaflet formation in pea (as lamina development is a late-stage developmental process). Whatever the reason, we should not be surprised at different nuances of genetic control of leaf dissection in different species, particularly as compound leaves in tomato (euasterid 1) and pea (eurosid 1) represent independent evolutionary events in very different lineages.

4.19 Pseudocompound leaves

There is a limit to how big a simple leaf can be, set by four physical problems of a massive continuous lamina. These are irrigating with water, supporting the weight, protecting from wind damage and shedding heat from the lamina. The largest simple leaves are therefore usually found in the understory of tropical rain forest where there is no problem supporting the lamina against wind, and the absence of direct sunlight prevents overheating. Examples include Araceae (*Alocasia robusta* has simple leaves up to 3 m long) and simple-leaf palms. Other big simple leaves are found in the giant water lilies (such as *Victoria* in the Nymphaeaceae), where the lamina is supported by water. Compound leaves solve these problems by having an increased amount of rachis to lamina to provide water and support, and by offering less wind resistance and being more efficient at convective cooling. For this reason most massive leaves, such as those of the palm *Corypha*, which may be up to 5 m across, are highly divided.

As we have seen, the majority of compound leaves are formed from branching of the leaf primordium by meristematic activity. Certain clades, however, have not evolved branched compound leaves but have instead achieved a compound-leaf phenotype by the physical disruption of a simple lamina. The simplest form of this is wind-tatter as seen in banana (*Musa* spp). When growing deep in the forest, bananas have simple leaves. In contrast, when growing in exposed areas, banana leaves tatter into strips as a result of wind action. Such leaves are still photosynthetically functional but offer less wind resistance and cool much more efficiently. Banana leaves that have not tattered, but which are exposed to the full tropical sun, often show brown patches indicating heat damage where the leaf lamina has heated up over a critical threshold.

Other plants have a much more controlled version of this process involving a type of **programmed cell death** (**PCD**) similar to **apoptosis** in animal cells. Populations of lamina cells die at an early developmental stage to create holes or cuts in the leaf lamina. An example of this is found in the swiss-cheese plant (*Monstera deliciosa*), which forms multiple holes in the lamina and at the lamina margin (Gunawardena et al., 2005). PCD is marked by the degradation of nuclear DNA into random small fragments by endonucleases and this process has been observed in *Monstera* (Gunawardena et al., 2005). There are numerous pathways and triggers for PCD in plants (Greenberg, 1996), but it is not yet known which of these are active in leaf morphogenesis.

The pinnacle of pseudocompound leaves is reached in the palms (Arecaceae). Palms have what is probably the weirdest of all types of leaf development (Corner, 1966b). Here the **primordium** starts out, as normal, as a rod-shaped structure. However, at a very early stage a series of tightly folded pleats form in the middle of the primordium. These pleats mark the places where the leaf will fracture to generate the pseudoleaflets. The fracturing is assisted by an abcission-like process (Nowak, Dengler & Posluszny, 2007). The rupturing may take place along the top fold of the pleat, giving rise to **leaflets** (**pinnae**) that are v-shaped in cross section (**reduplicate** leaflets). Alternatively, it may take place along the bottom fold of the pleat, giving rise to Λ-shaped leaflets (**induplicate** leaflets). It is possible that, at least in some species, PCD occurs along the rupture zone, but this has not been demonstrated. Alternatively it may be that a simple process of cell separation in files of cells is at work. Certainly mechanical rupture is completed by means of special-purpose **expansion cells**.

If no subsequent elongation of the **rachis** follows, then the leaflets are left in a fan-like arrangement and these species are called **fan palms**. However, if the rachis subsequently elongates the leaflets are carried apart in a feather-like arrangement, and such species are known as **feather palms**. A few palm species have a branched rachis and these palms are known as **fish-tail palms** because in species such as *Caryota* the leaflets resemble fish tails.

4.20 Stipules

Stipules are paired leaflet-like outgrowths on either side of the insertion of the leaf on the stem. Stipules are extremely important in plant taxonomy particularly at higher levels, as the occurrence of stipules is very constant within families. Plant families that do not have stipules are termed exstipulate. Organographically, stipules are part of the leaf even though a few stipules appear to originate in stem tissues. However, they are all associated with the leaf base and even when appearing to arise in the stem, they originate within the same population of meristematic cells as the leaf primordium although they may become separated by subsequent development of the shoot. The only exception to the rule that stipules are all associated with the leaf base is with **stipellas**. These are rare organs that appear to be developmentally equivalent with stipules but are found at the bases of leaflets on a compound leaf.

How, then, do we interpret stipules developmentally? In stipulate compound leaves, stipules are usually associated with acropetal leaflet development, as in pea (the first leaflets form at the base followed by progressively more apical leaflets). Other plants are basipetal (tip to base) in leaflet development, or centrifugal (developing first in the middle of the leaf and spreading to the apex and the base). Stipules are far more common in strongly acropetal rather than strongly basipetal types. Thus the acropetal pea has strongly developed stipules and the basipetal tomato is exstipulate. In this sense they can be viewed as the first leaflets of the leaf, extremes of an acropetal developmental tendency.

Stipules are significant for their effect on nodal anatomy. As they form so early, these basal branches of the primordium can affect the subsequent formation of procambial strands and hence stem vasculature. They are thus commonly irrigated by parts of the leaf traces that branch off while still in the stem. It is common for leaves to be provided with three traces from the stem and this character has been correlated with the possession of stipules (Sinnott & Bailey, 1914). When the stipules are big the whole of the lateral two traces may be used to irrigate the stipules. However, irrigation of smaller stipules is generally accomplished by branches from those lateral traces, and if the stipules are very small they may have no traces at all.

Stipules have a range of apparent functions. When large they may be important as extra leaf area for photosynthesis. More usually they serve to protect the developing leaf and the apical meristem as they often develop precociously.

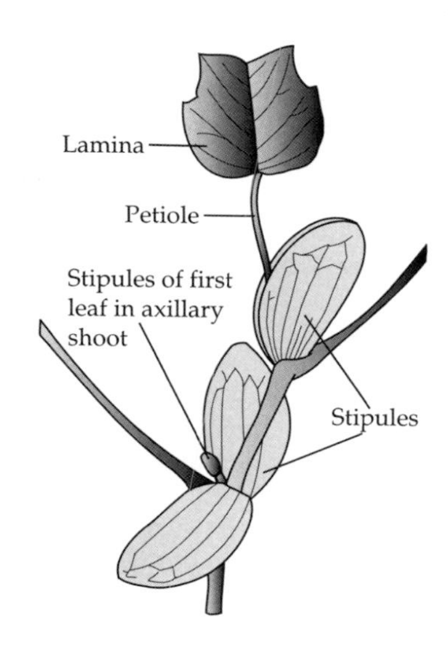

Fig. 4.18. Stipules (outgrowths of the leaf base) in *Liriodendron*. Their protective function in enclosing buds and young leaves is evident.

Associated with their wide range of apparent function is a wide range of morphology that derives primarily from variation in two processes: (1) various patterns of adnation with the petiole and (2) various patterns of congenital fusion with each other (see Box 4.7).

Stipules also differ in their **persistence** and this is related to their ecological function. If the stipule is adnate to the petiole the fate of the stipule is clearly tied to the fate of the leaf. However, free stipules can have an independent fate. They may be **caducous**, dropping soon after leaf expansion. **Caducous stipules** are usually a form of protection for the bud they therefore quickly become redundant after the shoot expands. Examples are *Ulmus campestris* and *F. sylvatica*. In these temperate trees with a spring flush of growth involving the expansion of leaves formed in bud, the stipules are shed soon after the winter bud opens. Unless plants with caducous stipules are examined at early stages they may be mistakenly thought to be completely lacking in stipules (**exstipulate**). **Deciduous stipules** are shed either around the same time the parent leaf is shed as in *L. tulipifera* (Fig. 4.18) and *Viola tricolor*. **Persistent stipules**, on the other hand,

Box 4.7 Types of stipules: patterns of adnation and congenital fusion.

Stipules are classified according to the degree of **fusion to the petiole** (**adnation**) and **fusion to each other** (**connation**) and by their **size, shape** and **persistence**. If stipules are not fused to the petiole or with each other they are known as **free stipules**. Examples are *L. aphaca* and *V. tricolor*. Alternatively they may be adnate to the petiole as in roses (*Rosa*) or clovers (*Trifolium*). In these plants the stipules, called *adnate stipules*, appear as a basal wing to the petiole. This happens when stipular initial cells divide along with the basal petiole cells so the stipules are carried along with growth of the petiole.

Stipules may be joined with each other. **Interpetiolar stipules** are the result of stipules of opposite leaves ontogenetically fusing. Neighboring stipules from different leaves may fuse to produce a continuous membrane of tissue between the opposing petioles. This is particularly characteristic of Rubiaceae but is also found in *Humulus* (hop).

In **axillary stipules** a pair of stipules from one leaf fuses together on the axillary side of the stem. The opposite is the **antidromous stipule** in which the two members of a pair of stipules fuse on the opposite side of the stem to the leaf. The most extreme form of stipular fusion is found in **ochreate stipules**. In these organs the stipules are joined on both sides of the stem forming a sheath around the stem often protecting the young shoot. It is considered as a single organ and is called an **ochrea**. It is characteristic of the Polygonaceae, Platanaceae (the plane tree, *Platanus*) and figs (*Ficus* in the Moraceae).

last much longer than their parent leaf. This is so for **stipular spines** such as those of *R. pseudacacia*. Stipular spines are usually strongly lignified and may remain on the plant long after the rest of the leaf has dropped.

A pteridophyte organ analogous to the stipule is the **ligule** of the lycophytes (e.g. *Selaginella* and *Isoetes*). Ligule is a very vague word in morphology meaning a small flap of tissue. It is also used for a flap on the leaves of grasses at the junction of sheath and blade, as well as for a small flap at the junction of claw and blade in certain petals of certain Caryophyllaceae (e.g. *Lychnis flos-cuculi*). These latter uses are not homologous with each other, or with the pteridophyte ligule.

4.21 Leaves in unusual places

Leaves are almost always formed as lateral organs of stems, formed on the flanks of the SAM as primordia. However, it is not uncommon for leaf-like organs, at least in mutants, to form on other leaves. Thus we can conclude that the leaf developmental pathway, although usually initiated from cells of the SAM, is not completely tied to the SAM and under certain conditions can be initiated on the leaf. By contrast, it is exceedingly uncommon for leaves to form on roots. The reverse is not true as ectopic roots form readily on many leaves and shoots.

An example of leaf-borne leaves is found in maize plants mutant for the gene Zm*LAX MIDRIB1* (*LXM1*; Schichnes, Schneeberger & Freeling, 1997). In these mutants leaf-like flaps arise at the base of the leaf, often paired on each side of the midrib. Interestingly, these do not appear to initiate from an ectopic meristem on the leaf, but instead from leaf cells that appear to have retained leaf-making competency in the mutant. The *LXM1* gene has not yet been cloned, but it is possible that it is an upstream regulator of *KNOX* genes as the phenotype shares some features with the knotted mutants of maize. The involvement of *KNOX* genes would suggest some similarity with the process of leaf formation (on other leaves) that is characteristic of rosulate species of *Streptocarpus* (Harrison et al., 2005b). In these species of *Streptocarpus* this is the only means of leaf production, as there is no vegetative stem or SAM.

A rather different situation is found in *B. hispida* var. *cucullifera*. In this plant, patches of abaxial cells form on the adaxial surface, promoting ectopic lamina growth at the boundaries of the adaxial–abaxial cell identity (Sattler & Maier, 1977). This phenomenon is further discussed in the section on the **bifacial theory of lamina development**.

Just as it is possible to find leaves developing on leaves, it is also possible to find shoots developing on leaves. These are often at the margin as in the reproductive propagules (plantlets) of *Kalanchoe pinnata*. In *Kalanchoe*, the molecular developmental genes *LEAFY COTYLEDON1* (*LEC1*) and *SHOOTMERISTEMLESS* (*STM*), characteristic of organogenesis (*STM*) and embryogenesis (*LEC1*) have been co-opted into leaf development (Garcês et al., 2007).

Shoots may arise from the leaf midrib, as in the inflorescences of unifoliate and rosulate *Streptocarpus* species. These do not have a conventional SAM but instead develop an inflorescence-producing **groove meristem** on the leaf midrib (Möller & Cronk, 2001).

However, in most cases where flowers or inflorescences appear to arise on the leaves, and are said to be **epiphyllous**, as in *Helwingia japonica*, this is a superficial connection only. The epiphyllous flowers of *Helwingia* (Fig. 4.19) are in fact axillary and are carried onto the leaf blade ontogenetically by meristematic growth of the leaf below the axillary flowers (Dickinson, 1978).

Fig. 4.19. Epiphyllous flowers in *Helwingia*.

4.22 Special forms of leaves

The leaf has proved to be an enormously important organ in the diversification of seed plants. Leaves can take on a huge number of forms besides the familiar **foliage leaves**. Floral leaves such as the **integuments** of the ovule, **carpels**, **stamens** and other **sporophylls**, **nectarophylls**, **petals**, **sepals, bracts** and **bracteoles** will be dealt with in other sections. Here the range of leaves not associated with reproduction will be examined. These are **twining leaves**, **leaf tendrils**, **cataphylls** (**scale leaves**), **ascidiate leaves** (**pitchers** and **bladders**), **spines** and **cotyledons**. Cataphylls and cotyledons often have important functions as **storage leaves**, examples being **bulb scales** and **hypogeous cotyledons**. Cotyledons in other species, such as many palms, do not function as storage leaves or photosynthetic leaves. Instead they are **absorption leaves** for absorbing the nutrition of the endosperm.

Stems can become twining organs or tendrils, and this is also true of leaves (Fig. 4.20). In this similarity, leaves and stem show their morphologically similar nature. When a leaf is a twining

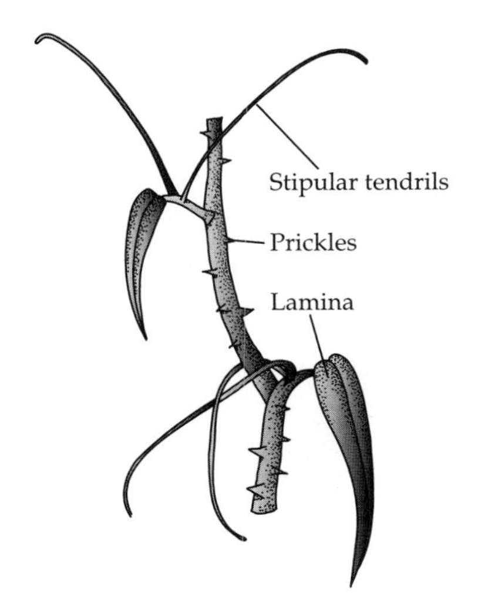

Fig. 4.20. Stipular tendrils in *Smilax*.

organ, it is usually the petiole that twines, or in the case of compound leaves, the rachis. A good example of a **twining leaf** is the fern *Lygodium*. This is a compound leaf with a twining rachis (Fig. 4.21). Another way leaves can be used for climbing is by means of **leaf tendrils**. The whole leaf may be converted to a tendril, or a tendril can be formed as a prolongation of the midrib or the rachis. Leaf tendrils are stem-like, without lamina and essentially similar to stem tendrils. Examples are the tendrils of the Cucurbitaceae and tendrils in Leguminosae such as the pea (*P. sativum*; Fig. 4.22) and *Lathyrus aphaca*.

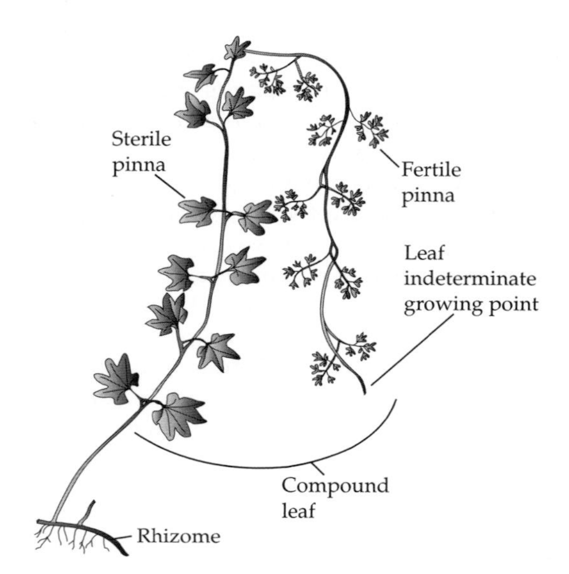

Fig. 4.21. Evergrowing climbing leaf of *Lygodium* with an indeterminate growing point.

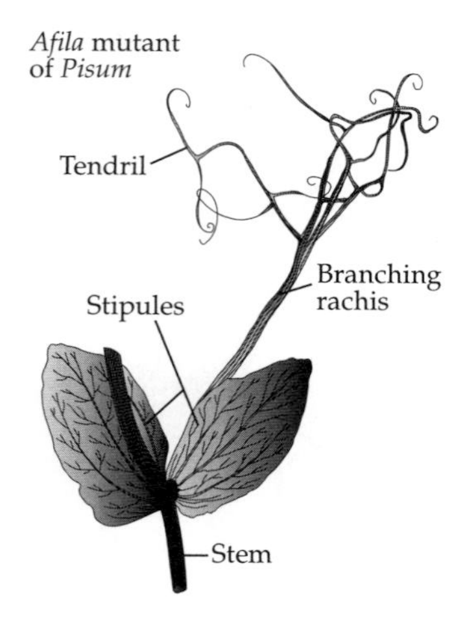

Fig. 4.22. The *afila* mutant of pea which has normal stipules but instead of leaflets a branching rachis terminating in tendrils.

Leaves frequently form **spines,** the leaf equivalent of **thorns** (stem organs). These may represent whole leaves reduced to a sharp pointed midrib as in Cactaceae and *Berberis*. The leaf-derived nature of such thorns can often be determined from the presence of shoots developing in the axils of the thorns as in *Berberis*. Alternatively, they may represent stipules as in the paired **stipular spines** of *Robinia*.

Some *Acacia* species, such as the bull-horn acacia (*Acacia cornigera*), have hugely enlarged stipular **ant-spines,** which function as homes for ants beneficial to the plant. The ants perforate the enlarged spines and hollow out the contents. In addition there are two other leaf adaptations relevant to this mutualism: the acacia provides food for the ants from **extrafloral nectaries** on the petiole and **beltian bodies** at the ends of the leaflets (rich in lipids and proteins). Curiously, the sugar solution produced by the extrafloral nectaries consists almost entirely of invert sugar (sucrose cleaved postsecretion by plant invertase enzymes). This is not favored by generalist ants but is apparently required by the specialist ants, which have very low levels of invertase in their guts (Heil, Rattke & Boland, 2005).

4.23 Epipeltate and ascidiate leaves

The phenomenon of **peltation** is an extremely important one in plant morphology. It results in the junction between the petiole and lamina being in the center of the underside of the leaf rather than at the lamina margin. As the surface of the petiole is continuous with the abaxial (lower) surface of the lamina, the petiole is unifacial with only abaxial surface. Examples are leaves of *Tropaeolum* and *Hydrocotyle*.

Peltation follows naturally from any weakening of adaxial identity at the base of the leaf primordium as a consequence of the **bifacial theory of lamina extension**. As the entire petiole surface in peltate leaves has abaxial identity, the boundary between adaxial and abaxial identity must run across the primordium (in a so-called **cross-zone**). This naturally results in lamina extension across the petiole as well as to the sides resulting in a **peltate leaf** (Fig. 4.23).

As the boundary between abaxial and adaxial identity, so important in peltation, is set by the differential expression of abaxial *YABBY* genes and adaxial *MYB* genes, one would expect to see in peltate leaves an expansion in the domain of leaf *YABBY* genes. This appears to be the case in *Tropaeolum* (Gleissberg et al., 2005). The *Tropaeolum majus* homologue of the *Arabidopsis* gene *Ath FILAMENTOUS FLOWER* (*TmFIL*) is widely expressed in the developing leaf primordium leaving only a small region of apparent adaxial identity.

The common form of peltation is known as **epipeltation** as it results from a weakening of adaxial identity and a consequent extension of abaxial identity. Following the same reasoning it is easy to see how the opposite, namely a weakening of abaxial identity and extension of adaxial identity, could give rise to an **inverse peltate** or **hypopeltate** leaf with an adaxially surfaced leaf stalk running into the middle of the adaxial surface of the leaf (Fig. 4.24). However, this is exceedingly rare. Well-studied examples include certain bracts of *Pelargonium* (Dupuy & Guédès, 1979) and the sepals of *Viola arvensis* (Jaeger, 1963). A curiosity occurs in the leaves

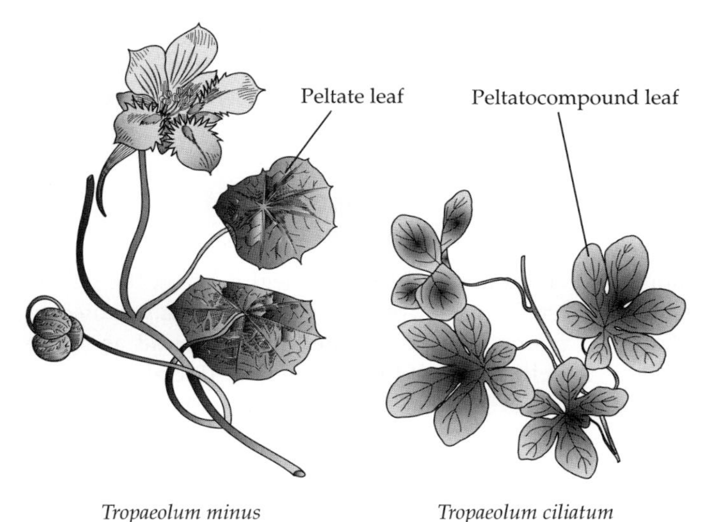

Fig. 4.23. Peltate and peltatocompound leaves in *Tropaeolum*.

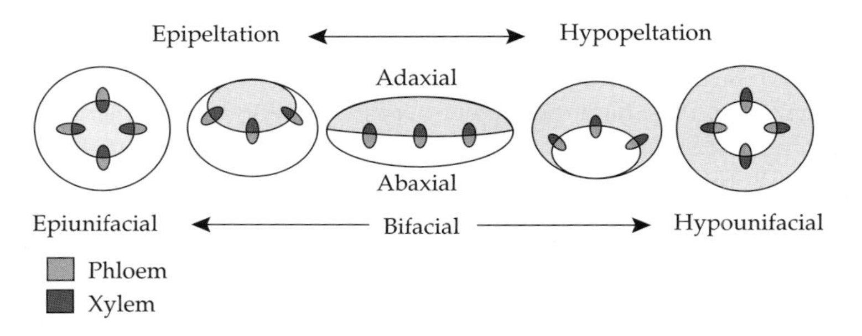

Fig. 4.24. Scheme of epipeltation and hypopeltation in relation to a normal bifacial leaf (centre). *Tropaeolum* leaves are epipeltate (the commonest sort) and their petioles resemble the figure at left in cross-section.

on *Codiaeum variegatum* cv. "Appendiculatum," which has a hypopeltate appendage growing out of the leaves (Baum, 1952b).

Some leaves, however, are completely **inverse unifacial** or **hypounifacial** (i.e. the top surface has spread under the leaf, which therefore has only adaxial surface). Cactaceae leaves (for instance *Opuntia*) are reportedly hypounifacial (Guédès, 1979), at least in some species (they are possibly best considered homologous to petioles and therefore are technically phyllodes). Cactus leaves, however, are not peltate as there is no second surface to promote laminal outgrowth.

It may be significant that these examples of **hypopeltation** and **hypounifaciality** have no evident leaf base, which is the part of the leaf in which the adaxial surface is likely to be most weakly expressed.

When peltate leaves have differential growth across the lamina they may deform into a cup, or, in extreme cases, **sac-like** or **ascidiate** leaves. **Ascidiate leaves** take the form of deep pitchers for trapping insects in *Cephalotus, Darlingtonia, Sarracenia* and *Nepenthes.* All these are examples of **epiascidiate** leaves. Occasionally, mutants or cultivars of noncarnivorous plants produce ascidiate leaves, as in the *Codiaeum* mentioned above. In *Codiaeum* the hypounifacial outgrowth often develops a **hypoascidiate** tip (Baum, 1952b). These varieties are interesting as they potentially give some idea of how ascidiate leaves might evolve.

The ovule of angiosperms has been ingeniously interpreted (Guédès & Dupuy, 1970) as a megasporangium-containing ascidiate "leaflet" at the margin of the carpel (a modified leaf). The evidence for this is almost entirely teratological. It is fairly common in developmentally abnormal plants for ovules to be folarized (becoming leaf-like). Strongly foliarized ovules have no outer integument and the inner integument is opened into a leaf-like blade with the inner surface of the integument equivalent to the adaxial surface of the leaf-like growth. The interpretation of teratologic observations is problematic, as mentioned earlier. However, in support of this interpretation is the observation that ascidiate leaflets and leaf margins are known as nonteratologic features of leaves. For instance, the glands at the top of the petiole in *Prunus padus* are ascidiate microleaflets. This interpretation of the ovule of angiosperms is at variance with interpretations of the seed-plant ovule from the fossil record of pteridosperms. However, it is not impossible that the ovule in angiosperm ancestors had a separate origin to ovules in other groups of seed plants.

Another ascidiate structure that should be mentioned is the carpel itself. The classical theory considers the carpel to be derived evolutionarily from a folded lamina. Many carpels, in early development, do indeed seem to illustrate a folded stucture. Many carpels, however, including those of many relatively primitive plants such as *Drimys* (Winteraceae) and *Persea* (Lauraceae), exhibit sac-like (ascidiate) growth at early stages of development. Thus the same processes that are active in producing the ascidiate structure of leaves may have been present in the evolution of some carpels.

4.24 Peltatocompound and diplophyllous leaves

A variant of the peltate leaf is when the peltate leaf is itself compound, and the same mechanisms apply to the **peltatocompound** as to the **peltate** leaf. If the petiole is **unifacial** and adaxial surface therefore occurs only at the end of the leaf, then the **adaxial–abaxial junction** occurs all around the leaf tip. Thus leaflets are produced symmetrically around 360° (Fig. 4.25). This is well seen in lupins (*Lupinus* spp., Leguminosae). In general with peltate leaves peltatocompound leaves may occur as well. An example of such a genus is *Tropaeolum*. In addition to peltate species such as *Tropaeolum major* and *Tropaeolum minus* there are numerous peltatocompound species such as *Tropaeolum edule* and *Tropaeolum polyphyllum*. Thus, in a genus with a common peltate ground plan the difference between peltate and peltatocompound species is determined by whether the lamina production is continuous or segmented.

An interesting variant on the peltatocompound leaf is the *diplophyllous* leaf. In this, the petiole is unifacial but instead of lamina production being circular as in *Tropaeolum*, there is strong proximodistal bias in lamina production with one "flap" of lamina being formed distally and one flap being formed proximally. Thus there is the appearance of two leaves (hence "diplophyllous"). Alternatively, it can be thought of as a peltatocompound leaf with two leaflets, one distal and one proximal. Diplophyllous leaves are very rare but exceedingly curious when they occur. They are present in monophyletic group of sub-Antarctic marsh marigold species such as *Caltha dionaeifolia* from Tierra del Fuego and *Caltha novae-zelandiae* from New Zealand (Fig. 4.26; Schuettpelz & Hoot, 2004).

The extensive studies of Baum and Leinfellner (Baum & Leinfellner, 1953) are suggestive that

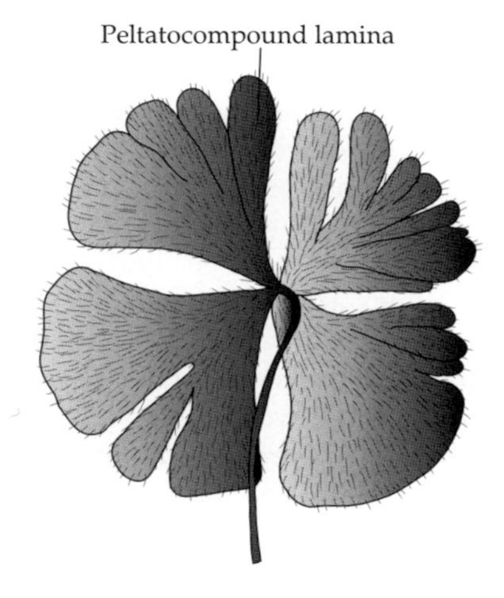

Fig. 4.25. Peltatocompound leaf in a fern (*Marsilea drummondii*).

stamens are equivalent to diplophyllous leaves. In this case the **filament** would be equivalent to the unifacial petiole and the **anther** is assumed to be composed of the two parts of the stamen folded inwards each bearing a marginal sporangium on each side of the lamina. Opposing **sporangia** on each of the leaflets then are united into **synangia** to give the **bilocular tetrasporangiate** condition. Thus the **adaxial** part of the anther is equivalent to one leaflet and the **abaxial** to another. Not only does this very neatly explain all the major features of the stamen, there is evidence for the diplophyllous stamen from organs transitional between stamens and petals that frequently occur in some species (Baum, 1952a). It follows that, if the stamen is correctly interpreted as a diplophyllous organ, diplophylly may be considered one of the most important conditions in angiosperm evolution.

4.25 Cotyledons

Cotyledons are merely the first leaves of the plants but as they develop in the unusual environment of the **seed** they are usually distinctive and therefore dignified with their own name. Some morphologists have considered them equivalent to **prophylls** (see below) in which case the **hypocotyl** (the stem between the radical and the cotyledons) is equivalent to the **hypopodium** of **sylleptic new shoots**. However, there is no evidence for any meaningful comparison with prophylls apart from, in some species, the reduced complexity and altered phyllotaxy. Cotyledons are thus best considered as specifically modified

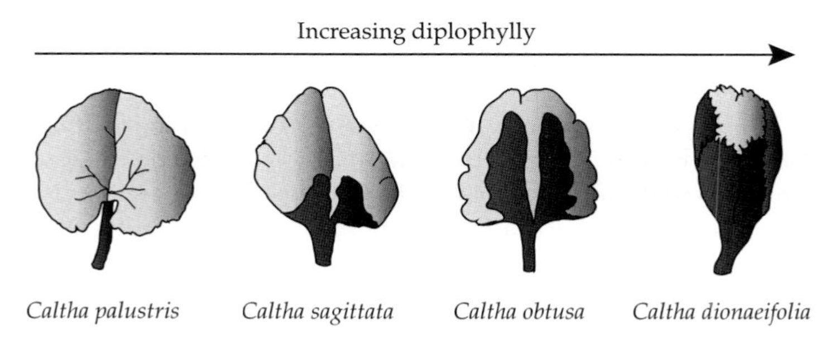

Fig. 4.26. Diplophylly in *Caltha* (marsh marigolds) (after Schuettpelz & Hoot, 2004).

foliage leaves, or even as the starting point of a heteroblastic series from juvenile leaves to adult (Tsukaya et al., 2000).

In gymnosperms there may be numerous cotyledons (up to 24 in *Pinus maximartinezii*) but they are green, photosynthetic and borne aboveground. In the angiosperms the number of cotyledons is more stable at one or two (very rarely three or more). Furthermore, a major distinction occurs between those cotyledons that are pushed aboveground by the elongation of the hypocotyls (**epigeous**) and those that remain underground as storage organs (**hypogeous)**. The evolution of cotyledons as hypogeous storage organs has had a major effect on the human diet. Hypogeous cotyledons of peas and beans are extremely protein rich and nutritious.

A third type of cotyledon is used for neither storage nor photosynthesis but for absorption. This is the absorption cotyledon found in palms (Arecacae) and in the water chestnut (*Trapa natans*). In these species the main seed nutrient reserves are in the form of endosperm. The cotyledons are therefore modified into absorptive and transport organs in order to convey endosperm nutrition to the rest of the embryo. They thus remain embedded in the seed on germination.

A curious cotyledon anomaly occurs in **unifoliate** (single leaf) members of the genus *Streptocarpus*. These plants are monophyllous because they develop no apical meristem. Instead one cotyledon aborts and one cotyledon continues growing with a basal meristem. It grows and grows, reaching 75 cm in some species. A groove meristem subsequently forms on the midrib of the cotyledon allowing inflorescence stalks to sprout directly from the cotyledon. This cotyledon is so distinctive and unusual, with so many stem-like characters that it has been given the name **phyllomorph**. The molecular mechanism of this unorthodox system is still unknown. However, what is known is that some species upregulate *KNOX* gene expression to produce further phyllomorphs from the primary phyllomorph. This habit is called **rosulate**, in contrast to the strictly unifoliate species (Harrison et al, 2005b). Curiously there are also perfectly normal or **caulescent** species within the genus, which produce a normal apical meristem and opposite leaves on stems. Evolutionary transitions appear to be possible between all three types (Moller & Cronk, 2001).

4.26 Scale leaves

Scale leaves, also known as **cataphylls**, are leaves that are generally reduced in both size and complexity compared to **foliage leaves**. They may, however, be conspicuously swollen if their function is storage. Developmentally scale leaves appear to be **heterochronic**, maturing at what would be an early stage of development in foliage leaves. Therefore scale leaves are usually morphologically equivalent to the basal parts of the fully developed foliage leaf. When a series of intermediates between scale leaves and foliage leaves is found, it is easy to discern their developmental nature.

Scale leaves are usually small, of simple structure and very often without chlorophyll. Rather than green they are often colorless, brown or yellow. They lack prominent veins and according to their function they may be chaffy, leathery or succulent. Scales are also found on stems and leaves (particularly in ferns). However, these are outgrowths of the epidermis and not morphologically related to leaves.

Shoots often have a pronounced scale-leaf zone (**cataphyllary region)** at the base (as in *Pinus* and *Quercus*), and this may be well demarcated or may gradually transition into foliage leaves (Fig. 4.27). In leaves with pronounced dormant buds, the most obvious cataphylls at the base of the incipient shoot are the **bud scales**, which are usually caducous when the shoot begins extension growth. The cataphyllary regions may be associated with stem adaptation. For instance, certain orchids, such as *Dendrobium*, have a pronounced cataphyllary region at the base where the stems are swollen into pseudobulbs consisting of several internodes with no foliage leaves. In extreme cases the entire plant may be cataphyllary. Many achlorophyllous plants have only scale leaves. Their leaves, no longer constrained by a photosynthetic function, are thus highly

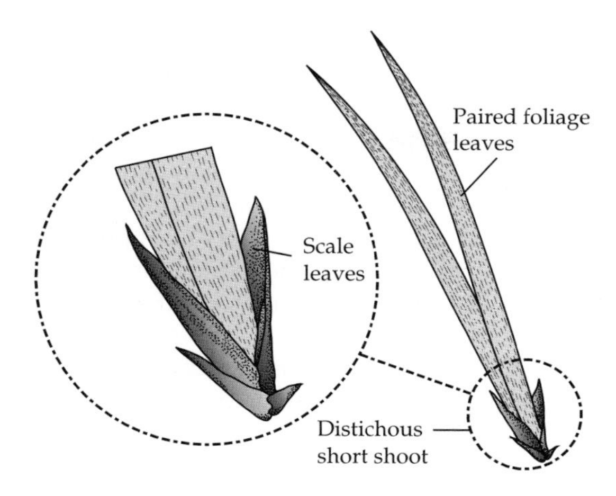

Fig. 4.27. Short shhots of a tropical pine (*Ducampopinus krempfi*) showing the scale leaves in two ranks and two flattened foliage leaves.

reduced throughout the plant (e.g. *Monotropa, Neottia, Orobanche*).

The evolutionary developmental problems posed by cataphylls come in two parts. First there is the question of how the developmental programs responsible for foliage leaves are truncated, or otherwise modified, to give scale leaves. The second question is how cataphyllary regions are demarcated on the developing shoot. The existence of such basal cataphyllary regions implies a phase change in the SAM.

4.27 Prophylls

Prophyll is a name given to a special type of scale leaf or cataphyll. They are the first leaves of new shoots (excepting bud scales). They are only worth distinguishing from foliage leaves or basal cataphylls if they have a distinctive placement (**phyllotaxy**) and are reduced in form compared to normal foliage leaves. They are morphologically interesting as they often show a definite phyllotactic relationship relative to each other or to the parent stem of the branch, rather than falling into the phyllotactic series of the subsequent foliage leaves of the shoot.

Prophylls are most commonly distinguished in inflorescences where they are also referred to as **bracts** or **bracteoles**. However, some vegetative stems exhibit prophylls, but more rarely. The cotyledons have occasionally been considered as the prophylls of the primary shoot. However, there is little justification or utility in this. The cataphylls of the inflorescence and flower, particularly bract and bracteole, will be dealt with in other sections.

Monocots usually have a single prophyll whereas dicots usually have two, although many plants do not have anything worth distinguishing as a prophyll. In monocots they are usually placed on the adaxial side of the base of the branch. These are called **adaxial prophylls**. Dicots generally have **transverse prophylls**, in which the paired prophylls are placed on a branch transversely, relative to the parent stem. Dicot prophylls are often **opposite** or **subopposite**. There is usually a different phyllotactic **divergence angle** between prophyll and the first foliage leaf, compared to the divergence angle between foliage leaves.

In **sylleptic branches** the prophylls are often carried a little away from the parent branch by the growth of the shoot. In this case the basal stem under a prophyll is called a **hypopodium**. If the prophyll is basal there is correspondingly no hypopodium and this is characteristic of **proleptic branches**.

The existence of prophylls raises the question of how they are differentiated, at a developmental level, from subsequent foliage leaves. The first leaves of a lateral SAM do not have previous leaf primordia to instruct their phyllotactic pattern by acting as auxin sinks (see Section 4.32). Instead they may be influenced by the SAM of the parent axis (still in close proximity when the prophylls form), and by the subtending leaf. This **external influence theory** (**EIT**) predicts that prophylls should be absent from lateral shoots formed *de novo* on older stems as a result of wounding or other stimulation of the secondary cambium (or at least different in form).

4.28 Scale leaves underground

Scale leaves are often the only leaves formed on underground stems. Underground they are known as **rhizome scales**, **corm scales** or **bulb scales**.

Scales on rhizomes may be merely vestigial versions of leaves. In bulbs they have a storage or protective function. An important morphological distinction between bulbous types hinges on the extent of the **cataphyllary region** in bulbs. In *Lilium* and *Fritillaria* the entire bulb is composed of bulb scales. However, in *Galanthus* and *Gagea* the cataphylls are only the outermost leaves, forming a sheath around the bulb. In these plants the storage function is primarily taken by the swollen bases of the normal foliage leaves which rise above the ground. When the above-ground parts of the leaves die, the swollen bases are often persistent, as in *Galanthus*. Thus what may appear to be bulb scales are in fact **swollen leaf bases** of normal leaves.

Another important distinction to make about bulbs is whether they are tunicated or naked. Bulbs with an outer cataphyll that wholly surrounds the bulb as a sheath are called **tunicated bulbs** and the cataphyllary sheath is called a tunic. Bulbs in which each outer cataphyll only partially surrounds the bulb, as in *Lilium*, are called **naked bulbs**.

Corms (i.e. tubers) also usually have a cataphyllary region at their base to provide a tunic of scale leaves as in *Colchicum*, *Malaxis* and *Crocus*. Often these tunicating scale leaves are very persistent so they contain the original tubers and subsequent ones produced. In *Arum maculatum*, however, they are less persistent so the older tubers are exposed and only the younger ones tunicated.

4.29 Phyllotaxy: orthostichies and parastichies

Phyllotaxy is the arrangement of leaves on a stem relative to each other. It is dependent on the position on the flanks of the **SAM** in which successive leaf **primordia** develop. This is evolutionarily constrained by two main factors. First, there are developmental constraints on the successful packing of leaf primordia onto the **apical dome** without interference. Second, ecological constraints may influence phyllotactic patterns of how mature leaves are stacked on top of one another. This is something that has adaptive consequences for light interception and efficient photosynthesis, as it will influence the amount of **self-shading** in the canopy. Many plants can escape self-shading by petiolar bending and torsion to carry out fine adjustment of leaf position. However, phyllotaxy is the primary means of building an efficient leaf canopy and is of great adaptive importance.

Primordia are successively initiated around the perimeter of the SAM. The helix joining successive primordia is called the **genetic helix**. There is an angle, called a **divergence angle**, between successive primordia. If that angle divides exactly into 360° (or multiples of 360°) then rows of leaves up the stem will result. For instance in the case of *Carex*, which has leaves in three ranks, the divergence angle is 120°. By imagining that leaf primordia produced successively 120° apart, it is easy to see that three ranks of leaves will be produced. The fourth leaf primordium will be positioned exactly in line with the first and with others in that rank. These straight **vertical ranks** of leaves are called **orthostichies**. *Carex* is therefore **tristichous** (**three-ranked**) with three orthostichies (three straight ranks). If the divergence angle is 180°, two orthostichies will result and the leaf arrangement is called **distichous** (**two-ranked**). This is found in some grasses (e.g. *Zea mays*) and other monocots and in some dicots such as certain shoots of elms (*Ulmus*).

A potential adaptive problem with orthostichies is that as the ranks are vertical some self-shading is likely to occur, at least when the sun is overhead. There are two ways in which phyllotaxy can depart from the rigid orthostichy. One is to have poor developmental control of divergence angle so that the orthostichies, although statistically present, are irregular. This appears to happen in some *Carex* species in which the tristichous arrangement is sometimes difficult to discern. However, the divergence angle of phyllotaxy is amazingly constant in most plants, and appears to be under quite tight developmental control. Neatly ranked orthostichies can be broken, however, depending on the divergence angle. If the divergence angle does not divide

PLATE I. Forms of the stem. (A) Cormous stem of a banana species (*Musa balbisiana*). This photograph shews an old corm (i.e. a short, erect, swollen underground stem, see section 2.18.2) from which the pseudostem of leaves (see section 4.6.1) has rotted away, revealing the circular leaf scars. (B) Filiform, functionally leafless yellow stems of the parasitic member of the Lauraceae, *Cassytha filiformis* (Lauraceae). The only leaves are the cotyledons, which remain in the seed coat (germination hypogeal) and the minute alternate scale leaves, in the axils of which the inflorescences are borne. The superficially similar genus *Cuscuta* (Convolvulaceae) scarcely forms leaves (see section 2.16). (C) Stems can be modified into thorns and hooks (see section 2.22.1). The widespread liana *Pisonia aculeata* (Nyctaginaceae) has axillary shoots in the form of thorns, produced in grapple-like pairs at the nodes. (D) The inflorescence hook of *Artabotrys hexapetalus* (Annonaceae) is a modification of the inflorescence stem that assists climbing.

PLATE II. Forms of the root. (A) Pendulous prop roots (section 3.15) of *Ficus benghalensis*. Note the proliferation of root branches where the roots have been cut. (B) Coralloid roots of a cycad (*Bowenia* sp). These surface roots have a distinctive coralloid form (3.17.3), and are negatively geotropic, normally providing a home to photosynthesizing colonies of nitrogen fixing cyanobacteria. (C) Vining, thigmotropic root tendrils (section 3.15) of *Philodendron* sp. (D) Transverse section of velamen of a *Dendrobium* orchid. The velamen is a multiple-layered epidermis (rhizodermis) of dead cells that fill passively with water during wet periods (section 3.14). The pink staining indicates suberin deposition and the green cells are the photosynthetic chloroplast-containing cells of the cortex below. The outermost velamenous layer has very little suberin deposition.

PLATE III. Forms of the leaf. (A) Fenestrate leaf of *Aponogeton madagascariensis (A. fenestralis).* The holes in the leaves are caused by precisely controlled apoptosis (programmed cell death) (section 4.19). (B) Fenestrate leaves of a palm *Reinhardtia gracilis* (Arecaceae). The fenestration is caused by the incomplete separation of leaflets. Separation of leaflets in palms appears to be caused by an abscission-like process (section 4.19). (C) Heteroblastic leaves (section 4.11) of the West Indian shingle plant, *Marcgravia rectiflora* (Marcgraviaceae). (D) Small brown scale leaves (section 4.26) of the Canary Island butcher's broom *Semele androgyna* (Ruscaceae). The large green leaf-like organs are cladodes (flattened stems, section 2.22.3) borne in the axils of the scale leaves. At a later stage flowers are borne along the edges of the cladodes (from "nodes" within the cladode).

PLATE IV. Sporangium and seed. (A) Naked ovules on the female strobilus of *Gnetum gnemon.* (B) Stamen-like microsporophylls in the male strobili of *Gnetum gnemon.* Rudimentary ovules are present in the male strobilus which function as nectarial organs and assist in insect pollination. One of these "false pollination drops" is arrowed. (C) Dissection of a *Zamia* ovule, showing a "nest" of seven male gametophytes (arrowed) growing down from the pollen chamber and protruding through the nucellus towards the female prothallus (cut away) (cf. Figure 5.21). Microgametophytes of a cycad (*Zamia* sp.). The microgametophytes will eventually burst to release motile sperm. (D) A bisected coconut (*Cocos nucifera*) showing germination. The transfer cotyledon (section 5.20) embedded in the coconut flesh (endosperm). Emerging from the "eye" is a structure comprising the radicle, hypocotyl and cotyledonary petiole. The plumule is encased in the tubular cotyledonary petiole, from which it will later break out.

PLATE V. Sporophylls and sporangia in Cycas. (A) A female *Cycas* showing indeterminate "cone" of young megasporophylls with the apex reverting to the production of foliage leaves above (*Cycas clivicola*) (see section 6.4). (B) A female *Cycas* at a later stage shewing the elongated megasporophylls and the enlarged egg-like ovules/seeds (*Cycas rumphii*). (C) A young *Cycas* megasporophyll showing six marginal megasporangia. (D) A male plant of *Cycas* showing the determinate male cone (*Cycas maconochiei*). The microsporophylls are subtended by transitional, somewhat spinous, leaves.

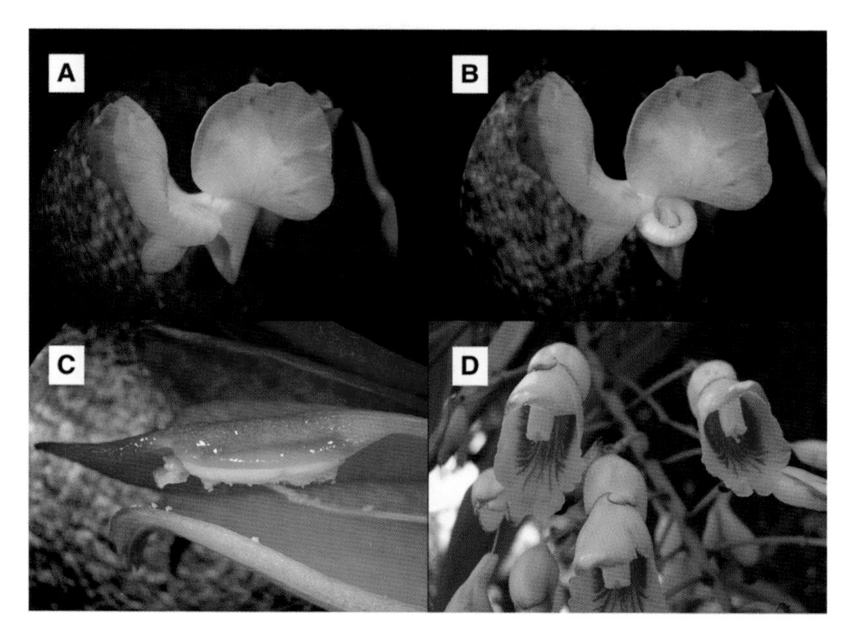

PLATE VI. Specialized flowers in the Zingiberales. (A) and (B). Tripped and untripped flowers in *Marantochloa* (Marantaceae) shewing the fundamental asymmetry of these flowers (section 6.15). (A) shows the flower in untripped state and (B) shows the same flower having being artificially "tripped" as it would be by a pollinating insect. (C) Longitudinal section through a flower of *Costus sp.* (Costaceae), showing the single stamen adnate to the roof of the flower, and the style threaded through the stamen. (D) Flowers of *Alpinia zerumbet* (Zingiberaceae) showing the prominent labellum (lip) which is of staminodial origin (section 6.21).

PLATE VII. Specialized flowers in the Leguminosae. (A) The bird pollinated *Delonix regia* (Caesalpinioideae) with 10 stamens. (B) *Bauhinia monandra* (Caesalpinioideae) with a similar floral form but showing extreme staminal reduction (section 6.30) with the androecium reduced to a single stamen. The flower in the centre is pollinator-receptive. The flower to the left is post-pollination, showing the bent back (retroflexed) adaxial petal and a colour change, features also shown by *Delonix regia*. (C) The inflorescence of African sicklebush, *Dichrostachys cinerea* (Mimosoideae) shows floral dimorphism. The upper (pink flowers) are sterile, serving to increase the attractiveness of the inflorescence to pollinators. The lower (yellow) flowers are fertile. (D) The regular flower of *Cadia purpurea* (Papilionoideae) has evolved from typically monosymmetric legume flowers by means of a change in expression of floral symmetry genes (see section 6.16). A change of pollination syndromes, from bee pollination to a bird pollination syndrome is associated with this change in symmetry.

PLATE VIII. Fruit and seed. (A) The pineapple, *Ananas comosus*, an indeterminate syncarp (section 6.27). The meristem of the fleshy inflorescence has reverted to the production of foliage leaves at the top, but has stopped growing. Growth will only resume if the top of the syncarp is cut off and planted, which is a method of propagation. (B) The syconium (fig) of *Ficus dammaropsis*, a syncarp that is effectively determinate. This fig is unusual in that numerous inflorescence bracts are prominent, showing the derivation of the fruit wall from peduncle tissue. (C) Excavated seed of a germinated double coconut (*Lodoicea maldivica*) showing the seedling a short distance away. Not visible is the cotyledonary tube (sheath) that connects the haustorial cotyledon within the seed to the rest of the seedling. *Lodoicea* has the distinction of bearing the world's largest seed. (D) The leafy fruit of *Pereskia grandifolia* (Cactaceae), demonstrating that the fruit wall is derived from stem tissue in the epigynous flowers of the Cactaceae (section 6.26).

exactly into 360° (or into low multiples of 360°) then leaves will not be positioned in vertical ranks.

It is easy to see the effect of a divergence angle of 120°, as noted above. However, if the divergence angle is slightly more, say 125°, the fourth primordium, instead of coming to be positioned exactly over the first, will be offset by 15°, and so on. This progressive offsetting causes the ranks to take the form of a helix. These helical ranks are called **parastichies**. In the case of the 125° divergence angle, there would be three parastichies each with an **offset angle** of 15°.

Similarly, a divergence angle of 144° gives five orthostichies. This arrangement is known as a **pentastichous** or **quincuncial phyllotaxy**. A divergence angle of 1° more, 145°, would give five parastichies with a 5° offset angle. The potential adaptive advantage of offset phyllotaxy is that leaves are only rarely exactly over another.

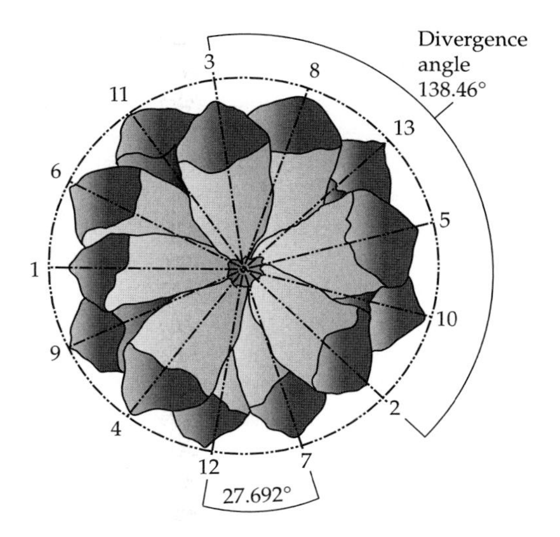

Fig. 4.28. Spiral phyllotaxis of cone scales in a pine (*Pinus strobus*).

4.30 Helical phyllotaxy

The phyllotaxy literature is extensive going back hundreds of years. Generations of botanists have become fascinated by the mathematical patterns that emerge from the phyllotactic helices. This has tended to cause some mystification of the subject. It should be borne in mind that the patterns and spirals have no intrinsic geometric or biological significance. They are merely consequences of natural selection for different **divergence angles** between primordia. These divergence angles have evolved because (1) they are developmentally possible on the SAM and (2) they create leaf arrangements that are adaptively viable in the ecosystem. The divergence angle is thus the fundamental basis of plant phyllotaxy. It should be noted that in the phyllotaxy literature the divergence angle is often given as a fraction of 360°. Under this system, 120° (120/360) is denoted as 1/3 and 144° (144/360) as 2/5.

In the cone of *Pinus strobus* the divergence angle of the cone scales is approximately 138.5° or 5/13 (Fig. 4.28). This can be visualized as 13 orthostichies with every thirteenth primordium perfectly aligned with the primordium 13 before it. This is not very easy to see or very useful as a model. Alternatively, this phyllotaxy can be modeled as a deviation from the pentastichous divergence angle of 144° (2/5), giving rise to five parastichies with a 27.5° offset angle. It is easy to see the five helices in this sort of pinecone. However, it could also be modeled as a deviation on tristichy with three parastichies with an offset of 55.5° (a very shallow helix), but these three shallow helices are more difficult to see. Of course it is completely arbitrary whether we consider a pinecone to be made up of 13 orthostichies or 5 parastichies or 3 parastichies. The counting of helices is just a way of modeling the placement of primordia on the shoot and they do not have any intrinsic biological significance.

Almost all **divergence angles**, in the production of successive leaves, are between 90° and 180°. An angle of 90° is the **tetrastichous** (four-ranked) arrangement. At angles less than this, the successive primordia are produced very near to each other on the same side of the apical dome. The developing primordia are therefore very likely to interfere with each other's development. Probably in consequence these cases are very rare. Nevertheless even **one-ranked** or

monostichous (divergence angle 0°) or offset-monostichous arrangements (divergence angle small) are found occasionally. The reason that divergence angles greater than 180° are not found is that these would be equivalent to their complement which would be less that 180°, and by convention the smaller angle is taken as the divergence angle. Thus a divergence angle of 240° can be treated as a divergence angle of 120°, but going the other way.

In **helical phyllotaxy** most divergence angles are between 128° and 140°. Arguably the most efficient divergence angles are between 135° and 140°, as these tend to produce large numbers of orthostichies or highly offset parastichies. There is consequently very little overlap of successive organs and so little self-shading and efficient **space packing**. Space packing is important if large numbers of substantial organs are produced with short internodes, as in the cone scales of a pine (*Pinus*) or fir (*Abies*).

4.31 Whorled and decussate phyllotaxy

Rather than single primordia being initiated at any one time, it is common for sets of primordia to be initiated at once. If pairs of leaves are initiated simultaneously this is known as **opposite phyllotaxy**. If more than two leaves are initiated it is known as **whorled phyllotaxy** and these sets are called **leaf whorls (verticils)** at maturity. If whorls are of three leaves, they are called **ternate** or **trimerous** and if of five, **pentamerous**. Sometimes paired leaves can mimic whorled as in certain Rubiaceae (e.g. *Galium cruciata*) which appear to have four leaves in a whorl but two of these leaves are thought to derive from the **interpetiolar stipules** of what is only a single leaf pair. Examples of true whorls are found in *Nerium oleander* and *J. communis*. They are also found in many water plants such as *Hippuris* and *Elodea canadensis*.

There are two ways of arranging whorls. One is so that the leaves are always directly above a member of the preceding whorl. This arrangement produces the same number of ranks as there are leaves in a whorl. This is rare but is found in a floral condition known as **obdiplostemony**. In this the outer whorl of stamens is on the same radii as the preceding whorl of petals. It is also found in the fossil sphenopsid *Archaeocalamites*†. In this the lateral branch structures (protomegaphylls) are directly above each other in successive whorls. In angioperms with many leaves, however, the presence of whorl-members directly above each other means that successive primordia are very close on apical dome and likely to interfere. In *Archaeocalamites*† therefore this system is only possible because the lateral branches are widely separated by internodal tissue, even at a young stage.

The other way is to offset each successive whorl so that each leaf is produced between two leaves of the preceding, and succeeding, whorls. This has the effect of keeping the primordia apart on the apical dome and lessens the degree of potential self-shading. It produces twice as many orthostices as there are members of the whorl. This is the commonest form (and it predominates in most flowers in which the floral leaves are in whorls).

Opposite leaves (in pairs) are extremely common and although distichous arrangements occur, successive pairs are usually positioned at right angles. This arrangement is called **decussate** in which the leaves are **opposite and alternate**. It is the characteristic arrangement of most Lamiaceae and in the genus *Acer*. The orientation of primordia pairs at right angles to the preceding primordia allows them to initiate as far as possible from existing primordia. When stems bearing decussate leaves are oriented horizontally the leaves are usually brought into one plane by twisting of the internodes or petioles or both. These **pseudodistichous** horizontal shoots are well seen in many decussate shrubs producing arching stems, such as *Deutzia*, *Philadelphus* and the larger species of *Hebe*.

In cases where the pairs of leaves in decussate phyllotaxy are unequal, either by a slight difference in insertion, in axillary bud development or in leaf size, two arrangements are possible. Leaves of a particular type can be helically arranged down the stem. This is the usual pattern in the Caryophyllaceae, in which this helical-decussate phyllotaxy occurs frequently.

However, in *Fraxinus* and *Cuphea* leaves of a particular type occur in two ranks only, zigzagging down the stem. The shoot is thus, remarkably, asymmetrical. On one side of the stem leaves are of one sort, and on the other side of the stem they are of the other sort.

4.32 Phyllotaxy: molecular control of divergence angle in the SAM

Divergence angle is the determinant of **phyllotaxy**. It is therefore of great interest to know what determines divergence angle and what processes are responsible for the constancy of divergence angle in individual shoots and in taxonomic groups.

The place in which a new primordium initiates is very constant with respect to preexisting primordia. New primordia generally initiate as far as possible from existing primordia, where there is **available space** (Snow & Snow, 1952). Microsurgical experiments using lasers show that, in tomato, the two youngest primordia (P1 and P2) are sufficient to determine the position of the incipient primordium (Reinhardt et al., 2005). P3 and P4 are responsible for fine tuning, determining the size and boundaries of the primordium. An obvious explanation for this would be the diffusion of a chemical signal. Alan Turing modeled this as his well-known **diffusion–reaction hypothesis** (Wardlaw, 1953). Under this model it was assumed that the source of the diffusible signal was the primordia themselves. It is now known that the opposite is true. Primordia are sinks for a promoting substance, not sources of an inhibitor.

The diffusible signal is **auxin** (Reinhardt, Mandel & Kuhlemeier, 2000; Reinhardt et al., 2003), which is necessary for leaf primordium initiation. The auxin originates elsewhere in the plant and is transported, under polar transport, toward the apical meristem through the outer cell layers of the shoot apices. It is in these layers that the auxin efflux transporter, **AraPINFORMED1 (PIN1)**, and the auxin influx transporter, **AUXIN RESISTANT1 (AUX1)** are known to be coexpressed (Reinhardt et al., 2003). This might produce a uniform auxin field in the surface layers of the apical meristem were it not for the effect of the existing leaf primordia, which act as powerful sinks for auxin, locally depleting auxin from their vicinity. It appears that only in regions of the apical dome that are away from primordia can a critical threshold of auxin be reached to promote primordium initiation (Fig. 4.29).

It appears that auxin acts by repressing ***KNOX1*** genes, which in turn repress the ***ARP*** genes that are vital for leaf formation. Thus, locally high concentrations of auxin will lead to the expression of *ARP* genes and the commencement of leaf formation. Clearly differences in the amount of auxin in the system, differences in the strength of the primordium auxin sink, and variation in the critical auxin threshold for leaf formation could all potentially lead to variation in phyllotaxy. However, the size of the apical dome will also be a factor.

This appears to be the case in the **abphyl1** mutant of maize (*Zea*), which has opposite rather than alternate leaves (Jackson & Hake, 1999). ***KNOX1*** genes stimulate the production of **cytokinin** and thus fuel increased cell division. Increased cell division would, in the absence of negative regulators, lead to an expansion of the apical meristem. The ***ZmABPHYL1*** (***ABPHY1***)

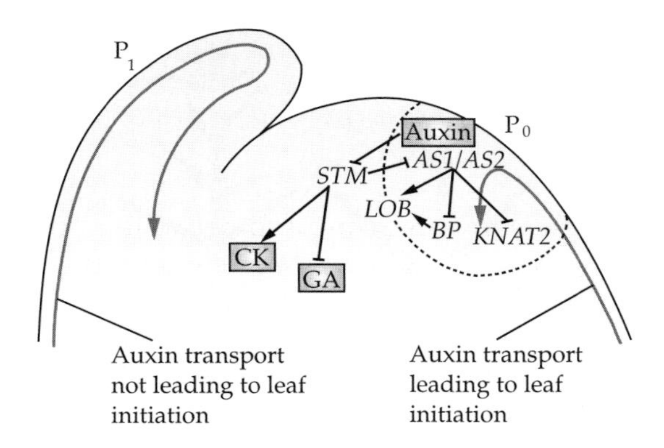

Fig. 4.29. The control of phyllotaxis at the level of the SAM. Existing primordia keep auxin away from the SAM. Auxin represses SHOOTMERISTEMLESS (STM) triggering the gene cascade leading to primordium development (after Champagne & Sinha, 2004). AS = ASYMMETRIC LEAVES, BP = BREVIPEDICELLUS, KNAT2 = Arabidopsis KNOTTED-LIKE2, LOB = LATERAL ORGAN BOUNDARIES.

gene is cytokinin inducible and probably negatively regulates this cytokinin-induced expansion of the meristem (Giulini, Wang & Jackson, 2004). When *ABPHY1* action is lacking, as in the **abphy** mutant, a larger meristem results, which is apparently permissive to a change in phyllotaxy (Fig. 4.30).

It would be extremely instructive to examine cases in which phyllotactic patterns change within a plant, for instance associated with **heterophylly** and to correlate this with hormonally driven phase changes in plant development. It would be instructive too to consider how general the primordium auxin sink model is. The early microsurgical experiments (Wardlaw, 1950) were done with *ferns*, particularly *Dryopteris*, and the results are consistent with this model. However, the **microphylls** of *lycophytes* and the **phyllidia** of *bryophytes* also have precisely determined phyllotaxy and it would be interesting to know how similar the determining system is.

It has been suggested that divergence angle may be affected by **mechanical signaling** due to the deformation on the apical dome produced by developing primordia (Green, Steele & Rennich, 1996). This might create stresses in the sheet of cells flanking the apical meristem and it has been suggested that physical buckling might be a cause of primordium initiation. Alternatively, cells have signal transduction mechanisms responsive to mechanical stress that might be involved in stimulating meristematic activities at certain positions. There is some experimental evidence for this from experiments on artificially changing the mechanical stresses within sunflower (*Helianthus*) receptacles and noting the change in floral initiation (Hernandez & Green, 1993). The sunflower receptacle initiates flower buds in a helical pattern similar to the helical initiation of leaves in phyllotaxy. However, it could be that mechanical stress in some way changes auxin flux through the apical meristem.

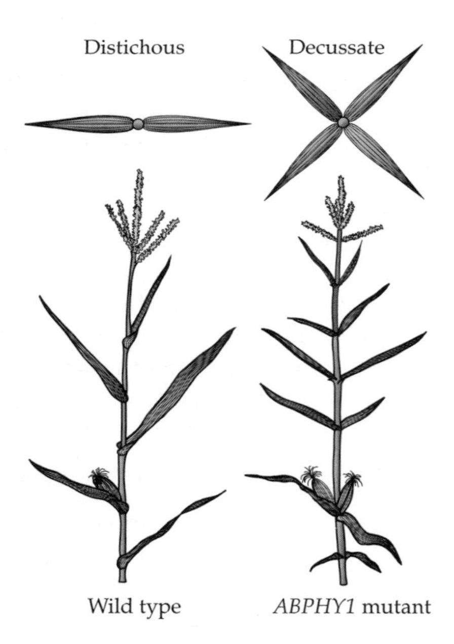

Fig. 4.30. Transition the distichous-alternate phyllotaxy in wild-type maize and the decussate arrangement of the *ABPHY1* mutant (after Jackson & Hake, 1999).

4.33 Leaf rolling and folding: ptyxis

Linnaeus first introduced the terms **vernation** and **aestivation** to denote how foliage leaves and perianth leaves are packed into vegetative buds or flower buds respectively. However, the terms are not completely equivalent. The central problem of packing in the flower bud is that many flowers have perianth parts in **whorls** so they must overlap in bud. It is this overlap that is centrally treated in the terminology of aestivation. The packing problem of a leaf bud is different, the relatively large and stout young leaves have to be rolled and folded as well as to have patterns of overlap. For precision therefore, the pattern of folding and rolling of an individual organ is distinguished by the term **ptyxis** (Fig. 4.31; Cullen, 1978). Vernation is therefore either used as a more general term, referring to both the folding of individual leaves and their overlap in bud, or, in the strict sense, only to the pattern of overlap between different leaves in bud.

These patterns are generally fairly constant and evolutionarily stable. They are consequently useful in distinguishing plant families. The patterns derive from morphogenetic processes (tissue bending) in the primordium that are rather poorly understood. On the expansion of

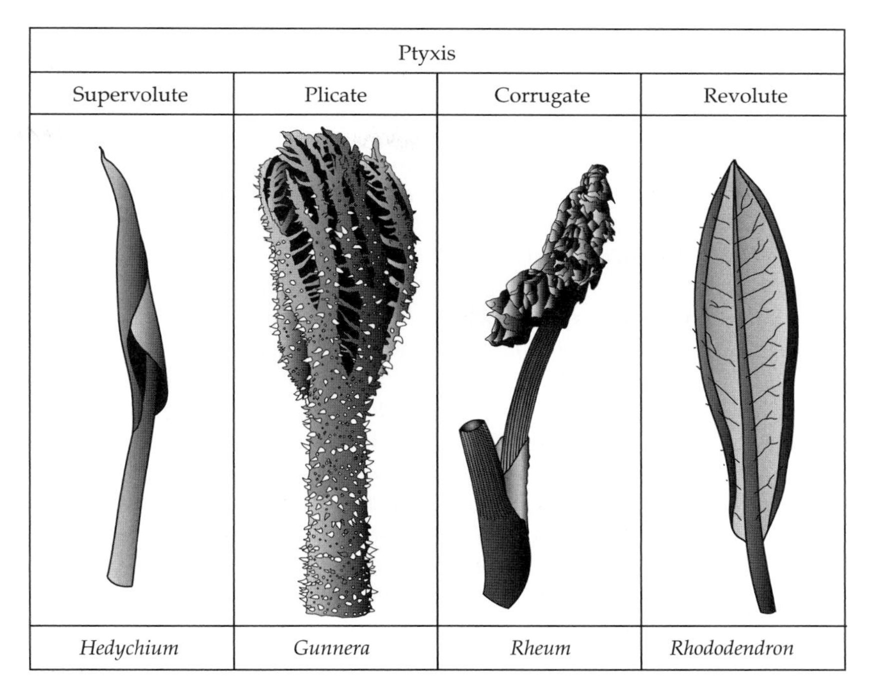

Fig. 4.31. Some patterns of ptyxis in leaves.

the mature lamina, ptyxis is normally lost and the leaves become planar. However, in certain cases vestiges remain at maturity, for instance in the pleated leaves of palms and in leaves with rolled margins.

The primary division is between **rolled** (**volute**) and **folded** ptyxis. In rolled ptyxis the immature leaves may be rolled on a **mediolateral axis**, that is, inwardly rolled from margin toward the midrib. Alternatively, they may be rolled back from the apex, that is, on a **proximodistal axis**, which is known as **circinate**. Circinate ptyxis is found in ferns, in the cycad *Bowenia* in the cotyledons of *Zannichellia*. The leaves are usually rolled back along the **adaxial** side of the leaf (antrorsely circinate), as in ferns. However, a few plants have the leaves rolled back along the **abaxial** side (retrorsely circinate). An interesting pair in this regard is *Drosera* (**antrorsely circinate**) and *Drosophyllum* (**retrorsely circinate**). Thus, types of ptyxis are differentiated, inter alia, according to their behavior relative to the **dorsiventral axis**.

Laterally rolled leaves may be rolled up on both sides toward the midrib on the adaxial side (**inrolled** or **involute**) or the abaxial side (**downrolled** or **revolute**). Involute plants include poplar, *Pyrus*, *Nymphaea*, *Viola*, *Sambucus*, *Lonicera*, *Weigela*, *Alisma*. Revolute leaves include those of *Salix*, *Rhododendron*, *Cycas revoluta*, Polygonaceae, *Nerium* and *Clematis*.

An extreme type of rolled leaves is the **tube-rolled** leaf (**supervolute** leaf). This sort of leaf is rolled up from one side only so that one margin is inside the roll and the other outside the roll. A well-known example of this is found in the banana (*Musa*), and it is characteristic of the whole of the Zingiberales as well as other monocots such as *Convallaria*. However, it is also found in dicots such as *Prunus* and *Berberis*.

Folded leaves may be crumpled (corrugate), pleated (plicate), infolded (conduplicate), marginally infolded (induplicate), marginally downfolded (reduplicate) or inflexed (reclinate). Crumpled leaves are scrunched up without any definite pattern like a handkerchief stuffed into a pocket. Examples include rhubarb leaves (*Rheum*) or the petals of poppy (*Papaver*). Pleated leaves are folded like a fan along the primary veins. This is a form found in many very large leaves, for instance in palms (Arecaceae), as it is an effective means of compaction. However, it is found in a surprising number of smaller leaves such as *Ribes, Vitis, Fagus, Alchemilla* and in some petals, as in the Solanaceae.

Inflexed leaves, in which the lamina is bent back along the top of the petiole, occur in *Liriodendron, Anemone* and *Aconitum*. In *Liriodendron* this has the result of making the leaves fit easily within the protective envelope formed by the stipules (Fig. 4.18), even though the petiole is relatively long. It is often combined with conduplicate ptyxis of the lamina. Conduplicate ptyxis is probably the commonest type. In this the two halves of the lamina are folded up along the midrib, rather in the manner of a closed book, with the adaxial side innermost. This type includes *Liriodendron, Syringa, Quercus* and many others.

As noted above, ptyxis is generally an evolutionary stable character. However, different patterns do sometimes occur in related species making for interesting comparisons. In *Primula,* involute and revolute ptyxis both occur (Conti et al., 2000) while in *Nothofagus* four types occur: conduplicate, revolute, plicate and flat (plane; Philipson & Philipson, 1979). Sometimes the orientation of the rolling or folding changes between abaxial to adaxial as in the related genera *Populus* versus *Salix* or *Drosera* versus *Drosophyllum*. In other cases, rolling, as opposed to folding, may be matters of degree. An extreme of variation is found in the compound leaves of cycads (Cycadales), in which several types of ptyxis can be found in the same leaf. For instance in species of *Bowenia* the **longitudinal ptyxis** (i.e folding along the **proximodistal axis**) is circinate while the **horizontal ptyxis** (i.e. folding along the **mediolateral axis**) is involute. The pinnae repeat the circinate ptyxis of the whole leaf while the pinnules have plane (flat) ptyxis (Stevenson, 1981).

CHAPTER 5

Sporangium to seed

5.1 Alternation of generations

A very remarkable feature of all land plants (and some algae) is that they have two distinct generations in their life cycle: a haploid **gametophyte** (gamete-bearer) that produces gametes and a diploid **sporophyte** (spore-bearer) that produces haploid spores by a reduction division (meiosis). The fusion of gametes at **fertilization** to form a diploid zygote is the gateway from the gametophyte to the sporophyte. **Sporogenesis**, the formation of haploid spores from the diploid sporophyte, is the gateway from the sporophyte to the gametophyte.

This contrasts with animals. Generally animals are diploid and the reduction division produces not spores but gametes directly, without the interposition of a gametophyte organism. Some algae, such as the bladderwrack (*Fucus*), also have this animal-like or **diplontic** life cycle (Fig. 5.1).

Other algae, including those in the lineage that includes land plants, are the reverse: **haplontic**. In haplontic algae the adult plant is haploid and produces gametes, which on fertilization form a zygote. The zygote undergoes a reduction division immediately after germination to produce another haploid generation. Charophytes (such as *Chara* and *Nitella*) and *Coleochaete* (Fig. 5.2) are the algae most closely related to land plants, and they are all haplontic.

Coleochaete thalli produce attached eggs and dispersed sperm. Fertilization gives rise to a **zygospore** (**oospore**), which then undergoes a combination of somatic and reduction divisions to form up to 32 haploid zoospores, each one of which is capable of producing a new thallus (Graham & Wilcox, 1983; Wesley, 1930). There is good evidence that the developing zygote is fed by the mother (**matrotrophy**) probably by the provision of hexose sugars through the maternal transfer cells abutting the zygote (Graham & Wilcox, 1983, 2000). From a life cycle point of view the production of eggs, the retention of eggs on the mother and the maternal feeding of the next generation (even if it is only a zygote and its few descendant cells) are important preadaptations in algae for the subsequent evolution of the land plants.

The land plants have further delayed the reduction division of the descendant cells of the zygote and **interpolated** further somatic cell divisions at the diploid level to form a complex organism: the sporophyte generation. They therefore have a **diplohaplontic** life cycle and so display **alternation of generations**. Seed plants typically have very large sporophytes and gametophytes that are evolutionarily reduced. The result is that in large trees the interpolated sporophyte generation may occupy 100 trillion times the volume of the minute gametophytes.

In bryophytes such as mosses the alternation of generations is evident, as the gametophyte (moss, i.e. protonema and gametophore) is very different from the sporophyte (a stalked spore capsule borne on the moss plant). In ferns the alternation of generations is also evident, as both the gametophyte and sporophyte are free living. The sporophyte is familiar as the commonplace fern. The gametophyte is less familiar as it is smaller. However, any close examination of suitable habitat, such as shady clay banks in moist temperate or tropical regions, will reveal the small green thalli of fern gametophytes (sometimes they are filamentous, as in *Schizaea*).

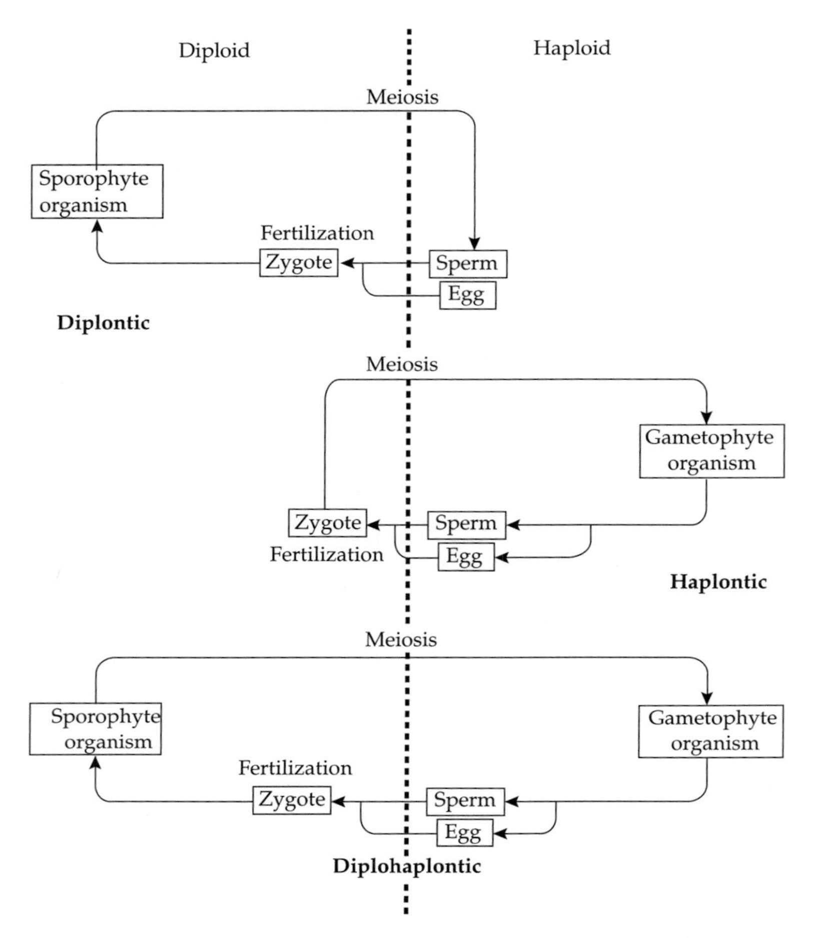

Fig. 5.1. Life cycles of plants. The diplohaplontic life cycle is found in all land plants, while the haplontic life cycle is found in those algae most closely related to land plants. Diplontic life cycles are found only in a few algae and in animals.

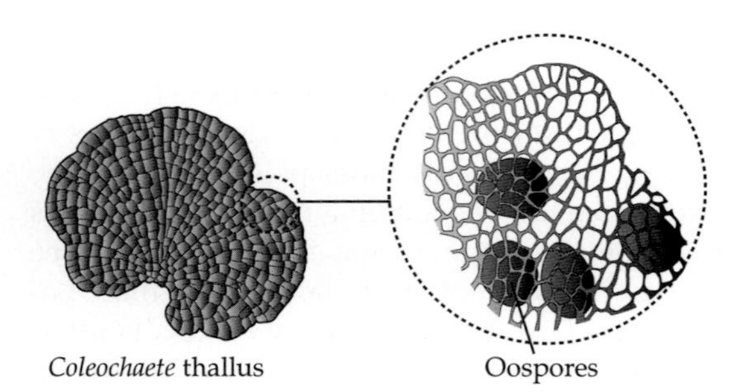

Fig. 5.2. Plate-like thallus of *Coleochaete* showing parenchymatous cellularity and the presence of oospores in the thallus.

In seed plants, however, alternation of generations is not obvious; the gametophyte is microscopic or hidden. The discovery that all land plants show alternation of generations therefore revolutionized botany, and the story of its discovery is one of the most brilliant pieces of detective work in the history of science. Most of the credit is due to Wilhelm Hofmeister (1824–1877), who first laid out the life cycle homologies of all land plants in his magisterially detailed "Comparative researches" (Hofmeister, 1851; Kaplan & Cooke, 1996).

Before Hofmeister, the gulf between the Linnaean groups of Cryptogamia and Phanerogamia[1] seemed unbridgeable. Attempts at homologies were wide of the mark. For instance, the fern prothallus was widely assumed to be the fern's "cotyledon." By showing the universal fertilization of eggs by sperm in free-living plant gametophytes and finding the homologous processes in seed plants, Hofmeister built bridges between the previously unbridgeable.

5.2 Evolutionary significance of the diplohaplontic life cycle

Two very important questions are raised by the universality of the diplohaplontic life cycle in land plants. First, what were the selective forces that drove the origin of the diplohaplontic life cycle from haplontic algal ancestors? Second, what are the evolutionary consequences of this life cycle for the land plants?

The green plant clade to which green algae and land plants belong (Viridiplantae) is largely haplontic but has evolved alternation of generations independently in the sea lettuces (*Ulva* and relatives)[2] and in the land plants. The two cases appear to be rather different. In the ulvophytes the two generations are usually identical (isomorphic alternation of generations). A complete transfer of developmental information has therefore taken place between diploid and haploid. Defining homology as a "continuity of information," the two phases are "homologous." However, in extant land plants, the alternation of generations is always heteromorphic and so requires a different developmental trajectory for each generation.

[1] Cryptogamia, Linnaeus' term for "lower" plants, derives literally meaning "secret marriage" from the Greek, *cryptos*, secret, *gamein*, to marry) and derives from their absence of pollination, which Linnaeus correctly associated with sex. Phanerogamia, Linnaeus' term for seed plants, derives from the Greek, *phaneros*, manifest or apparent. J. Hedwig (1730–1799) made a major contribution to breaking down this system by discovering the sexual organs of mosses in 1774.

[2] The ulvophyte lineage consists of six groups (the Ulvales, Ulotrichales, Trentepohliales, Cladophorales, Caulerpales and Dasycladales). Alternation of generations is recorded in all of these except the Ulotrichales.

In the ulvophytes, in which both generations are independent and free living it is possible to postulate an ecological explanation for the origin of the haplodiplontic life cycle. Slight differences in ecological preferences between phases may allow a more comprehensive exploitation of the environment than one phase alone (Hughes & Otto, 1999). In the land plants there is dependence of the sporophyte on the gametophyte. For instance a fern sporophyte can grow only in a position previously occupied by a gametophyte. This argues against the **ecological hypothesis**. However, some fern gametophytes do have a wider ecological range than the sporophytes, maintaining themselves by asexual reproduction (e.g. Rumsey et al., 1998, 1999).

Another explanation is the **dispersal hypothesis**. The diplohaplontic life cycle in land plants provides, at its most basic level, two dispersal phases in the life cycle, sperm and spore. Sperms have very limited dispersal capacity, mainly through rain splash. Spores, however, can travel thousands of miles to colonize remote oceanic islands. The interpolation of a sporophyte generation, capable of making innumerable spores, therefore had a profound effect on plant distribution. Heterosporous plants largely threw away that advantage by making spores large enough to provide a home for the female gametophyte. Seed plants have partially regained the advantage by wrapping their fertilized female gametophytes in structures suitable for dispersal (i.e. seeds and fruits).

A third explanation is that a sporophyte phase is advantageous as it can mask the effects of deleterious alleles. The **masking hypothesis** has been examined theoretically and experimentally (Mable & Otto, 1998, 2001; Otto & Goldstein, 1992) and found to be plausible under relatively high rates of recombination. However, if linkage is not broken up by recombination the haploid phase is favored.

The masking hypothesis works only if the diploid phase is more elaborate than the haploid, and there are consequently many genes that are not expressed in the haploid. If the two phases are isomorphic all the alleles will be exposed anyway during the gametophyte stage. However, the masking hypothesis may well explain why large organisms such as the giant Redwoods of California are diploid not haploid. Quite apart from the difficulty of sperm swimming from tree to tree, long-lived organisms will accumulate somatic mutations, which will be masked only in diploids. The need for masking may also explain the widespread occurrence, in algae, of **coenocytic cells** (cells with multiple nuclei). Such cells allow for internucleus masking of somatic mutations, even in haploid organisms. A similar logic applies to the **defense hypothesis** (Nuismer & Otto, 2004). Sporophytes are likely to be more able to defend themselves from pathogenic organisms as they may have allelic diversity in parasite recognition and defense genes, making it harder for parasites to evade the host defense.

Another attractive hypothesis is the **delayed sex hypothesis** (Richerd, Couvet & Valero, 1993). Sex is a feature of gametophytes as only gametophytes produce sex cells (sperm and eggs). However, sex may have a cost, for example, the waste of producing numerous male gametes for every successful fertilization event, or the difficulty of achieving fertilization in dry terrestrial conditions, and in particular cross-fertilization between suitable individuals. If sex is delayed by interpolating a sporophyte generation, then the costs of sex are reduced. One problem with this theory, however, is that there is an alternative, simpler, response to high-cost sex: vegetative reproduction.

Perhaps the most intriguing hypothesis of all is the **paternal forcing hypothesis** (Haig & Wilczek, 2006). This posits that the paternally and maternally derived complements in the genome of the zygote would be in conflict. It might be in the evolutionary interests of the maternal genome to limit allocation of maternal resources to the zygote. However, the paternal genome would benefit from the extraction of extra resources from the mother to interpolate extra cell divisions between zygote and meiosis if this were to result in the formation of more spores and more new gametophytes carrying the paternal genome. Thus paternally expressed genes may have been responsible for initiating mitotic divisions of the zygote. However, long-term evolution of the sporophyte would be impossible without mutual advantages for both maternal and paternal genomes.

5.3 Gene expression and the diplohaplontic life cycle

Extant land plants have strongly heteromorphic alternation of generations implying very different patterns of gene expression in each. There thus appears to be extensive repatterning of gene expression as plants pass through life cycle checkpoints (zygote and spore). The repatterning may be in part epigenetic through, for instance, differences in chromatin distribution. Whatever the mechanism, the set of expressed genes (**transcriptome**) of *Arabidopsis* is very different between gametophyte and sporophyte. In the male gametophyte (at the pollen stage) a recent study found that 40% of the 992 expressed genes detected were expressed in the gametophyte only (Honys & Twell, 2004). However, for the 60% of genes that are expressed in both gametophyte and sporophyte, gametophyte selection (which exposes all alleles to selection) can potentially purge recessive mutations deleterious in the sporophyte.

In contrast, the female gametophyte of *Arabidopsis*, being enclosed in an indehiscent megasporangium of sporophyte origin (the ovule), is very difficult to isolate, and therefore its transcriptome is difficult to assess. However, the mutant sporocyteless (*spl*, also called nozzle, *nzz*), that produces nearly normal ovules but no gametophytes, can be used to subtract sporophyte gene expression from an ovular pool (Yu, Hogan & Sundaresan, 2005). The *SPL* gene is a transcriptional regulator required for both microsporocyte and megasporocyte development (Yang et al., 1999). The mutant therefore has no megasporocytes and therefore no megagametophytes. Using the mutant a significant number of genes have been discovered with expression confined to the female gametophyte.

5.4 Epigenetics and parental conflict

The nourishment of the developing sporophyte phase by the female gametophyte means that a haploid organism may be investing in an organism that carries an alien gene set (the paternal contribution to the genome). There is thus scope for **sexual conflict** between the two genomes, maternal and paternal (Haig & Wilczek, 2006). Only in cases where the gametophyte makes more than one sporophyte, or engages in vegetative reproduction (and so has the ability to invest in its own genome alone) are these effects expected to be significant. In these circumstances it may be advantageous for the paternal genome to increase the maternal investment in the sporophyte and for the maternal genome to limit it. Conflict should be particularly strong in bryophytes where the gametophyte makes a substantial investment in the sporophyte (matrotrophy), and has the capacity for vegetative reproduction.

A consequence of sexual conflict might be parent-of-origin-based **gene imprinting** (Haig & Wilczek, 2006). This is the biased expression of allelic copies deriving from a particular parental genome, and results from the mitotically stable **epigenetic silencing** of one genome by another. It would be extremely fruitful to search for such imprinting in "bryophytes."

Another approach would be to look for the signature of sexual conflict in the physiology of matrotrophy. Of particular interest are the transfer cells responsible for passing food from gametophyte to sporophyte. These are anticipated to be the major battleground of sexual conflict. In bryophytes dead and collapsed cells are found at this junction (Ligrone, Duckett & Renzaglia, 1993), which may represent the collateral damage from the sex war (Haig & Wilczek, 2006).

In flowering plants, the situation is somewhat different as there are two "mothers." There is the maternal gametophyte and the maternal sporophyte on which the gametophyte is parasitic. Much of the nutrition comes from the maternal sporophyte, so the conflict is between the differing sporophyte genotypes (mother and embryo). Thus the miniaturization of the female gametophyte has led to a situation akin to that found in diplontic organisms that nourish their young, like mammals. Mammals have extensive imprinting (for instance, of genes expressed in the placenta) that is consistent with parental conflict theory, and so too do flowering plants. The placenta of mammals is foetal tissue that is invasive into maternal tissue therefore providing a conflict. The flowering plant endosperm is in some ways analogous to the mammalian placenta as it extracts food from the maternal sporophyte for the nourishment of the embryo. It has the same genotype as the embryo but with the addition of an extra haploid copy of the maternal genotype (i.e. it is triploid).

Gymnosperms may be somewhat different as the nutrition of the embryo is provided for by relatively large gametophytes that in some species, such as *Ginkgo*, grow even if there is no embryo. So there should be little conflict between the sporophyte and the embryo. Furthermore, as the female gametophyte usually produces only a single embryo there should be no conflict between the gametophyte and the embryo. The circumstances in *Ginkgo* should therefore attenuate imprinting, and this would be interesting to test at the molecular level.

In angiosperms the gametic genomes are substantially nonequivalent. As a result, only the maternal copies of certain key genes control endosperm development. The evolution of this uniparental genomic imprinting, that is, switching off the paternal alleles, is predicted by theory, as a means for the mother to control selfish resource extraction, which benefits particular paternal genotypes. The main imprinted pathway is the *FIE/MEA* pathway (*FERTILIZATION INDEPENDENT ENDOSPERM/MEDEA*). The action of this gene pathway is essential for normal endosperm and embryo development. Two genes of this pathway, *FWA* and *FERTILIZATION INDEPENDENT SEED2* (*FIS2*), are known to be imprinted by DNA methylation (Jullien et al., 2006b). *MEA* itself is imprinted by histone methylation, and by an intriguing twist *MEA* is self-imprinted. In endosperm, the maternal *MEA* allele encodes an essential component of the imprinting machinery, maintaining the silenced state of the paternal *MEA* allele (Jullien et al., 2006a). Medea is an apt name for a gene involved in parental conflict because, in Greek tradition, Medea killed her two children as part of her

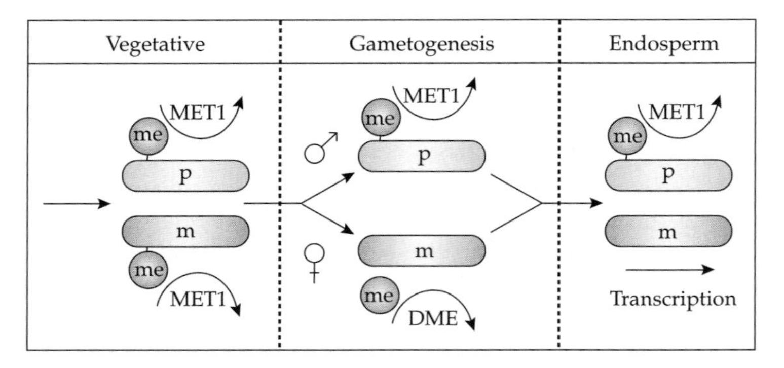

Fig. 5.3. Gene silencing by continuous MET1 action coupled with uniparental demethylation by DME (after Jullien et al., 2006b).

revenge on her errant lover, Jason (Euripides, 431 BCE).

An interesting aspect of uniparental gene silencing to be considered is that the gene is rendered functionally haploid, and therefore deleterious mutants can be purged even in the diploid phase of the life cycle. The maintenance of a haploid stage with significant transcriptome provides an important means of selection for genetic health and imprinting extends this process. Because uniparental gene silencing occurs in both plants and mammals it is tempting to see it as an ancient feature conserved between them. However, in plants and mammals imprinting has arisen independently driven by their individual physiological circumstances. This is clear when one considers that in mammals, it is the silenced state that is imposed *de novo*. However, in plant genes such as *FWA*, methylation is the ground state (Fig. 5.3), and it is the maternal allele that is released from silencing *de novo*, by the action of the DNA glycosylase *DEMETER* (*DME*; Kinoshita et al., 2004). *FWA* does not need to be resilenced as the endosperm does not contribute to the next generation.

5.5 The evolutionary history of the life cycle

The evolutionary history of the gametophyte in land plants is one of reduction, and indeed it is one of the best examples of prolonged directional reductive evolution. In angiosperms this reaches an extreme, with the female gametophyte consisting typically of eight nuclei, all contained within the megaspore cell wall (the embryo sac wall). The male gametophyte is smaller still and contained within the microspore cell wall. It consists typically of 1–2 vegetative nuclei and two sperm nuclei (one to fertilize the egg, the other to fertilize the endosperm).

The sporophyte, on the other hand is a new phase in the life cycle that evolved during the transition to land, probably to take advantage of aerial dispersal of sporopollenin-protected spores. From small beginnings the sporophyte has evolved with increasing complexity culminating in the relatively massive plant bodies of modern trees.

We are hindered by the lack of fossils of the gametophytes, or sporophytes, of the earliest land plants or "eoembryophytes." Our knowledge of the eoembryophytes comes solely from spore tetrads that first appeared in the mid-Ordovician, some 465 million years ago (Mya), along with fragmentary cellular remains that are liverwort-like. These eoembryophytes may plausibly have resembled liverworts (Edwards, Duckett & Richardson, 1995; Kenrick & Crane, 1997b; Kroken, Graham & Cook, 1996), maybe like the modern thalloid liverwort *Sphaerocarpos*, which also has spores in tetrads. The sporophyte of *Sphaerocarpos* is a small mass of sporogenous tissue that produces spores in tetrads much like the Ordovician fossils. The spores are released when the plant disintegrates and remain in the soil or blow around with soil particles during dry seasons. The resistant sporopollenin wall protects them from all these vicissitudes.

However, 65 million years later, by the time the Rhynie Chert in Scotland was laid down (c. 400 Mya) the situation is dramatically different, with both sporophytes and gametophytes that are complex tracheophytic organisms. The Rhynie gametophytes (Fig. 5.4 and Box 5.1) consist of small axes, about 2 mm wide and 1–2 cm long, supporting cup- or disc-shaped gametangiophores up to 1 cm across, bearing either archegonia or antheridia (they are unisexual structures). The axes have vascular conducting tissue (protosteles) and arise from small basal stem masses (protocorms) anchored in the ground by means of rhizoids and penetrated by mycorrhizal fungi (Remy, Gensel & Hass, 1993; Taylor, Kerp & Hass, 2005). The sporophytes of the Rhynie plants are generally larger and more robust. This is especially true in *Aglaophyton*†, which has sporangia borne terminally on axes that may be up to 18 cm high.

The gametophyte protocorms may conceivably be reduced thalli retained from a thalloid ancestry. It is very interesting that the gametangia are borne aloft on stems, as are the gametangia of certain extant liverworts such a *Marchantia*. In *Marchantia* the lofted gametangia serve to facilitate cross-fertilization by rain splash of the sperm, and this may be the purpose of the lofted gametangia of the Rhynie plants. Thus the evolution of the stem may initially have been a means of raising sperms and eggs into the air.

The parallel with *Marchantia* ends there. The gametangiophore in *Marchantia* is a hollow tube and not a protostelic stem. Furthermore *Marchantia* produces a sporophyte in the form of a spore capsule that does not become free living, but instead remains on the gametangiophore, which also therefore functions as a sporangiophore. In the Rhynie plants, however, the weight of the developing sporophyte must have collapsed the gametangiophore allowing the sporophyte to attain a free-living existence on the ground, where it is capable of vegetative spread and is the dominant part of the life cycle.

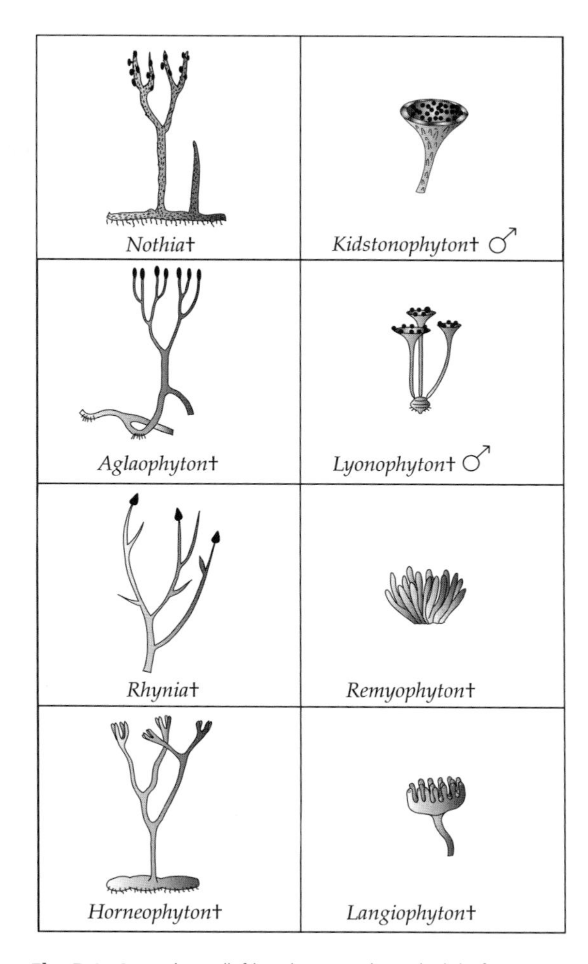

Fig. 5.4. Sporophytes (left) and gametophytes (right) of some early fossil land plants (after Remy et al., 1993; Taylor et al., 2005). The dagger indicates extinct taxa.

Gametophytes of the sort found in the Rhynie plants have entirely disappeared from the extant tracheophyte lineages. It may be that although they functioned well in the low competition environment of the lower Devonian, their small size and probable shade intolerance (their horizontal area to volume ratio is low, exactly opposite to the fern prothallus) made them very vulnerable to competition from their own larger and more vigorous sporophytes. In this scenario they, as it were, harbored the zygotes of their own destruction.

Unable to compete with the increasing vigor of Devonian sporophytes, there are three ways

Box 5.1 Gametophyte–sporophyte pairs in the Rhynie Chert (400 Mya).

1. *Aglaophyton major* (S)/*Lyonophyton rhyniensis* (G) (Rhyniopsida). Gametophyte unisexual (both sexes known)
2. *Rhynia gwynne-vaughnii* (S)/*Remyophyton delicatum* (G) (Rhyniopsida). Gametophyte unisexual (both sexes known)
3. *Horneophyton lignieri* (S)/*Langiophyton mackiei* (G) (Horneophytopsida). Gametophyte unisexual (both sexes known)
4. *Nothia aphylla* (S)/*Kidstonophyton discoides* (G) (Zosterophyllopsida?). Gametophyte unisexual (male only known)

out of the nutritional dilemma. One is for the gametophytes to reduce still further producing gametangia directly on the protocorm, which then may rely on the food reserves of the megaspore or become wholly mycotrophic (the situation in lycophytes and ophioglossoids). Another is for the protocorm to become a shade-tolerant foliose structure, as in the fern prothallus. Finally, and most spectacularly, the gametophyte may become parasitic on the sporophyte, as in the seed plants. All these options may be considered means of escaping competition.

5.6 The sporangium

A **sporangium** is an organ that produces **spores**. During development a sporangium produces **archesporial cells** at the center. These develop into **spore mother cells**, which undergo meiosis to form four haploid daughter cells. These then develop into spores, typically with a tough outer wall made of the highly resistant polymer **sporopollenin**.

In bryophytes, the sporophytes have but a single sporangium, as part of a specialized organism called a **sporogonium**. In other land plants (the **polysporangiophytes**) a sporophyte may produce many sporangia. In *lycophytes* the sporangia are borne in the axils of **microphylls**. In other extant plants the sporangium is generally interpreted as leaf borne, although this has been controversial. Leaves bearing sporangia are termed **sporophylls**.

The position in which sporangia are borne is of evolutionary interest. In early land plant fossils such as *Cooksonia*† and *Rhynia*†, the sporangia are borne terminally on naked stem axes. Therefore in these early land plants there was a developmental mechanism for phase change at the apical meristem. Stem apices become determinate and differentiate into a sporangium in an analogous way to how the stem apex of a flowering plant can become determinate and form a terminal flower. In *Zosterophyllum*† the sporangia are lateral on the stems, presumably by highly unequal dichotomous branching. From this rather irregular arrangement of lateral, short stalked sporangia, the later zosterophylls evolved several interesting patterns of sporangium arrangement (**sporangiotaxis**). These include helical arrangement and arrangement in rows (monostichous and distichous). Remnants of these patterns are seen in the extant relatives of the zosterophylls, the modern lycophytes. In modern lycophytes the sporangia are generally associated with microphylls. Indeed microphylls themselves have been interpreted as sterilized sporangia (see Chapter 4, Section 4.2.1).

The lateralization of sporangia in lycophytes may be interpreted as solving a growth problem. Terminal sporangia terminate the growth of axes and therefore of the plant. In nonlycophyte vascular plants this problem appears to have been solved by producing the sporangia on lateral organs that evolved from stems: megaphylls. Sporangia in monilophytes and most seed plants may be interpreted as borne on megaphylls, or on structures derived from megaphylls.

If this is so, it is curious that few plants have retained stem borne sporangia. It may be that genes involved in terminal sporangium formation, such as genes for determinacy, were

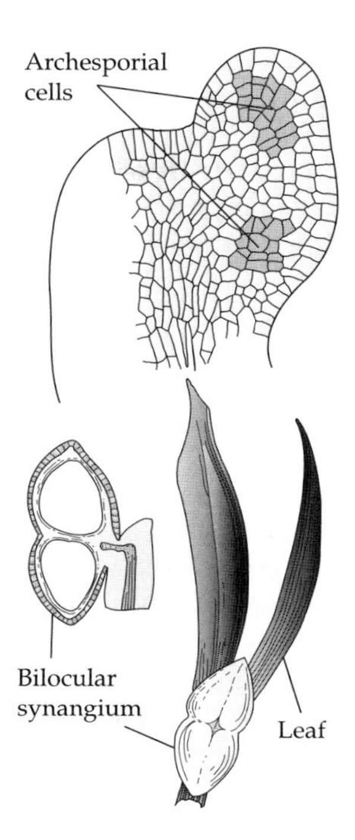

Fig. 5.5. Synangium of *Tmesipteris*, an ophioglossoid fern.

appropriated to leaf developmental pathways early in leaf evolution thus establishing a developmental link between leaf and sporangium. In this view a leaf becomes seen as a sterilized reproductive branch.

The attachment and the arrangement of sporangia vary. In some ferns there is a **sporangial stalk**, but most other sporangia are **sessile** (sitting directly on the lamina). Some sporangia are solitary, but others are joined into **synangia** (composite organs formed of many sporangia). Examples of synangia are found in the apparently terminal trilocular synangia of *Psilotum*, the leaf-associated synangia of *Tmesipteris* (Fig. 5.5) and the synangia of marattialean ferns.

5.7 The sporogonium

In bryophytes the sporangium develops as a complex organ, which represents the entire sporophyte generation. This is the **sporogonium**, or **footed sporangium**, an organ that differs greatly in size and robustness. It generally consists of two parts: a foot and a capsule (**theca**) containing the spores. The foot abuts the gametophyte tissues and contains transfer cell through which nutrition passes. The foot may be extended into a long stalk (**seta**). The capsule is equivalent to the sporangium and often bears specialized tissues for spore dispersal. The homologies of the sporogonium are uncertain. The stalk has been interpreted as homologous to the stem of polysporangiophytes. However, the development of the stalk is rather different to that of a stem, and it is best regarded as a bryophyte specific development of tissue between foot and capsule (Kato & Akiyama, 2005), hence the use of the term sporogonium for the entire organ.

As the sporogonium represents the entire sporophyte generation of the plant, it begins its development in the **archegonium** (the usually flask-shaped structure that contains the egg and the developing sporogonial embryo). At first, the archegonial wall expands with the developing sporogonium allowing it to complete much of its development while protected by the archegonial wall. However, when the stalk starts to elongate, the archegonial wall ruptures freeing the sporogonium. The pattern of archegonial rupture is highly characteristic. It may rupture at the base, in which case the top of the archegonium is carried up by the developing sporogonium as a cap, or **calyptra**, which covers the capsule. This is characteristic of most mosses. Alternatively it may rupture at the top, in which case the sporogonium grows through the archegonial wall leaving it as a **sheath** around the base of the seta, as happens in most liverworts and *Sphagnum*.

The stalk (seta) of the sporogonium consists of a foot at the base, and the main part, which may be very long or absent. Thus the stalk of the *Anthoceros* sporogonium consists of only a foot, which in this case is covered with rhizoid-like hairs. In other species such as the aquatic liverwort, *Riccia*, even the foot is lacking and there is no stalk at all. The foot may be swollen and nearly spherical as in *Sphagnum* or it may merely be tapered as in *Polytrichum*. At the top of the stalk is the **apophysis** where the stalk meets the theca. These often bear stomata and may be swollen into a ring as in *Polytrichum* or into a large plate as in *Splachnum*. The enlarged apophysis in *Splachnum* is often red colored and attracts flies which are thought to aid in spore dispersal between its specialized substrate patches, animal dung.

The capsule consists of a capsule wall, a central column (**columella**) and a spore sac. The capsule wall may be only a single cell layer as in the liverworts, *Riccia* and *Marchantia*. On the other hand it may be comparatively thick-walled and be provided with stomata as in many mosses and *Anthoceros*. The central column also varies. It is absent in most liverworts, and in *Anthoceros* it is present as a thin line of cells only. In mosses it is more massive, sometimes reaching the top of the capsule (*Funaria*) or just being present as a dome at the base (*Sphagnum*) or is ephemeral (*Takakia*). The spore sac is a membranous cell layer surrounding the spore mass. The spore mass itself is composed of spores and highly differentiated cellular structures called **elaters** that are thought to assist with spore dispersal. Elaters are absent in mosses and some liverworts, e.g. *Riccia*. Most liverworts, for example *Marchantia*, have unicellular elaters provided with spiral thickenings on the cell walls. In *Anthoceros* the elaters are usually pluricellular and often branched.

In some species, for example *Riccia, Sphaerocarpus* and *Pleuridium*, the capsule is **indehiscent** and the wall just decays when the spores are ripe. However, other species have special mechanisms, which involve either **valvate dehiscence** or **operculate dehiscence**. Valvate dehiscence is characteristic of liverworts (four valves) and anthocerotes (two valves). In valvate dehiscence the capsule wall splits into strips. If the valves are short, the strips form teeth over the opening of the capsule as in *Marchantia*. The moss *Andreaea* has valvate dehiscence with strips that are united at the top forming a cage through which the spores exit. More usual in mosses, however, is operculate dehiscence in which the top of the capsule falls off as a **lid** (**operculum**). This is usually due to the split at a **dehiscence ring** (**annulus**) of specially differentiated epidermal cells. When the operculum drops off, one or two sets of teeth called the **peristome** may be revealed underneath, although this is lacking in some mosses (*Sphagnum, Oedipodium*). The peristome teeth are hygroscopic. In dry weather they bend up, releasing the spores, while closing in wet weather. In cases where there is a **double peristome** the **outer peristome** is usually tooth-like and the **inner peristome** is usually hair-like.

5.8 The superficial sporangium

In most polysporangiophytes the sporangium is superficial, that is, it occurs on the surface of other organs. The superficial sporangium occurs in lycopods, ferns and in the pollen-producing sporangia of seed plants. There are some minor exceptions: in *Ophioglossum* the sporangia are deeply embedded within the rather thick fleshy tissue of the sporophyll. In seed plants the megasporangium is tightly wrapped in a covering of sporophyte tissue (the integument). However, sporangia are mostly completely exposed for an obvious reason: their spores must be released into the atmosphere.

The **sporangium wall** varies in thickness but is always initially several cell layers thick. In ferns, however, this is reduced to a single cell layer at maturity. Most ferns have a highly distinctive sporangium called a **leptosporangium**, which is small, stalked and thin-walled, but, most telling of all, develops from a single superficial cell of the fern leaf (Fig. 5.6). This superficial cell divides periclinally (parallel to the surface) giving a surface daughter cell and an inner daughter cell. The ferns develop their sporangia from the daughter surface cell. The marattioids and the ophioglossoids however, although also having a periclinal division, develop their sporangia (called **eusporangia**) from the inner daughter cell (and usually some surrounding cells too).

Cells of the outer layer of the sporangial wall usually have thickenings on the cell walls. The consequence of this fibrous layer is to set up differential stresses on the sporangial wall as it dries out. This ruptures the sporangial wall and liberates the spores. In the **anthers** (i.e. microsporangia) of angiosperms the fibrous layer is called the **endothecium**. Ferns have their thickened cells in the form of a ring, called the **annulus**, running over the top of the sporangium. Cells of

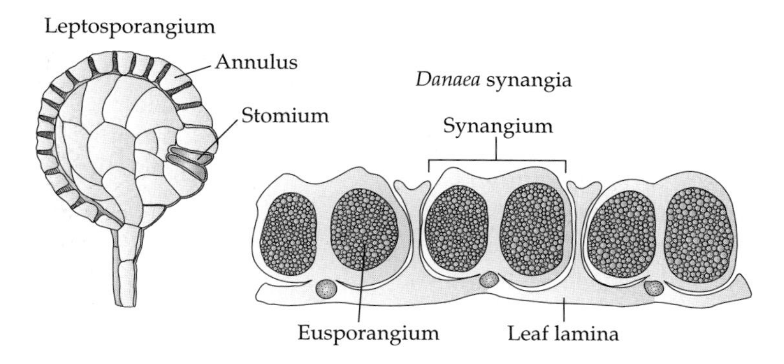

Fig. 5.6. The stalked, thin-walled leptosporangium of the filicoid ferns contrasted with the thick-walled eusporangia of the marattioid ferns.

this ring have their bottom walls and their radial sidewalls are thickened. The top wall is unthickened. When it dries out, the top of the annulus collapses ripping the annulus out of the sporangial wall and liberating the spores. Some ferns such as the Schizaeaceae have a **complete annulus**, which runs right over the sporangium. Other fern families, such as the Polypodiaceae have an **incomplete annulus** that runs part of the way over the sporangium, while in the Osmundaceae there is a **reduced annulus** which consists only of a group of cells with thickened walls positioned just below the apex of the sporangium.

The inner part of the sporangial wall is called the **tapetum**. The tapetum provides nutrition to the developing spores. The cells are very proteinaceous and although it is sometimes persistent, as in *Lycopodium*, it often breaks down to produce a mucilaginous **periplasm**, which bathes the developing spores, as in ferns and angiosperm pollen sacs. Fern and bryophyte spores are particularly interesting in this regard as they often have a well-defined outer skin called a **perine**, which is derived from the breakdown products of this sporangial tissue.

In its basic form the sporangium is a simple chamber filled with sporogenous tissue, all of which develops into spores. However, it may be divided into chambers by plates of cells called **trabeculae**. In some sporangia not all the contents become spores. In liverworts and hornworts some of the sporogenous cells differentiate as **elaters**, specialized elongated cells whose function is to assist spore dispersal. *Equisetum* also has elaters but in this case they are derived from outgrowths of the spore wall. In most sporangia the spores form as separate units. However, they can also occur as aggregations (**massulae**). Some spores such as the microspores of Ericaceae and the spores of *Sphaerocarpos* remain in **tetrads** (four cells that are the product of a single meiosis). However, spores can aggregate as larger masses or **polyads**. Spore masses that incorporate dispersal structures are known as **pollinia** (in the orchids and asclepiads).

5.9 Megasporangia and microsporangia

Most ferns and bryophytes are **homosporous**, producing only one type of spore. However, a few aquatic ferns, many lycophytes and all seed plants are **heterosporous** producing **megaspores** in **megasporangia** and **microspores** in **microsporangia**. Megaspores are large and germinate to produce **female prothalli**, and microspores are smaller and germinate to produce **male prothalli**. Heterospory is important as the pathway to the seed, in which the megasporangium is a wrapped and indehiscent structure that retains the megaspores (of which only one may be functional) on the parent sporophyte.

The lycophytes provide interesting variation as *Lycopodium* is homosporous and *Selaginella* heterosporous. *Selaginella* typically produces only four megaspores in the megasporangium and not all of these may be functional. The microsporangium on the other hand produces hundreds of microspores (Figs. 5.7; 5.8). The fossil record is replete with extinct examples of heterosporous plants.

An interesting intermediate between homosporous and heterosporous types is found in the dioecious and **anisosporous** fern, *Platyzoma*. Spores of different sizes produce gametophytes of differing sex. This has been interpreted as **incipient heterospory** (Duckett & Pang, 1984; Tryon, 1964). An earlier suggestion that spore size variation in *Equisetum arvense* is also linked to the sex of the gametophytes has been rejected (Duckett, 1970). Sexual behavior is indeed complex in *Equisetum*, but it is not linked to spore size (Duckett, 1977). There are also a few mosses that are interpreted as anisosporous and often the smaller spore is said to produce the male gametophore.

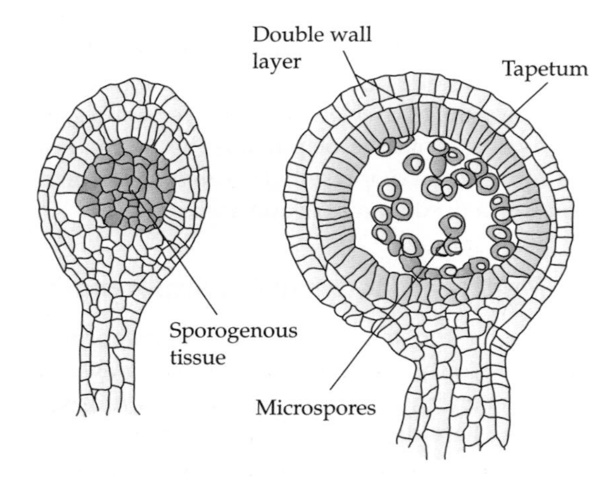

Fig. 5.7. Microsporangium of *Selaginella*, and its development (after Slagg).

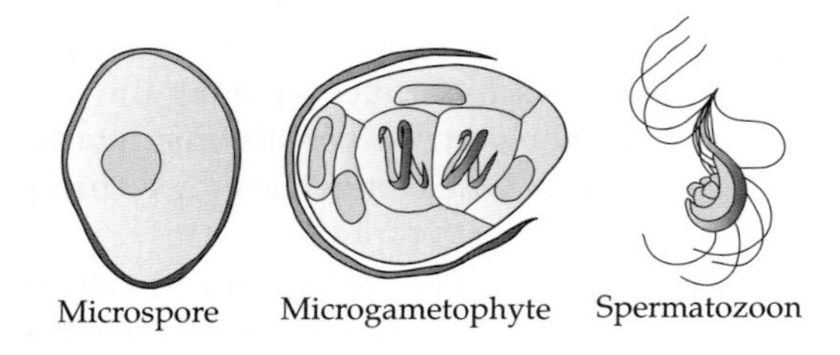

Fig. 5.8. Development of the microgametophyte of *Selaginella*. The microgametophyte is partially enclosed in the spore wall throughout its duration.

5.10 The tetrasporangiate anther

One remarkable feature of angiosperms is the uniformity and oddity of the microsporangial complex, called an anther. Angiosperm anthers are **tetrasporangiate**, that is composed of four microsporangia. These four sporangia are arranged in a particular pattern: in two pairs on either side of the **connective** (the tissue to which the microsporangia are attached). The shape of the connective determines whether the sporangia are directed out (**extrorse**), to the side (**latrorse**) or inward (**introrse**). Anther direction relative to the center of the flower is a relatively labile feature. In many Ericaceae, the anthers are extrorse in early development but later invert to become introrse (Fig. 5.9).

The wall dividing the two sporangia of a pair breaks down by **programmed cell death** (PCD, **apoptosis**) during development to give a single case or **theca**. The cavity inside a theca is called a **locule**. The thecae, being composed of two united sporangia, are thus comparable to the **synangia** of a eusporangiate fern. In gymnosperms the equivalent structure to the anther is usually **bisporangiate**, and these two sporangia remain separate.

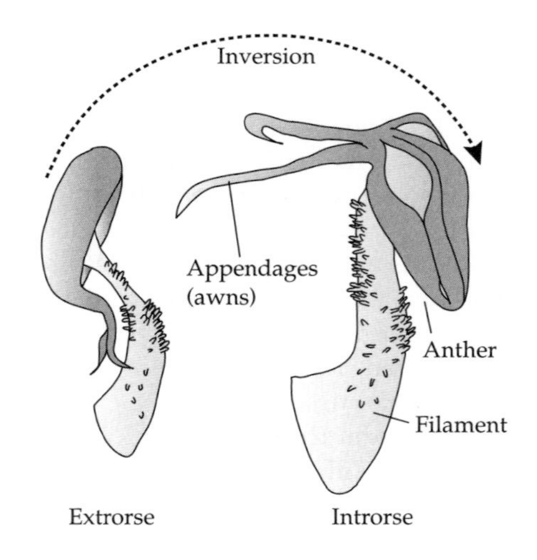

Fig. 5.9. Stamens of *Enkianthus* (Ericaceae) showing the initially extrorse orientation of the microsporangia (anthers dehiscing away from the floral axis) changes to introrse by developmental torsion (after Hermann & Palser, 2000).

Generally the basic tetrasporangiate/bithecate pattern is conservatively retained throughout the angiosperms, but some variation does occur. One theca can be reduced or lost, as in *Salvia* in which the connective stalk of the lost theca is retained as part of the lever mechanism of pollination found in this genus. Alternatively the two thecae may unite to form a **synthecous** anther. This may happen early in development as in *Scrophularia*, or late in development, just before flower opening (**anthesis**) as in *Digitalis*. In other plants both thecae are present but a sporangium has been lost from each theca. In consequence, each theca is formed from a single sporangium and the anther is bisporangiate (Fig. 5.10). An example is found in the composite *Microseris bigelovii* (bisporangiate) while related *Microseris* species such as *Microseris douglasii* are tetrasporangiate (Battjes, Chambers & Bachmann, 1994). As these two species are crossable, it has been possible to ascertain that a single major quantitative trait locus (QTL), although with several modifiers, is responsible for the difference (Gailing & Bachmann, 2003a). In *Senecio vulgaris*, there is variable reduction of microsporangia within the species, and thus the microsporangia appear to be in a state of active evolution (Gailing & Bachmann, 2003b).

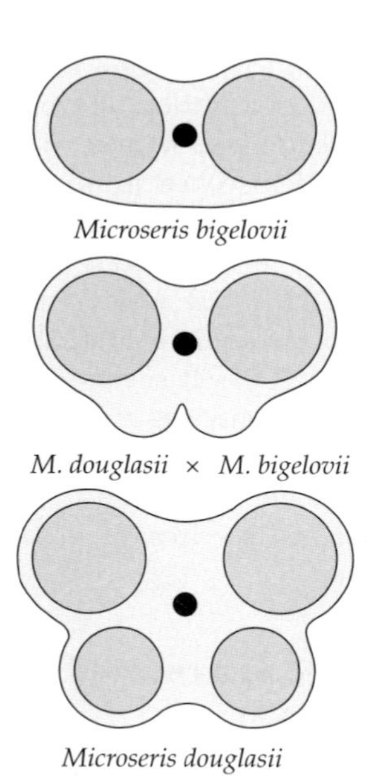

Fig. 5.10. The fundamentally tetrasporangiate nature of the angiosperm stamens can change to bisporangiate by reduction, as shown here in *Microseris* (Asteraceae) (after Battjes et al., 1997).

The sporangial wall (or theca wall if the synangial unit is being considered as a whole) is a bilayered structure. The sporangial epidermis (**exothecium**) is usually unspecialized. However, the inner layer or **endothecium** is composed of cells with differentially thickened walls, which on drying set up internal mechanical stresses that serve to rupture the thecae, creating slits or pores for the egress of pollen.

The thickening of the endothecium is related to the deposition of lignin in prescribed patterns in the cell walls. Thus mutants in genes controlling secondary thickening pathways in plants can disrupt the process, and lead to indehiscent anthers. An example is provided by the *NAC SECONDARY WALL THICKENING PROMOTING FACTORS* (*NST1* and *NST2*) in *Arabidopsis*. Double *NST1/NST2* mutants have an indehiscent anther phenotype (Mitsuda et al., 2005).

Generally there is only one slit per theca, allowing common dispersal of pollen derived from two sporangia. The rupturing is generally highly controlled, being along a weaker line of cells between the two microsporangia called the **stomium**.

The internal cells of the sporangium develop from mother cells known as the **archesporial** cells. The archesporial cell lineage differentiates into an inner and an outer layer. The inner cells are the **sporogenous** cells[3] that will undergo mitosis to form numerous **sporocytes**, and then meiosis and **sporogenesis** to form the **microspores**. The innermost cells of the outer layer differentiate as the **tapetum**. A **middle layer** separates the tapetum from the endothecium. Tapetal cells are essential for the transfer of nutrients and other

[3] Archesporium (i.e. the archesporial cells) has historically been used in varying senses. Some authors use this term to indicate the mother cells of just the sporocytes (i.e. the sporogenous cells). Other authors (as here) use the term to indicate the mother cells of all the internal cells of the sporangium, excluding only the wall.

substances required for the correct development of the microspores. The setting of the cell fate boundary between the tapetum and the sporogenous cells is a very important step in microsporangium development as it regulates the number of sporogenous cells and hence the number of microspores (Fig. 5.11).

In *Arabidopsis TAPETUM DETERMINANT1* (*TPD1*) gene is probably the master gene required to maintain tapetal identity (Yang et al., 2003b). The maintenance of tapetal cell fate probably requires a signal from the sporogenous tissue. The *TAPETUM DETERMINANT1* protein has the characteristics of a secreted protein and may be this signal. Furthermore, TPD1 interacts with a member of a putative signaling pathway to determine the cell fate boundary (Yang et al., 2005). This is the leucine rich receptor kinase *EXCESS MICROSPOROCYTES1* (*EMS1*; Zhao et al., 2002) also called *EXTRA SPOROGENOUS CELLS* (*EXS*; Canales et al., 2002).

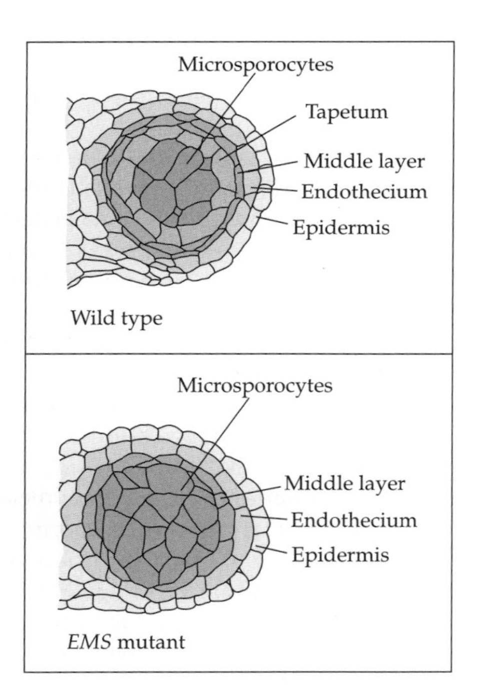

Fig. 5.11. Diagram showing the five cells layers of the wild type anther, and the deletion of the tapetum in the *EMS* mutant. The concentric cell layers of the sporangium are determined by a signaling network. Tapetal cells are replaced by extra sporogenous cells in the mutant, showing the important role of *EMS* in the determination of the tapetum (after Zhao et al., 2002).

Other members of this putative signaling pathway are now emerging, including the SOMATIC EMBRYOGENESIS RECEPTOR-LIKE KINASES1 and 2 (SERK1 and 2), which like EMS1, are leucine rich receptor kinases, probably forming signaling complexes in the secondary parietal cell ready to respond to a signal from the sporogenous cells and determine tapetal cell fate (Albrecht et al., 2005; Fig. 5.12).

Different gene systems probably specify the identities of the outer cells layers. The most promising candidates so far are the *CLAVATA1*-like receptor kinases, *BARELY ANY MERISTEM1* and *2* (*BAM1* and *BAM2*; Hord et al., 2006). These genes may be involved in the cell–cell signaling required for the exquisitely precise cell layer determination in theca development. The double mutants (*bam1/bam2*) fail to develop endothecium, middle layer and tapetum.

Much of the molecular control of anther development remains to be discovered, but the anther is emerging as a promising system for the study of cell fate specification and other biological processes. One such is PCD (apoptosis), which is involved in anther processes, including anther dehiscence and maturation of the tapetum (Ku et al., 2003; Schreiber, Bantin & Dresselhaus, 2004). Anther dehiscence is an active process that in *Arabidopsis* is synchronized with flower opening. The gene *DEFECTIVE IN ANTHER DEHISCENCE1* (*DAD1*) was found to encode an enzyme in jasmonic acid (JA) biosynthesis (Ishiguro et al., 2001). It seems that JA synthesized in the filaments regulates water flow in both petals and stamens to bring about petal expansion, filament elongation and anther dehiscence at the same time, in a remarkable economy of mechanism.

5.11 The ovule as a covered megasporangium

The ovule is an **integumented megasporangium**, and the provision of a coat (integument) for protection during dispersal of the whole megasporangium and to control access to the

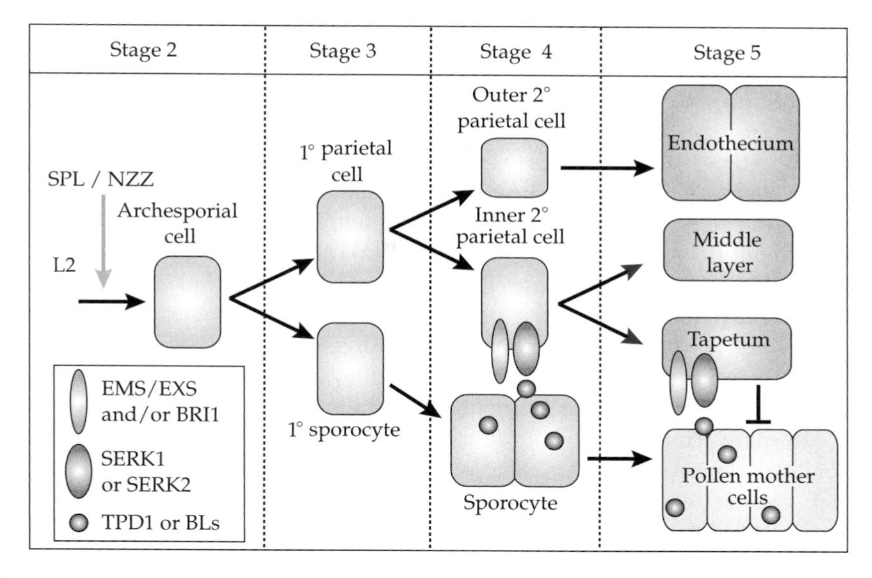

Fig. 5.12. Model showing the development of the four layers of the anther tissues from an archesporial cell. A model of signaling between the inner layers necessary for the correct specification of the tapetum is also indicated, involving numerous genes important in anther development and patterning (after Albrecht et al., 2005).

megasporangium by microgametophytes, was one of the most important events in land plant evolution, around 365 Mya (Gerrienne et al., 2004).

Some plants have ovules borne on sporophylls, and this is particularly evident in *Cycas* in which the sporophylls are obviously leaf-like (see Chapter 6). However, in conifers although the ovules are borne on leaf-like ovuliferous scales (Tomlinson & Takaso, 2002), these are interpreted as a reduction of a radial stem system on the basis of fossil evidence (Florin, 1951, 1954). In *Ginkgo* the ovule rests on a collar of tissue atop a radial stem system, and in the conifer *Taxus* (yew) the ovuliferous scale is lacking by reduction and the ovule is terminal on the shoot. In angiosperms the sporophyll is called a carpel, and this structure encloses the ovules it bears.

The **integument** is a covering wrapped around the megasporangium and constitutes the key character of the seed plants and the main difference between megasporangia and ovules. This structure is therefore of great morphological interest. The seed-plant integument is thought to arise from the **cupule**, a small cup-shaped structure surrounding the megasporangia seen in certain fossils.

Two main hypotheses are put forward for the origin of these cupules. The first is the **branch theory** (or telomic theory), in which the cupule is composed of a reduced and united ring of lateral branches subtending the megasporangium. The other hypothesis is the **synangial theory** (Benson, 1904; Kenrick & Crane, 1997a). In this the integument is formed from a ring of sterilized stalked sporangia surrounding the remaining fertile megasporangium. In the synangial theory the cupule is thus homologous to part of a progymnosperm sporophyll, for instance of the *Aneurophytales*†, with the distal parts becoming laminar. Thus it is reminiscent with the evolution of the bifacial megaphyll, which also involved distal parts of branch structures becoming laminar.

There is generally one integument in gymnosperms (which are therefore called **unitegmic**). However, in angiosperms there are two integuments (**bitegmic**). Although there are many unitegmic angiosperms, these appear to have evolved by the loss of one integument (McAbee,

Kuzoff & Gasser, 2005). In bitegmic ovules the integuments are differentiated as **inner integument** and **outer integument**.

As the second integument is an evolutionary innovation characteristic of the angiosperms, it is of considerable interest. There are two main hypotheses for the origin of the second integument of angiosperms. It may be derived from a leaflet of the sporophyll bearing the ovule: the **leaflet hypothesis**. This is suggested by the leaf-like **cupules** surrounding the ovules in some fossil pteridosperms, and some angiosperm teratomorphs in which the outer integument becomes leafy. Alternatively it may result from a reiteration of developmental program of the first integument: the **reiteration hypothesis**. Reiteration is supported by the comparative ease with which supernumerary integuments can evolve in angiosperms. The true **aril** (a fleshy third integument that is important in the seed dispersal of many groups of plants) is an example. In this case integuments may be viewed simply as protective or attractive outgrowths of the **ovule stalk** (**funicle**).

5.12 The morphology of the ovule

The integuments usually wrap completely around the megasporangium leaving only a small entrance pore called a **micropyle**, through which **pollen tubes** (angiosperms) or **pollen grains** (gymnosperms) gain access to the ovule. In gymnosperms the integuments may be drawn out so that the micropyle takes the form of a small tube, and there may also be a hollow chamber, called a **pollen chamber**, at the base of the micropyle to receive the pollen. In bitegmic ovules the mouth of the outer integument is called the **exostome** and the mouth of the inner integument is called the **endostome**. The exostome and endostome together make up the micropyle. In Euphorbiaceae, the micropyle is blocked by a plug of tissue formed by an outgrowth of the **placenta**, called an **obturator**. The pollen tube has to grow through the obturator, which thus represents an extra barrier to fertilization.

Inside the integuments is the **nucellus**. This tissue corresponds to the megasporangium wall. The tissue at the base of the ovule, where the integuments originate, is called the **chalaza**. Some angiosperms such as elm trees (*Ulmus*) display **chalazogamy**, or penetration of the pollen tube by invasive growth through the chalaza rather than through the micropyle. Some angiosperm seeds are **crassinucellate** and **pachychalazal**, with a thick nucellus and chalaza. These characteristics are usually thought to be primitive, with ovules evolving by reduction to become small and thin-walled: **tenuinucellate** and **leptochalazal**.

The ovule stalk (**funicle**) in angiosperms connects the ovule to the ovule-bearing part of the carpel, called a placenta. An ovule stalk is not present in many gymnosperms where the ovule is sessile on the sporophyll. However, the **ovuliferous scale** in conifers may well represent the fused stalks of ovules that have been lost in the reduction of the number of ovules from several to one per cone scale. The funicle can produce outgrowths of morphological importance. The **aril** represents a fleshy extra integument, and appears to derive from a funicular outgrowth rather than be of foliar origin. The aril is attractive to seed-dispersing animals. A familiar aril is the lychee (*Litchi chinensis*). Other examples include water lilies (*Nymphaea*) and yew (*Taxus*). If the outgrowth of the ovule stalk is small it is called a **strophiole**, as in *Viola*, *Chelidonium* and *Colchicum*. These often function as **elaiosomes** to attract ants for seed dispersal. If the funicle is adnate to the outer integument for some distance the resulting ridge of tissue is called a **raphe** (Fig. 5.13).

The orientation of the ovule, brought about by asymmetric growth along the adaxial–abaxial axis, is an interesting character in evolution. There is considerable variation of this character in angiosperms, but it is nevertheless conserved enough to provide a taxonomically useful character. Bending of the ovule is presumably important to bring the ovule into a suitable orientation for pollination or seed development. If there is no curvature of the ovule, it is known as **orthotropous**, whereas if it is bent over to be inverted it is **anatropous**. These are the commonest types in angiosperms. If the inversion is not complete the ovule is **hemianatropous**, whereas if the ovule is doubly inverted, as in *Opuntia*, it is known as **circinotropous**. In all these types the embryo sac is straight, as the bending is at the base of the ovule. However, the bending may occur in the middle of the ovule in which case the embryo sac itself is bent. If the bending is at the tip, it is known as **campylotropous**, and if the bending is in the middle it is known as **amphitropous**. The extent and precise location of asymmetric growth across the ovule therefore determines ovule type and it would be interesting to know the mechanism of this asymmetric growth at the molecular level (Fig. 5.14).

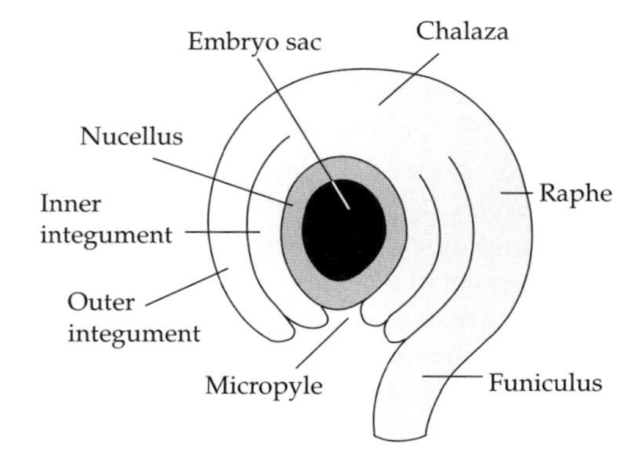

Fig. 5.13. The mature ovule of angiosperms is a megasporangium (nucellus) containing a germinated megaspore (embryo sac) surrounded by one or more (generally two) coats called integuments.

5.13 The molecular control of ovule development

The enigmatic nature of the ovule raises the question of whether a molecular understanding of ovule development will shed light on its morphological nature. Ovule development starts as a radially symmetrical primordium with patterning established on a proximodistal axis. The proximodistal patterning divides the ovule into three regions: funicle, chalaza and nucellus.

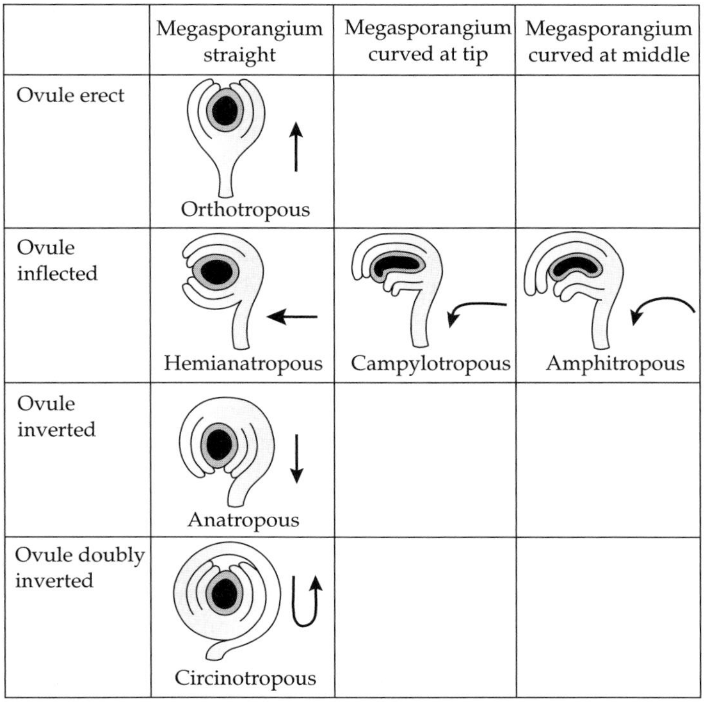

	Megasporangium straight	Megasporangium curved at tip	Megasporangium curved at middle
Ovule erect	Orthotropous		
Ovule inflected	Hemianatropous	Campylotropous	Amphitropous
Ovule inverted	Anatropous		
Ovule doubly inverted	Circinotropous		

Fig. 5.14. Diversity of ovule orientation.

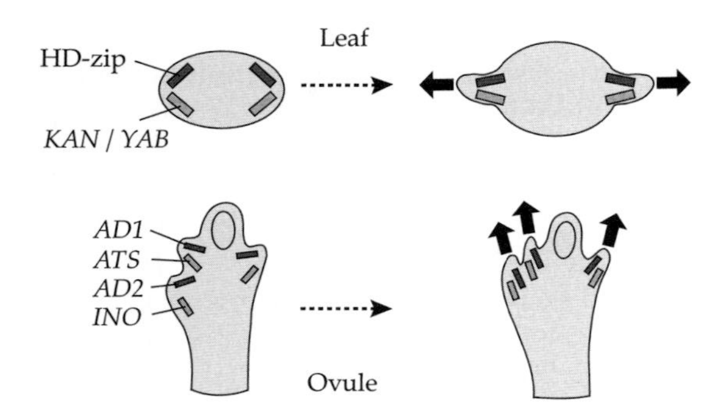

Fig. 5.15. Parallels between the adaxial–abaxial patterning of leaves (upper) and integuments (lower). At the molecular level there is plausible homology between integuments and leaves (after McAbee et al., 2006). See text for explanation; AD1 and AD2 refer to putative HD-zip adaxial factors.

There is general agreement as to the interpretation of the nucellus: it is the megasporangium.[4]

The chalaza is the region that bears the integuments (one in gymnosperms, often two or more in angiosperms). It is around these integuments that there is most debate as to their evolutionary origin, so it is particularly interesting to consider the development of these. There are immediate parallels with foliar organs. Integuments are obviously laminar structures, and their outgrowth seems to require the interaction of abaxial and adaxial factors, as is the case in leaves (McAbee et al., 2006). As in leaves, the abaxial factors are *KANADI* and *YABBY* family genes. Specifically, for the outer integument of *Arabidopsis*, *KANADI1* (*KAN1*) and *KANADI2* (*KAN2*) are implicated along with the *YABBY* gene *INNER NO OUTER* (*INO*). For the inner integument the *KANADI* gene *ABERRANT TESTA SHAPE* (*ATS*) is implicated. The putative adaxial factors are not yet known with certainty, but the HD-Zip gene *PHABULOSA* (*PHB*), important as a leaf adaxial factor, is expressed in the adaxial region of the inner integument (Sieber et al., 2004; Fig. 5.15).

The parallels between the integuments and leaves are consistent with the leaflet theory of the ovule (Guédès & Dupuy, 1970). Guédès was struck by the common morphological pattern displayed by teratologic ovules. Abnormal ovules in nature are foliarized. In ovule foliarization the outer integument disappears, the chalaza flattens to become blade-like and the inner integument becomes tube-like. The leafiness of the chalaza is often striking, even with the formation of palisade parenchyma (characteristic of foliage leaves). Of course foliarization of ovules is no surprise now that we know that elements of the genetic mechanism of leaf formation are expressed during the development of normal ovules. If these genes are misexpressed during teratologic development, leafy structures are to be expected.

However, the leafy nature of the integuments is not borne out by the fossil record. The pteridosperm fossil record can be interpreted as the evolution of integumented megasporangia from the progressive fusion of either branches (telomes) or sterile megasporangia stalks present around a central megasporangium: the branch and synangial theories, respectively (Kenrick & Crane, 1997a). For instance in *Hedeia corymbosa*† (Cookson, 1935) a series of sterile branches surround the megasporangium, whereas in *Elkinsia polymorpha*† (Rothwell, Scheckler & Gillespie, 1989) an integument of linear lobes surrounds the megasporangium. In *Eurystoma angulare*† the integumentary elements are nearly entirely fused (Camp & Hubbard, 1963).

The filamentous structures formed in the place of integuments in certain mutant backgrounds (for instance *INO* double knockout mutants with another gene *PRETTY FEW SEEDS* (*PFS*)) have even been interpreted, perhaps somewhat hopefully, as mimicking such original branches or telomes (Park, Hwang & Hauser, 2004). However, as leaves become filamentous as a result of loss of abaxial–adaxial polarity in *kan1/kan2/kan3* mutants, this observation might be considered equally consistent with a leaf interpretation of integuments.

At first sight, there appears to be a discrepancy between the telomic (branch or stalk) interpretation of the basic seed-plant integument, as gained from fossils, and the leafy interpretation of integuments emerging from the molecular

[4] Although it has been suggested, but not widely accepted, that it represents a highly reduced megasporangial stalk (Herr, 1995).

genetics. The cupule predates the advanced megaphyll, characteristic of pteridosperms and angiosperms. However, it may be that in the evolutionary transformation of fertile telome systems into megasporophylls, leaf-like characteristics were not only acquired by the sporophyll but by the integuments also. In the **synangial theory** the integument is interpreted as homologous to part of a progymnosperm fertile telome system. There is therefore potential for the evolution of the integument and the megasporophyll to be developmentally linked. Under this scenario of "developmental hitchhiking" it would not be surprising for genes characteristic of leaf development to be integrated into integument development.

On the other hand it may be that the stem-derived cupules surrounding the megasporangia of pteridosperms are simply not homologous to the potentially leaf or leaflet-derived integuments of angiosperms.

The outer integument appears to be a structure unique to angiosperms. It frequently displays asymmetric growth, thus bending the ovule and giving the ovule as a whole abaxial–adaxial polarity. Thus the outer integument can be seen to have two major roles: an early role of bringing the ovule into a precise orientation appropriate for fertilization within the carpel, and a late role of giving extra protection to the seed and extra anatomical complexity to the seed coat. However, in many angiosperms the outer integument has been lost. This derived unitegmy is characteristic of the euasterid clade and appears to result from the congenital fusion of the two integuments. Derived unitegmy also occurs in other groups, notably *Impatiens* (McAbee, Kuzoff & Gasser, 2005) in which both unitegmic and bitegmic species occur together with species having a bifid integument (bitegmic at the apex and unitegmic at the base). Such variation lends itself as a model system for studying the origin of derived unitegmy (McAbee, Kuzoff & Gasser, 2005; Fig. 5.16).

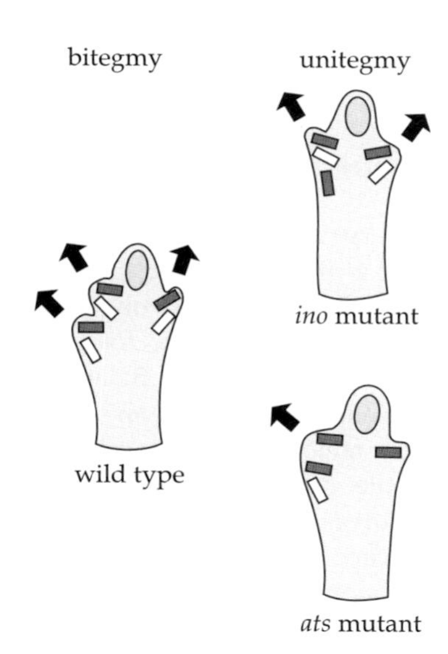

Fig. 5.16. Model of unitegmy in the *ino* and *ats* mutants. Active abaxial and adaxial identity is required for laminal outgrowth. The absence of functionality in the inner integument abaxial identity factor (*ats*) or the outer integument identity factor (*ino*) leads to a failure of integumentary outgrowth (after McAbee et al., 2005, 2006).

It is interesting to consider how the precise proximodistal ordering of funicle—chalaza (outer integument)—chalaza (inner integument)—nucellus arose. The gene *NOZZLE* (*NZZ*, also called *SPOROCYTELESS, SPL*) has emerged as a key regulator of this. Interestingly *NZZ* is involved in both anther and ovule patterning, indicating that these two organs are broadly equivalent at the level of megasporangial structures and have likely evolved from a common, homosporous, structure (Schiefthaler et al., 1999). This conclusion is further supported by evidence from the *MULTIPLE SPOROCYTE1* (*MSP1*) gene in rice (Nonomura et al., 2003). This gene too is active in both microsporangia and megasporangia. It encodes a leucine rich receptor kinase involved in cell fate specification within both sporangia.

In many *nzz* mutants the nucellus and the pollen sacs fail to form, suggesting that *NZZ* is involved in patterning. *NZZ* spatially and temporally regulates *INO* expression and thus controls the growth of the outer integument from the abaxial domain of the chalaza (Balasubramanian & Schneitz, 2002). Furthermore *NZZ* is required to restrict *PHB* to the region of initiation of the inner integument (the distal part of the chalaza; Sieber et al., 2004). The homeobox gene *WUSCHEL* (*WUS*) is expressed in the nucellus from where its expression is required noncell autonomously for integument development in the chalaza. It appears that *NZZ* is required for WUS expression in the nucellus too (Sieber et al., 2004).

A complex network of genes regulating ovule development is thus emerging (Skinner, Hill & Gasser, 2004). Other genes involved in this network are *BELL1* (essential for *INO* expression) and *AINTEGUMENTA* (*ANT*; Balasubramanian & Schneitz, 2002). ANT is an AP2 domain transcription factor that has a pleiotropic role in both ovule development and floral organ growth (Elliott et al., 1996). In the ovule it functions in concert with *NZZ* to regulate the temporal expression of *INO* (Balasubramanian & Schneitz, 2002). The ultimate specifiers of the sex organs are the floral homeotic genes. It is significant that, at least in the stamen, NZZ is activated by *AGAMOUS* (*AG*), thus linking sporangium development directly to the MADS-box floral identity genes (Ito et al., 2004).

5.14 Sporogenesis and spores

In the sporangium, the cells at the center assume an **archesporial cell** identity distinct from that of the sporangium **wall cells**. In bryophytes the archesporial cells give rise only to sporogenous cells (or to sporogenous cells and elaters in liverworts and hornworts). In pteridophytes, the archesporial cells give rise to sporogenous cells and tapetal cells, whereas in angiosperm microsporangia the archesporium gives rise to all of the following: sporogenous cells, tapetum, middle layer and endothecium. In the angiosperm ovule the archesporium gives rise to what is usually a single sporogenous cell, as well as inner parts of the nucellus. The inner wall of the nucellus is somewhat tapetum-like in some species. This is known as an **integumentary tapetum**, but it is unlikely to be homologous to the tapetum of microsporangia.

Considering the above, it is clear that there is evolutionary variation in the degree to which early divisions of the archesporium lead to sporogenous tissue, or to other tissue types that play an assisting role rather than a direct role in sporogenesis. In the angiosperm microsporangium the decision tree for the acquisition of cellular identity is as follows. Each archesporial cell divides periclinally to give a **primary sporogenous cell** and a **primary parietal cell**. The resultant primary parietal cell layer then divides into two **secondary parietal cell** layers. From these two secondary cell layers derive the endothecium, middle layer and tapetum by appropriate cell fate specification (Davis, 1966; Scott, Spielman & Dickinson, 2004).

The number of mitoses that the primary sporogenous cells undergo determines the number of sporocytes and hence the number of spores. Regulation of the number of mitoses is therefore an important step in the evolution of heterospory, and its effects may be seen to the fern *Platyzoma* and the moss *Archidium* in which sporangia with fewer mitoses produce fewer, larger spores (Tryon, 1964). Megasporangia generally produce fewer spores than microsporangia, culminating in the monomegaspory seen in *Marsilea* (Schneider & Pryer, 2002) and seed plants.

Within land plants the details of sporogenesis and meiosis are remarkably conserved (Renzaglia et al., 2000; Scott, Spielman & Dickinson, 2004). However, except in the basic processes of meiosis, plants differ markedly from animals as a consequence of the demands of alternation of generations. In plants long distance dispersal units are formed from the products of meiosis, and, in the case of heterosporous plants, these units may become homes to developing gametophytes.

A tetrad of four cells is generally formed from meiosis (Fig. 5.17). However, in the *Arabidopsis* mutant tetraspore (*tes*) the absence of a molecular motor-like kinesin protein results in a failure of the four products of meiosis to pull apart. Large "tetraspores" that contain all the products of meiosis are instead formed (Spielman et al., 1997; Yang et al., 2003a).

Developing spores are generally enveloped in a highly resistant coat formed of a remarkable polymer called sporopollenin. Sporopollenin has traditionally been regarded as a polymer of carotenoid esters. However, more recently this view has been rejected in favor of sporopollenin

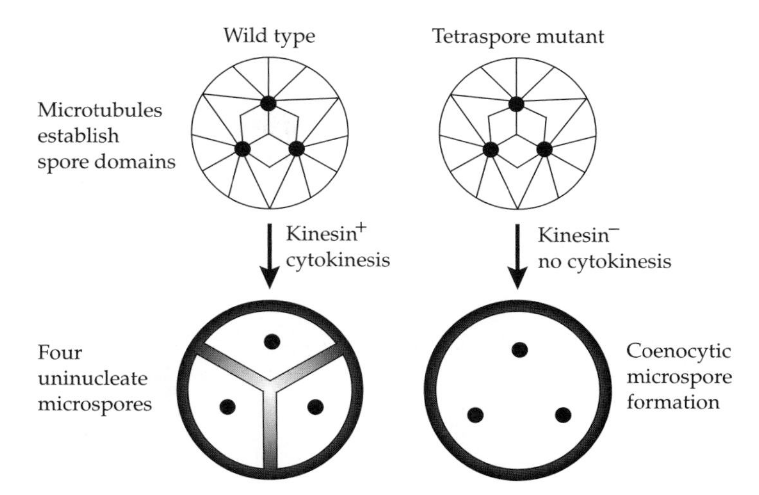

Fig. 5.17. Diagram illustrating the requirement for kinesin in the formation of microspore tetrads, as illustrated by the *tet* mutant. Only three spores are shown as in a tetrahedral arrangement, the fourth spore of the tetrad is in a different plane (after Spielman et al., 1997).

being a mixture of long chain fatty acids with ether-bonded polyphenolics (Ahlers et al., 1999, 2000; Dominguez et al., 1999). Whatever its make up this very effective protective material is vitally important in the function of the spores as dispersal units. The megaspore of seed plants, being permanently encapsulated in an indehiscent sporangium, does not have this protective wall, although some transient deposition of sporopollenin may occur (Pettitt, 1977).

The seed-plant megasporangium usually forms a single functioning sporocyte (spore mother cell), and of the four products of meiosis that this sporocyte produces, three generally abort leaving a single spore. This is not universally the case, however, as some angiosperms require two spores to form an embryo sac. Such embryo sacs are known as **disporic** as opposed to the usual **monosporic** ones. Nevertheless, the usual abortion of three products of meiosis to leave a single spore is an interesting apoptotic process, which must be under tight physiological control.[5] In linear tetrads the surviving spore is generally at the chalazal end, as in *Ginkgo* (Stewart & Gifford, 1967), suggesting that a tetrad polarity is established. In some angiosperms this polarity is reversed, but there are no reported cases of the surviving spore being the middle ones of a linear tetrad.

5.15 The gametophyte and prothallus

The gametophyte has evolved many forms characteristic of different groups of organisms. It ranges from chlorophyllous and foliose (as in mosses and leafy liverworts) or cylindrical, achlorophyllous and mycotrophic (as in lycophytes and ophioglossoids). The gametophyte must produce **gametangia** in which the processes of **oogenesis** and **spermatogenesis** occur, although in angiosperms these have been reduced away during evolution. The **archegonium** is the female gametangium of land plants. It is a relatively highly developed structure consisting of a stalk, a bag (**venter**), and a neck (through which the sperms gain entry). Within the venter are the unfertilized sex cell (the **oosphere**) and the **ventral canal cell**, and within the neck are the **neck canal cells**. The walls of the archegonium are generally a single cell layer. The male gametangium is called an **antheridium**, which when well-developed consists of a sac containing **spermatocytes** that develop into **spermatozoa**.

In all extant plants with a dominant sporophyte (i.e. all plants except mosses, liverworts

[5] It has been suggested that this process uses as a mechanism the segregation of a lethality factor at meiosis. This remarkable theory (Bell, 1996) is based solely on the results of a single piece of experimental work on *Mirabilis jalapa* described by Correns in 1900. No further evidence in support of the idea has yet been forthcoming.

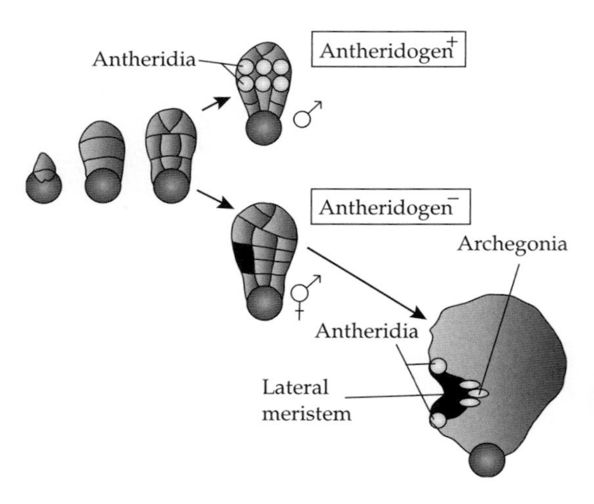

Fig. 5.18. Development of male and hermaphrodite prothalli in homosporous ferns (after Banks, 1999).

and anthocerotes) the spore germinates to produce a prothallus. Prothallus is a term for a gametophyte with a low order of complexity. In ferns and lycopods prothalli are free living. In ferns they constitute a small plate of green tissue (Fig. 5.18) formed from a single apical cell (e.g. *Dryopteris*) or a filamentous system (resembling moss protonema, as in *Schizaea* and *Trichomanes*). In lycophytes with a free-living prothallus, it is usually a small brown mycotrophic tuber. In *Equisetum* the prothalli are generally strongly lobed (Fig. 5.19).

In strongly heterosporous plants, however, the prothallus is usually small and contained entirely within the spore wall. In the case of the megaspore of *Selaginella*, the large unicellular megaspore germinates by forming cell walls within the spore. The spore wall splits open and archegonia form on top of the prothallial mass bursting from the spore. Sperm-producing microspores are carried to the megaspores in the surface water film.

5.16 The prothallus of seed plants

In seed plants this germination within the megaspore is taken to extremes. In the center of a mature ovule the single megaspore has germinated to produce a prothallus within the spore wall. As the megaspore wall is unthickened and elastic, the wall remained intact as a cellular prothallus grows inside the spore wall (gymnosperms) or a free nuclear prothallus forms inside the spore (angiosperms). The angiosperm prothallus is the most reduced in the plant kingdom.

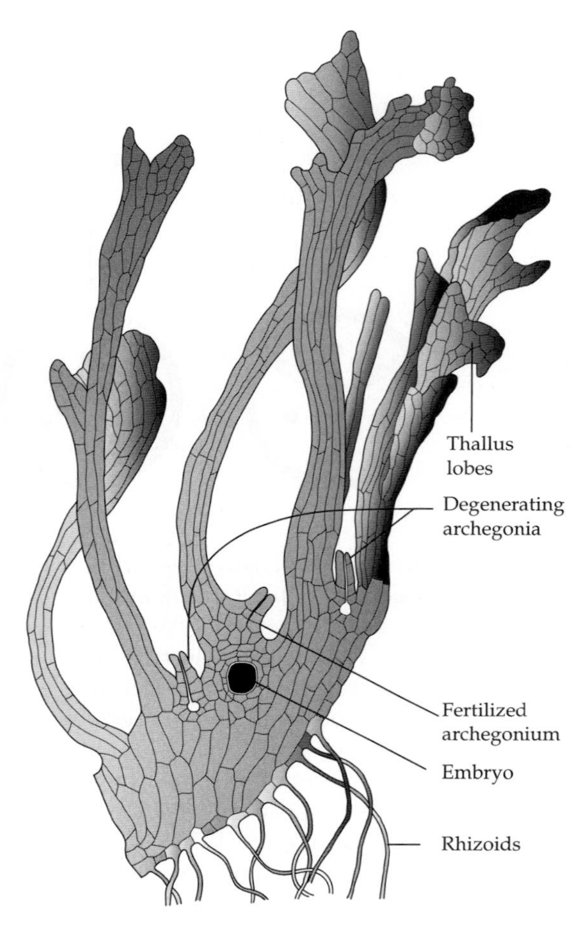

Fig. 5.19. Gametophyte of *Equisetum* showing fertilized and unfertilized archegonia (after Hofmeister).

A germinated megaspore in seed plants is called an **embryo sac**. An embryo sac is bounded by the unbroken spore wall and spore cell membrane. In plants with massive ovules, such as *cycads*, this leads to a single cell membrane being (in some species) big enough to cover the surface of a hen's egg and therefore perhaps one of the largest cells in existence. In gymnosperms, the prothallus within the megaspore is cellular and forms a number of rather reduced but still recognizable archegonia, each containing an egg. If more than one egg is fertilized, all but one of the resulting embryos usually abort.

In angiosperms, germination of the megaspore leads to a prothallus, very often of eight nuclei distributed in seven cells (Fig. 5.20). There are numerous patterns of prothallus development in angiosperms, and in some cases fewer or more nuclei are formed (Davis, 1966; Maheshwari, 1950; Williams & Friedman, 2004; Fig. 5.20).

The eight-nucleate "Polygonum" type is found in some 70% of angiosperms, however. In this type, at maturity, the female prothallus generally consists of (1) a large central cell with two centrally located "polar nuclei," (2) an egg apparatus consisting of an egg cell and two synergid cells at the micropylar end of the ovule and (3) three antipodal cells at the chalazal end of the ovule. In angiosperms both the egg and the central cell are fertilized by sperm nuclei. In *Arabidopsis* the type 1 MADS-box identity gene, *AGL80*, is essential for specifying the central cell in development (Portereiko et al., 2006).

The two synergids, companion cells adjacent to the egg, are thought to be homologous to the archegonium, mainly because of their proximity to the egg—in cell lineage and spatially. It is interesting that the early diverging angiosperm *Amborella* has three synergids and therefore appears to be less reduced (Friedman, 2006). However, it is an open question whether the cells of the bryophyte and pteridophyte prothallus can be homologized in any strict fashion with the cells of the angiosperm prothallus, as prothallial development has diverged so much.

It is easy to see how molecular organography could be interestingly used to test putative archegonial homology. If genes specific in expression to the archegonium of bryophytes and pteridophytes were found to have orthologues, which were specific in expression to the two egg companion cells, the hypothesis of homology would be greatly bolstered, and an extraordinary case of conservation of gene function would have been uncovered. Recently a gene apparently involved in specifying the development of the synergid

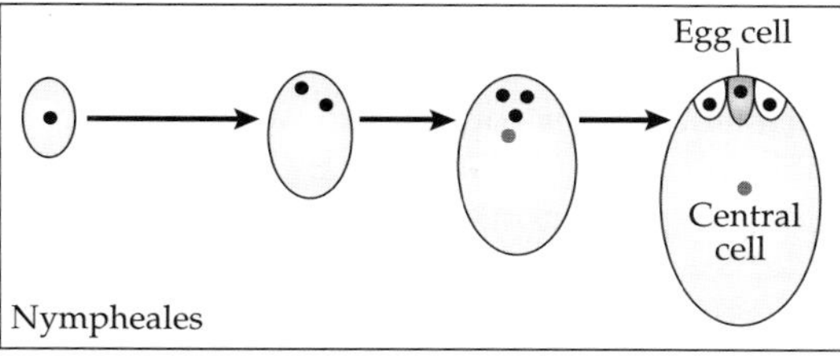

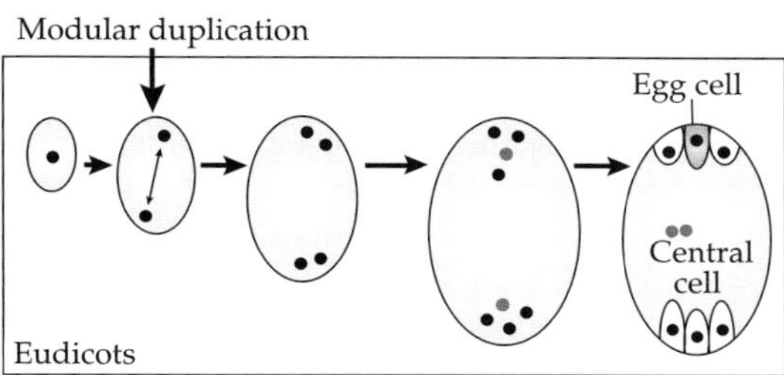

Fig. 5.20. Development of the four- and eight-nucleate embryo sac types in flowering plants. The eight-nucleate type may derive from a modular duplication of the four-nucleate type (after Friedman and Williams, 2004).

filiform apparatus has been discovered, *MYB98* (Kasahara et al., 2005). The filiform apparatus is part of the synergid that is a likely source of chemiattraction by the egg apparatus. Indeed mutants lacking this myb transcription factor could not attract pollen tubes to the egg. It would be of great interest to determine whether homologues of this gene are expressed in the archegonia of bryophytes and ferns. The filiform apparatus accumulates a receptor-like kinase protein, FERONIA, which is necessary for short distance signaling between synergids and pollen tube at fertilization. The specificity of this system is probably a major interspecies reproductive barrier (Escobar-Restrepo et al., 2007).

5.17 Gametogenesis and fertilization

"If two persons wish to find each other it is better for one to wait while the other searches" (Corner, 1964). This is one way of expressing the logic of sperm and egg (heterogamety), to which might be added "especially if she who waits has a powerful chemiattractive beacon." Land plants are heterogamous, with eggs and spermatozoa that clearly derive from algal precursors. The egg appears to be evolutionarily conservative, perhaps because, for waiting, a simple spheroidal shape is difficult to improve upon. The plant spermatozoon, in contrast, shows some considerable variation. This variation spans the range from complex motile cells to the two sperm nuclei of angiosperms. The helically coiled, biflagellate spermatozoon of bryophytes bears a strong resemblance to the helically coiled, biflagellate spermatozoon of the aquatic alga *Chara* (Renzaglia et al., 2000). This is one line of evidence that the Charales are related to the land plants (Graham & Kaneko, 1991). The conservation is hardly surprising as the bryophyte spermatozoon also swims through free water (in the surface film) to find the egg. There is one important difference however. Whereas the charalean spermatozoon has up to 30 mitochondria, the bryophyte spermatozoon is a stripped down version with a single mitochondrion.

Many lycophytes also have biflagellate spermatozoa, but some have multiflagellate ones, including *Isoetes* and *Phylloglossum* (Renzaglia & Garbary, 2001; Renzaglia & Maden, 2000). Ferns, horsetails and those gymnosperms with motile spermatozoa (cycads, *Ginkgo*) are all multiflagellate (Renzaglia & Garbary, 2001). The spermatozoon of cycads is particularly remarkable, commonly being a quarter of a millimeter in length and with a spiral row of very many flagella. Unlike the spermatozoa of bryophytes and pteridophytes, which swim in pure water, the cycad spermatozoon swims in an extruded pollen drop of very high osmotic pressure and hence has extreme adaptations to maintain mobility (Fig. 5.21).

Land plants are characterized by flask-shaped archegonia, their egg-containing organ. This is a unique and relatively conserved structure and one of the reasons why land plants are thought to be monophyletic. The archegonium has a single egg although it possibly is derived from an oogonial structure containing multiple eggs, remnants of which possibly persist in the form of the neck canal cells, which lyse to create a "transmitting mucilage" for the spermatozoon. On the spermatozoon reaching the mouth of the archegonium the neck cells lyse producing mucilage through which fertilization is effected.

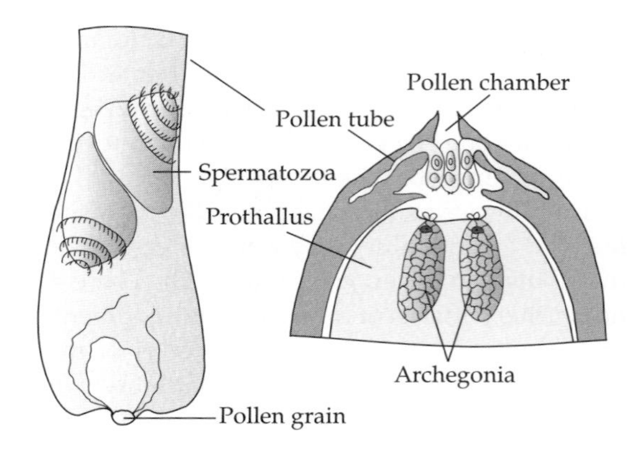

Fig. 5.21. Fertilization in a cycad (*Zamia*), according to Webber.

In angiosperms the archegonium is apparently represented (in a highly reduced state) by the two synergids together forming the "egg apparatus." The idea of the synergids as intrinsically associated with the egg is an old one, and the name itself derives from Greek *synergos*, "working together."

Spermatozoa are formed from spermatogenous cells in antheridia, or in the case of gymnosperms spermatogenous cells in the germinated pollen grain. In cycads and *Ginkgo* germination of the microspore produces a generative cell among the cells of the prothallus. This generative cell then divides to give a sporogenous cell and a sterile cell. The sterile cell usually takes no further part (although in *Microcycas* it too becomes sporogenous). The sporogenous cell forms the sperm. In *Ginkgo* spermatogenesis occurs after one further division so that two spermatozoa are formed (Friedman, 1986, 1987a,b; Friedman & Gifford, 1988).

In angiosperm pollen the process is generally simple. The single microspore nucleus divides once (pollen mitosis 1) to give a vegetative and generative nucleus. Then the generative nucleus divides again (pollen mitosis 2) to give the two sperm nuclei, which travel along the pollen tube to the ovule. These divisions are tightly controlled, apparently by homologues of the important animal/yeast gene *CELL DIVISION CONTROL PROTEIN 2* (*CDC2*). Cdc2 is known to be involved in the tight integration of cell division

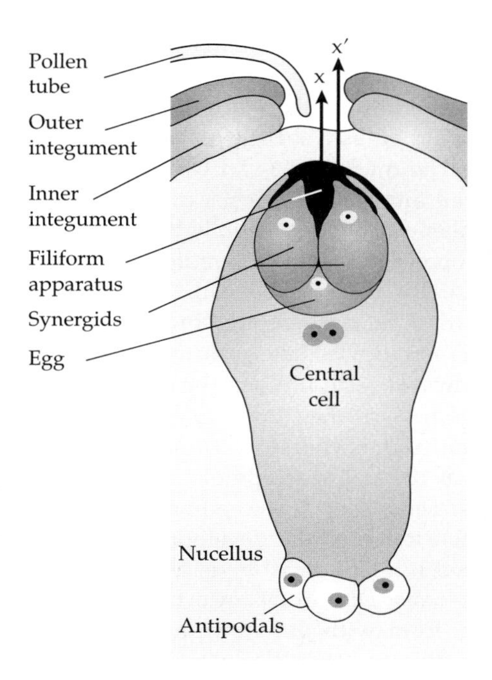

Fig. 5.22. Fertilization in an angiosperm. The arrows x and x′ indicate short- and long-range chemoattractive signaling from the filiform apparatus of the synergids (after Dresselhaus, 2005).

and cell fate determination implicit in animal and yeast developmental systems (Hayashi, 1996; Weigmann, Cohen & Lehner, 1997). In *Arabidopsis*, the plant homologue, *CYCLIN DEPENDENT KINASE A-TYPE* (*CDKA*), is required for the generative cell to divide in male gametogenesis (Iwakawa, Shinmyo & Sekine, 2006; Sekine, Iwakawa & Harashima, 2006). Knockout mutants therefore form one spermatozoon, not two. The single sperm fertilizes the egg, leaving the endosperm fertilization unaccomplished. The mutant, although having a perfectly viable sperm, is effectively male sterile. It is intriguing to speculate that the tight linkage between cell division and cell fate in the male gametophyte and its evolutionary variation may be controlled through developmental regulation of cyclin dependent kinases.

Motile spermatozoa travel chemotropically to the egg (Brownlee, 1994). In angiosperms much of the travel to the egg is effected by chemotropic growth of the pollen tube in the stylar tissue. In this case the guidance is performed both by signaling from the style and by the egg and synergids (Fig. 5.22). In the style, the guidance mechanism have still not been worked out, but it is known that an intriguing small basic blue protein (plantacyanin) found in the style attracts pollen tubes *in vitro* (Dong, Kim & Lord, 2005; Kim et al., 2003c), at least in lily (*Lilium*). Over shorter distances, a signal from the ovule is strongly implicated. There are *Arabidopsis* mutants known whose pollen tubes grow through the style but fail to find the micropyle (Johnson et al., 2004). There are also genotypes of *Titanotrichum* (Gesneriaceae) whose pollen tubes fail in the same way (Wang, Moller & Cronk, 2004b).

The synergids are now known to have a strong role in this pollen tube guidance (Higashiyama et al., 1998; Higashiyama, Kuroiwa & Kuroiwa, 2003). Laser ablation of the synergids has been shown to disrupt pollen tube guidance, implicating a diffusible substance produced by the synergids in this process (Higashiyama et al., 2001). Parallel work in maize has identified a protein "homing signal" produced by the egg apparatus (Marton et al., 2005). This homing signal is a small transmembrane protein of 94 amino acids called Zm*EGG APPARATUS1* (Zm*EA1*).

Furthermore, there are nonspecific signaling molecules such as Gamma-aminobutyric acid (GABA; Palanivelu et al., 2003; Palanivelu & Preuss, 2000). Curiously, in addition to its role in plants GABA functions as an inhibitory neurotransmitter in animals. Despite GABA's nonspecific role, there appears to be a strong specific component to the ovule signaling, as the process of pollen tube guidance has been shown to be highly species-specific in interspecies *in vitro* tests (Palanivelu & Preuss, 2006).

The key role of the synergids and egg cell (the egg apparatus) in this short range signaling is particularly interesting in the light of the likely homology between these cells and the egg and archegonium, which chemotropically attract motile spermatozoa. This raises the possibility that elements of gametic chemotropism may be derived from a common mechanism in all land plants.

Whether or not the mechanism shares common elements, the net result is the same: the fusion of gametes to form a zygote. Gametic interaction before fusion is almost certainly a highly controlled process involving cell–cell signaling of proteins on the membrane surfaces. Recently a particular protein has been implicated in this. GENERATIVE CELL SPECIFIC1 (GCS1) is required for gametic fusion in *Arabidopsis* (Mori et al., 2006). With fertilization effected, the ovule becomes a seed.

5.18 The seed

A seed is a fertilized integumented megasporangium, functioning as a propagule and is therefore the ultimate development of the ovule. It has a **seed coat**, or **testa**, formed by maturation of the integuments, and a **seed stalk** formed by maturation of the funicle. In bitegmic angiosperms the seed coat derived from the inner integument is sometimes distinguished as a **tegmen**. Under this scheme, when most of the seed coat is derived from the outer integument the seed is known as **testal**. Alternatively, if most of the seed coat derives from the inner integument it is referred to as a **tegmic** seed. It should be borne in mind however that parts of the **nucellus**, and even the **endosperm**, can contribute to the formation of the seed coat.

The seed coat can be many cell layers thick, or, as in orchids, reduced evolutionarily to a single cell layer. The outside of the seed coat can bear many features such as a **coma**, or **hair tuft**, for dispersal. Alternatively the seed coat may become **winged** (**alate**) by outgrowth of the testa. Other outgrowths of the testa occasionally occur and include a **caruncle**, an excrescence in the micropylar region as in *Ricinus*, usually connected to dispersal. An **arillode** is similar to an aril but is distinguished in being formed as an outgrowth primarily of the testa, rather than the funicle. A **raphe** (adnation of the funicle to the seed coat) may be present as a clearly visible feature of the mature seed, as in *Nympaea*. It represents a portion of the funicle adnate to the seed. The base of the ovule, the chalaza may also be visible as a surface feature of the mature seed, usually as a bump, as in *Vitis vinifera*. Finally the micropyle of the ovule can sometimes, although by no means always, be discerned as a surface feature of the mature seed. The micropyle is an important landmark in seed anatomy and if it cannot be discerned on the surface it can usually be located by cutting open the seed as the radicle of the embryo is usually directly beneath it and pointing toward it. It is therefore through the vestige of the micropyle that germination usually, but not always, occurs. When the seed is dispersed the vestiges of the funicle may be left as a scar (**hilum**) and, of course, by funicular outgrowths such as the aril or strophiole.

The seed coat is exceedingly complex and variable in its cellular development, a fact documented in great detail by E. J. H. Corner in his *The Seeds of Dicotyledons* (Corner, 1976). Various layers form with different physical properties. The main ones are a fleshy external layer called a **sarcotesta** to aid in dispersal when an aril is not present. There may be a very hard stony layer to protect the seed called a **sclerotesta**. Sometimes there is a membranous inner layer called an **endotesta**. Various cell layers of the integuments, or mixtures of them can contribute to these seed coat layers, so there is no necessary or simple correspondence between layers in the integuments and layers in the seed coat.

At the heart of the seed lies the embryo and its food reserves.

5.19 Formation of the embryo

In Embryophytes (land plants), the new sporophyte commences its development from a single diploid cell (zygote) within specialized nutritive and protective structures (either archegonia or embryo sacs).

The zygote results from the fusion of haploid gametes, fertilization of egg by spermatozoon. It proceeds to divide and so form the new diploid organism. In vascular plants the first cell divisions are interesting as they are filamentous. This produces a small pluricellular structure called a "suspensor" contained within the prothallus or the remains of the embryo sac.

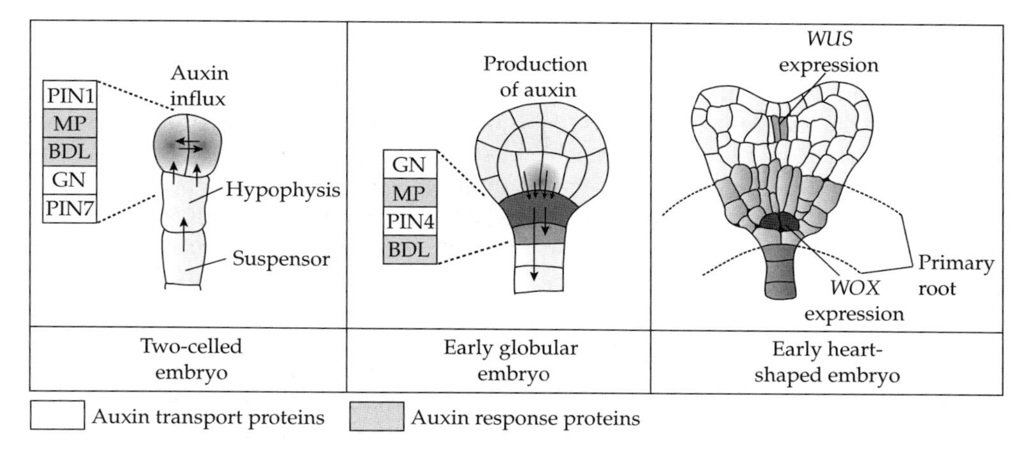

Fig. 5.23. The role of auxin and meristem determining genes (*WUS* and *WOX*) in determining the patterning of the angiosperm embryo. The left panel shows the strong expression of genes for auxin transport and auxin response in the two-celled embryo resulting in auxin-controlled embryogenesis. The middle panel shows changing auxin gene distribution at the globular stage. The right panel shows the location of organ initials at the heart stage. After Friml et al., (2003) and Scheres et al., (1994) (PIN = *PIN-FORMED*, MP = *MONOPTEROS*, BDL = *BODENLOS*, GN = *GNOM*, WUS = *WUSCHEL*, WOX = *WUSCHEL-LIKE HOMEO BOX*)

The production of the suspensor appears to be the result of a sort of "germination" of the zygote, and it is tempting, but speculative, to see this as a recapitulation of filamentous diploid generations of some algae, or as analogous to the filamentous protonema of *Chara* or of mosses. The protonema of mosses is the initial product of germination of the haploid spore and later gives rise to the more complex haploid moss organism (gametophore).

Many pteridophytes and all gymnosperms and angiosperms so far examined (with the possible exception of *Ginkgo*) have a recognizable suspensor. In mosses and liverworts there does not appear to be a suspensor, suggesting that this feature is a derived character and a probable synapomorphy for the tracheophytes. However, the significance of the suspensor is unlikely to be clear until its development, together with that of other filamentous structures in the green plants, is understood at the molecular level.

The suspensor has an apical cell from which the main embryo derives, although in seed plants the root derives from the uppermost cell of the suspensor (called a **hypophysis**). The suspensor plus apical cell is sometimes referred to as a **proembryo**. However, it is the division of the apical cell, as the main precursor of the embryo proper, that is the critical checkpoint in embryo development, setting in train the auxin signalling cascade that guides further organogenesis (Fig. 5.23; see also Chapter 3, Section 5).

As the embryo develops, the constituent parts form. In bryophytes these consist of the foot and the seta; in ferns the first leaf initials and the first lateral root; and in seed plants the stem (hypocotyl), cotyledons and primary root (see next section). These embryos either continue to develop *in situ* or, in the case of seed plants, are dispersed with the enclosing parental sporophyte tissues to establish the new sporophyte in a different location.

5.20 Structure of the seed-plant embryo

The embryo itself is made up of **radicle**, **hypocotyl** and **cotyledons**. They also contain the root and shoot primary meristems. Embryos that are well developed may have adventitious root buds at the base of the hypocotyl and a **plumule** (i.e. a bud with the makings of a shoot) with primordia of the first **cauline leaves** (i.e. noncotyledon leaves). Cauline leaves formed in embryo are referred to as primary cauline leaves. Many plants have two of these, but more may be formed as in the almond (*Prunus dulcis*), which may have up to eight. The embryo is very often oriented in a u-shape so that both the cotyledons and the radicle point toward the **hilum** or seed scar (marking the point of attachment of the seed stalk or funicle). The hilum is the usual point of germination for the embryo. However, this is not always the case. In walnut (*Juglans regia*), for instance, the radical is turned away from the hilum.

The **cotyledons** are the first leaves of the plant, remarkable because they form in embryo. In gymnosperms they are often very numerous, with up to 12 or 15 in *Pinus*. However, in angiosperms there are one or two (three occurs as an aberration). There is one in monocotyledons and two in other angiosperms (dicotyledons). They are usually strongly differentiated in morphology from subsequent leaves as a consequence of their unique environment of formation. They are usually of a simplified shape and smaller size, but are mostly green and leaf-like, carried aboveground by the elongation of the **hypocotyl (epigeous cotyledons)**.

However, in some plants the cotyledons are grossly swollen with food reserves and are not carried aloft to photosynthesize but remain underground to supply nutrients to the seedling while only the **plumule** expands aboveground. These are termed **hypogaeus cotyledons** (e.g. *Aesculus*, *Phaseolus*, *Quercus*, *Vicia*). In many palms the cotyledons are organs of transfer for nutrients from the endosperm to the developing plumule. These are called **transfer cotyledons**. They are well exemplified by the date palm (*Phoenix dactylifera*) and the Double coconut (*Lodoicea*). Transfer cotyledons are also found in *Rhizophora* and *Trapa*.

The cotyledons are folded up in the embryo in various ways. The commonest pattern is that the cotyledons are flat and are folded back to lie beside the radicle with either the edges against the radicle or the flat face against the radicle. Alternatively the cotyledons, especially when they are relatively large, may be folded. Folding may occur at the midrib (*Raphanus*), or by spirally rolling (*Bunias*) or plicate folding, as in *Acer pseudoplatanus*.

In certain Gesneriaceae, such as *Streptocarpus* and *Monophyllaea*, growth of the **cotyledons** is asymmetric. In extreme cases one cotyledon aborts and the other, the **macrocotyledon**, enlarges massively to become the only leaf of the plant. In *Streptocarpus* the apical meristem also aborts and inflorescences form adventitiously on the midrib of the macrocotyledon. This macrocotyledon is therefore called a **phyllomorph** as it has stem-like and leaf-like properties. It is highly meristematic, having a **basal meristem** (for continuous, indeterminate basal lamina growth) and a **groove meristem** (to produce inflorescences). *Streptocarpus* is a morphologically interesting genus, as plants with ordinary stem morphology occur (**caulescent** *Streptocarpus*), as well as plants consisting of a single phyllomorph only (**unifoliate** *Streptocarpus*) and plants which develop further phyllomorphs from the groove meristem (**rosulate** *Streptocarpus*).

5.21 The grass embryo

The grass embryo is particularly complex, often having adventitious root buds and extra organs of as yet unknown homology. One of these is the **coleorhiza**. This is a pouch that covers the root, and, in some cases, the developing adventitious root buds. On germination of the embryo the radicle punches through the coleorhiza, which may remain for some time as a ring around the base of the radicle.

As the coleorhiza surrounds the radicle, the plumule is surrounded by the **coleoptile**. This is a sheathing leaf that completely covers the developing cauline shoot. It may be homologous to a cotyledon or to a specialized prophyll.

There are two other grass-specific organs in the embryo: the **scutellum** and the **epiblast**. The scutellum is a shield-shaped outgrowth that abuts the endosperm and has a function in the absorption of nutrition. It probably represents an outgrowth of the cotyledon, either all of it or part of it. If the coleoptile is also homologous to the cotyledon, the scutellum may represent the bottom half of the cotyledon. A small scale called the epiblast is usually found opposite the scutellum, although this is not present in all grasses.

The epiblast is another candidate for homology with the cotyledon. If the epiblast is indeed homologous to the cotyledon then the scutellum and the coleoptile must represent one or more specialized prophylls.

5.22 Food reserves of the embryo

The mature embryo varies considerably in development. In orchids it is minute, while in other groups root and shoot are well-developed and very frequently completely fills the interior, or kernel, of the seed to the exclusion of other tissues. This is the case in hypogeous species like oaks and horsechestnuts, but also in epigeous species such as peas and cucurbits. In these cases the food reserves of the embryo are often stored in the **cotyledons** (as in broad bean, *Vicia faba*) or the **hypocotyl** and **radicle** in the case of the brazil-nut (Lecythidaceae). However, food reserves are often found in other tissues. These may be (1) prothallus, (2) endosperm and (3) nucellus (perisperm).

Prothallial reserves are only found in seed plants with a well-developed prothallus such as pines (*Pinus*) and other gymnosperms. In angiosperms the prothallus is reduced to a few cells, so it can only form reserves if there is secondary development by means of **double fertilization** (a second fertilization at the same time as the fertilization to form the zygote). The tissue resulting from double fertilization is called **endosperm**. However, in some angiosperms the nucellus develops into storage tissue called **perisperm**. In plants with a perisperm, there is usually endosperm too. However, in some plants (e.g. *Piper*) the perisperm forms the bulk of the food reserves.

Double fertilization is an extremely significant morphological innovation and a **synapomorphy** shared by all angiosperms. In many gymnosperms prothallial food reserves are formed by continued development of the seed, regardless of whether there is an embryo. This is wasteful of resources, often producing large amounts of sterile seed. In angiosperms the food reserves (endosperm) usually only form when an embryo is formed, as both are dependent on spermatozoa from the male prothallus (pollen tube). This is generally much more efficient as sterile and wasteful seeds and fruits are therefore not formed in the absence of fertilization.

However, occasionally endosperm development may be autonomous, occurring without double fertilization. Normally endosperm is **triploid**, formed from a **triple fusion** of two **polar nuclei** of the prothallus with a sperm from the male prothallus. However, this may differ depending on the number of polar nuclei. Consequently it is diploid in *Oenothera* and pentaploid in *Plumbago*. However, on occasion, such as in some elms, endosperm development is triggered merely by the fusion of the two polar nuclei in the absence of a sperm. This **diploid endosperm** often develops for long enough to trigger the development of a fruit: the phenomenon of **parthenocarpy**.

The endosperm has its own morphology both in terms of shape orientation and physical properties. It may be straight or curved or ruminate. **Ruminate endosperm** is formed when outgrowths of the testa divide the endosperm in an irregular pattern as in some Annonaceae and Myristicaceae. Ruminate endosperm is most familiar in nutmeg (*Myristica fragrans*). The physical nature of the endosperm at maturity varies from floury (farinose) as in cereals, fleshy (carnose) as in *Ricinus* (castor oil plant) or stony as in many palms, such as the date palm (*Phoenix dactylifera*).

5.23 Molecular control of endosperm development

Because of its vast economic importance and structural simplicity the molecular aspects of endosperm development have recently received much attention (Berger, 2003; Olsen, 2004). The endosperm generally begins life with the fertilization of the polar nuclei (usually two) in the large central cell of the embryo sac by a spermatozoon. However, in *Arabidopsis* the fertilization of the egg cell, even without the second fertilization, is sufficient to trigger endosperm development, at least for a time (Iwakawa, Shinmyo

& Sekine, 2006; Sekine, Iwakawa & Harashima, 2006). This implies that there is a signal produced as a result of egg cell fertilization that promotes endosperm formation.

The events allowing endosperm development to proceed are under tight control by imprinted polycomb-type genes that are specific to the central cell. These include *FERTILIZATION INDEPENDENT ENDOSPERM* (*FIE*; Ohad et al., 1999), *MEDEA* (*MEA*; Kiyosue et al., 1999), *FERTILIZATION INDEPENDENT SEED2* (*FIS2*; Luo et al., 1999) and *MULTICOPY SUPPRESSOR OF IRA* (*MSI1*; Guitton & Berger, 2005; Guitton et al., 2004). These genes all play a controlling role in normal endosperm development, and knockout mutants of any one of them will allow endosperm to form in the absence of fertilization. A likely explanation is that these proteins form a complex that represses genes involved in endosperm development (Guitton et al., 2004). The existence of these repressor genes should be no surprise. Endosperm development is an exponential cell-proliferative process, and the consequences of failure to control it may be severe.

There are two main forms of endosperm development, cellular and free nuclear. Most angiosperms have a free nuclear stage in their endosperm development, under which a syncitium develops in the central cell. In other species endosperm is cellular from the beginning, and in a rather odd type (found in some monocots) there is a cellular division of the central cell into two compartments followed by a reversal to free nuclear development in one or both compartments. This is the **helobial type** of endosperm development. Endosperm is absent in orchids (Orchidaceae) and podostems (Podostemaceae) although in some orchids (*Bletilla* is an example) a few cells do form and persist.

Arabidopsis has a free nuclear endosperm, and the syncitial stage lasts for eight rounds of mitosis (up to 256 nuclei), whereas it lasts nine rounds in maize (*Zea*, up to 512 nuclei). When the syncitial stage comes to an end, cellularization begins, generally on the outside first progressing toward the middle. The control of the onset of cellularization is of great developmental interest. It is so tightly coupled to the eighth round of mitosis that its control probably involves the same genes that control somatic cytokinesis (Sorensen et al., 2002), such as the cell-division-specific syntaxin gene *KNOLLE*, the Sec1 protein gene *KEULE* and the kinesin-like protein gene *HINKEL*. There is an intriguing parallel to the syncitial-cellular transition in the *Drosophila* embryo, which is also linked to the machinery of cytokinesis and tripped by a critical nucleo-cytoplasmic ratio (Edgar & Lehner, 1996).

In cereals cellularization is eventually completed throughout the endosperm, which is solid at maturity. However, sometimes cellularization is not completed and the endosperm remains liquid. This is clearly seen in the coconut (*Cocos nucifera*). The outer endosperm is cellularized (the coconut "flesh") while the inner layer is syncitial (the coconut "milk"). In coconut milk the mitoses are often incomplete leading to highly polyploidy (superploid) nuclei in the syncitial fluid (liquid endosperm; Abraham & Mathew, 1963). In coconut the nuclei may reach dodecaploidy (12*n*). It seems that maintenance of normal mitosis in such a giant syncitium is difficult. This type of endosperm superploidy (above 3*n*) is a phenocopy of endosperm superploidy that can be induced mutationally, as in the very numerous titan-class mutants (Tzafrir et al., 2002). The failure of many different components can cause endosperm superploidy, such as knockouts of the tubulin folding complex *PILZ* genes (Steinborn et al., 2002) and knockouts of the condensin gene *TITAN3* (*TTN3*; Liu et al., 2002).

After cellularization comes the patterning phase. Many endosperms have a threefold differentiation along the proximodistal axis of the seed. At the micropylar end is the **micropylar endosperm** (called **embryo surrounding region**, or ESR, in cereals). In the middle is the **starchy endosperm** (the main source of stored carbohydrate). At the base is the **chalazal endosperm** (called the **transfer layer** or **basal transfer layer** in cereals and involved in reserve acquisition). In wheat, and some other cereals, there is another endosperm region, forming an outer layer, called the **aleurone layer**. This layer contains storage proteins that are essential to bread-making (*aleuron* is Greek for wheat flour). It also shows the existence of radial patterning as well as a proximodistal patterning.

The existence of proximodistal patterning implies the action of identity genes capable of responding to spatial context information to specify endosperm regions. A candidate is the maize transfer layer specific MYB transcription factor ZmMYB RELATED PROTEIN1 (Zm*MRP1*; Gomez et al., 2002). In *Arabidopsis* there is evidence that this basal endosperm identity is restricted to the chalazal end by the FIE/MEA endosperm repression pathway. Mutants in genes associated with this pathway show an expanded distribution of chalazal endosperm (Sorensen et al., 2001).

Candidates have also been identified for radial patterning (i.e. specifying the outer or aleurone layer). In fact a large number of mutants are known from maize that fail to specify or properly develop an aleurone layer (Becraft & Asuncion-Crabb, 2000). In addition to these, a gene of particular interest is the *OUTER CELL LAYER4* (*OCL4*), encoding an HD-ZipIV homeodomain protein (Ingram et al., 2000).

5.24 The origin of the endosperm

An important question in plant evolution has been the origin and nature of endosperm. There are two scenarios. First the endosperm may be a development of the prothallus thus making it directly comparable with the prothallial reserves of gymnosperms, but with a different timing and the intercalation of an addition genome through double fertilization. Alternatively, as it results from a fertilization event it may be homologous to a separate organism: an altruistic second embryo that becomes food for the first embryo (Friedman, 1992). It would therefore be a case of differentiated multiple embryos or **polyembryony**. The chief evidence for this is that endosperm formation requires fertilization like an embryo and that some gymnosperms, including members of the Gnetales, form multiple embryos.

The gymnosperm prothallus may produce multiple archegonia, which may be fertilized by different sperm. It is thus possible to find dispersed gymnosperm seeds with twin embryos (Buchholz, 1920; Clare & Johnston, 1931). However, it is also probably common for all but one of the resultant embryos to abort, leaving a single embryo in the seed at maturity. This could be seen as an intermediate step in the formation of endosperm with a differentiation between successful and "suppressed" embryos in the same seed.

In 1990 Friedman put forward a bold hypothesis for endosperm origin (Friedman, 1990) based on a study of fertilization in the gnetophyte, *Ephedra*. In *Ephedra*, a sister nucleus of the egg nucleus is fertilized by a second sperm nucleus. Suggestively, the first divisions are reminiscent of the endosperm development. There is an initial process of free nuclear division and then a period of cellularization (Friedman, 1992). Eventually, however, these cells lead to the production of additional embryos. As gymnosperms are only distantly related to angiosperms this double fertilization process is probably not homologous between *Ephedra* and angiosperms. However, it may provide an excellent analogue model for early evolutionary stages of endosperm formation.

If a similar mechanism is responsible for endosperm in angiosperms, we are led to the conclusion that the endosperm is homologous to a supernumerary embryo. The conclusion is bolstered by the discovery that in some early divergent groups of angiosperms, such as the Nymphaeales (but not *Amborella*), the endosperm is diploid, involving a single polar nucleus (Williams & Friedman, 2002), and thus more reminiscent of an embryogenic fertilization. It is possible that **triploid endosperm**, involving fusion of two polar nuclei before fertilization, may have evolved more than once and may postdate the evolution of endosperm.

5.25 Polyembryony

Polyembryony is the presence of multiple embryos in a single seed. Polyembryony from multiple fertilization events is called **simple polyembryony**. Simple polyembryony is relatively common in *Ginkgo* and cycads and has even been recorded in the fossil record. Fossils of a Permian seed from Antarctica have been found with two archegonia, each one containing an embryo with suspensor (Smoot & Taylor, 1986).

Another form of polyembryony that should be mentioned is **cleavage polyembryony** in which the first divisions of a zygote give rise to two embryos rather in the manner of the formation of identical twins in mammals. The first division of the zygote is asymmetric, giving rise to an apical cell and a basal cell. The apical cell forms an embryo and the basal cell forms a suspensor. The suspensor is highly embryogenic, but its embryogenic potential is suppressed, possibly by the signaling from the apical cell (Vernon et al., 2001; Vernon & Meinke, 1994). If the apical cell is compromised, as it is by mutations at the TWIN2 (TWN2) locus, the embryogenic potential of the suspensor is released and multiple embryos can form. TWN2 encodes a valine-tRNA ligase, which appears to be needed for the correct functioning of the apical cell (Zhang & Somerville, 1997).

A third means of generating multiple embryos is **adventitious polyembryony**. This is a form of vegetative reproduction called **adventitious embryony**. In this, the embryo is not formed by sexual means but rather apomictically from diploid nucellar cells. If the nucellus is highly embryogenic, then several embryos may form in the ripe seed as in some cultivars of *Citrus*. This is one of the forms of agamospermy, which will be further discussed below.

5.26 Agamospermy

Agamospermy or **seed apomixes** is the asexual formation of seed. In seed plants it is a means of using the apparatus of alternation of generations to produce a new sporophyte while bypassing the genetic consequences of alternation of generations. Agamospermy produces new sporophytes that are identical copies of the parent sporophyte, and it is thus a form of asexual reproduction or **apomixis**.[6]

There are different ways of circumventing alternation of generations depending on the stage at which the process is "hijacked." The three main mechanisms are (1) diplospory, (2) apospory and (3) adventitious embryony (Fig. 5.24).

In **diplospory**, the process of meiosis is hijacked so than diploid spores are produced. In this process the reduction division converts to a mitosis so forming an unreduced **restitution nucleus**. The female gametophyte produced from spore germination is therefore diploid and identical to the parent sporophyte. Diplospory (diploid spores) is sometimes referred to as **apomeiosis** (without meiosis). The egg nucleus from the diploid female prothallus can therefore become a diploid embryo without fertilization. The process by which a female gamete develops into an embryo without fertilization is termed **parthenogenesis**. There are thus two separate processes involved: apomeiosis and parthenogenesis. These two processes may have separate genetic control as in *Taraxacum* (Van Dijk et al., 1999) and *Erigeron* (Noyes & Rieseberg, 2000). Alternatively both processes may apparently fall under the control of a single locus.

In **apospory** (without a spore) the process of meiosis and sporogenesis proceeds as normal except that all four spores abort (Fig. 5.25) and the spore function is assumed by a somatic cell (**apospore**). The female gametophyte is made from a diploid sporophytic cell (from the megasporangial wall), and the female prothallus is genetically identical to the parent. Again, a nucleus from the diploid female prothallus becomes an embryo without fertilization by parthenogenesis.

[6] The term apomixis is sometimes used synonymously with agamospermy (seed apomixis; Bicknell and Koltunow, 2004). However, it is convenient to call apomixis all those processes involving the conversion of sexual reproductive structures to asexual reproduction, including the conversion of flowers to bulbils as in some *Allium* species. The term agamospermy is then used to indicate seed apomixis specifically. Winkler originally introduced the term apomixis in 1908 to mean "substitution of sexual reproduction by an asexual multiplication process."

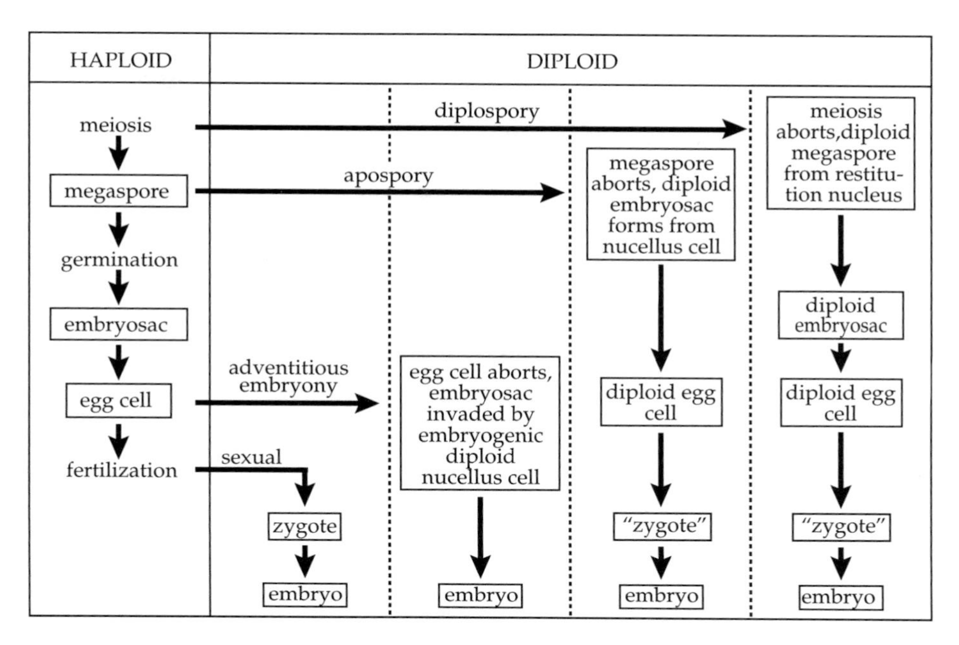

Fig. 5.24. Summary diagram of different forms of agamospermous reproduction, contrasted with normal sexual reproduction (left). Each form of agamospermy involves a departure from the wild type pattern at a different stage. (Not shown, apospory, like diplospory except that the spore originates from a diploid nucellar cell.)

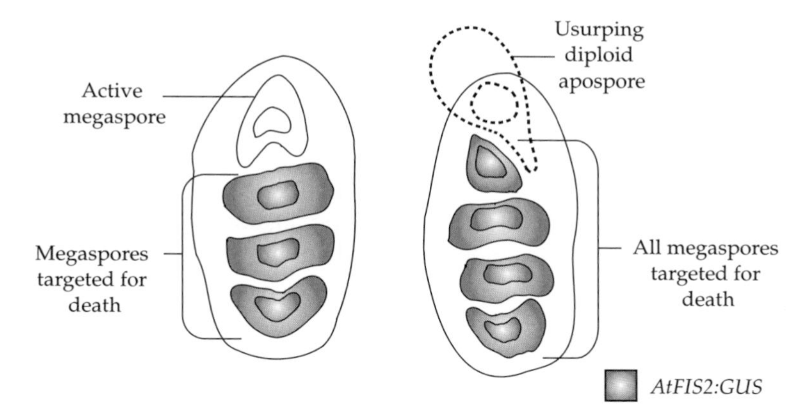

Fig. 5.25. The targeting of three of four megaspore for apoptotic death is correlated with *FIS2* expression. In apospory all four megaspores are targeted for death (after Bicknell & Koltunow, 2004).

The third mechanism, **adventitious embryony**, hijacks the process of alternation of generations right at the end. In adventitious embryony, spores are made and a haploid female prothallus forms. However, the new embryo is not made by fertilization of the egg but from invasion of the embryo sac by an embryo formed from a diploid cell of the nucellus. The new embryo is therefore genetically identical to the parental sporophyte. This process is known as **somatic**

embryogenesis, a common feature of some plant tissues under some conditions. Very often not one but two or more embryos form in the seeds of plant with adventitious embryony. An example is *Hosta ventricosa*, an apomict very often with multiple embryos. Adventitious embryony has a footnote in history as it is the first form of agamospermy to be discovered, over 150 years ago, in *Alchornea* (Euphorbiaceae; Smith, 1841).

Just circumventing the sexual process of embryogenesis from egg and sperm is not enough. The endosperm is also the product of a fertilization event and is essential for proper seed development. Normally the endosperm will not develop in the absence of fertilization, as a diploid endosperm is repressed by the *FIE/MEA* repression pathway. The normal requirement for endosperm is that it should be $2n$ (maternal) plus $1n$ (paternal). Many apomictic plants get around this by having normal fertilization of the endosperm but not of an egg. This is called **pseudogamy**. As there is a requirement for pollination in pseudogamous apomicts, apomixes can be difficult to establish in such plants. Usual tests for apomixes such as seed set with the exclusion of pollen, fail in pseudogamous apomicts which still require pollen. Genetic tests such as fixed heterozygosity at multiple loci are therefore applied (without recombination, no segregation of alleles is to be expected in apomictic populations).

Other apomictic plants are, however, able to produce a functional diploid endosperm, without a paternal genome. Such plants are said to have **autonomous endosperm** development. Normal endosperm regulation by imprinted genes and methylation can therefore no longer apply. One possible mechanism for this that has been proposed is to downregulate the imprinted *FIE/MEA* pathway and to have DNA that is less methylated (**genomic hypomethylation**; Vinkenoog & Scott, 2001; Vinkenoog et al., 2000; Box 5.2).

5.27 The biological context of agamospermy

The molecular basis of agamospermy has not yet been worked out in any system, even though it is usually found, on genetic analysis, to result from one or a few major loci. For instance, in *Taraxacum* a single dominant locus on the satellite-bearing chromosome is responsible for diplospory (van Dijk & Bakx-Schotman, 2004). A dominant major locus has been implicated in both apospory and diplospory in diverse systems (Bicknell, Borst & Koltunow, 2000; Nogler, 1984; Sherwood, Berg & Young, 1994). Although agamospermy comprises several different mechanisms, there are a number of generalizations about agamospermy that suggest that these phenomena may be related to some underlying biological principles and these generalizations may even point the way to the molecular basis of the mechanisms (Bicknell & Koltunow, 2004).

First, agamospermy is not randomly distributed in the angiosperms but is characteristic of

Box 5.2 Examples of genera containing apomictic species.

1. Diplospory with pseudogamy

Alliaceae: *Allium*. Brassicaceae: *Boechera holboellii*. Poaceae: *Tripsacum*

2. Diplospory with autonomous endosperm

Asteraceae: *Erigeron, Taraxacum, Eupatorium*

3. Apospory with pseudogamy

Poaceae: *Brachiaria, Cenchrus, Panicum, Paspalum, Pennisetum, Poa*. Hypericaceae: *Hypericum*. Rosaceae: *Amelanchier, Crataegus, Potentilla, Rubus*. Ranunculaceae: *Ranunculus*

4. Apospory with autonomous endosperm

Asteraceae: *Hieracium*

5. Adventitious embryony

Euphorbiaceae: *Alchornea*. Rutaceae: *Citrus*

certain clades suggesting that there may be preadaptations that are required.

Second, agamospermy is strongly associated with hybridity and polyploidy (and hence especially with allopolyploidy). In *Taraxacum* there are sexual diploids and apomictic polyploids. Residual sexuality in apomicts (particularly tetraploids) allows the production of new diploids and triploids. The diploids are always strictly sexual and the triploids always obligate apomicts. Thus there is microevolutionary cycling between agamosperms and sexuals (the sexual–asexual cycle) that is tightly associated with ploidy level (Verduijn, Van Dijk & Van Damme, 2004).

Third, there seems to be some association between types. Although the mechanisms are different, there are many cases of diplospory and apospory occurring in the same genus, in the same species, or even in the same ovule (Bonilla & Quarin, 1997; Smith, 1963), implying that the evolutionary pathways may share common elements.

Fourth, although generally under clear genetic control, agamospermy is often facultative and behaves as a quantitative trait (Barcaccia et al., 2006; Houliston & Chapman, 2004). Some plants are strictly agamospermous, but the trait can be variable and apparently affected by environmental factors. For instance, early studies on *Hieracium flagellare* revealed that different florets in the same capitulum could vary between sexual and agamospermous (Rosenberg, 1906).

The association with polyploidy is particularly intriguing. It suggests that there is a feature of polyploid biology that is intrinsically suitable for agamospermy (Quarin et al., 2001). It has been suggested that agamospermy depends on disparate gene copies from different genomes having subtly different expression domains or timing (Carman, 1997).

5.28 The mechanism of agamospermy

The hijacking of the normal developmental sequence, for instance spore mother cell function being taken over by a nucellar cell (aposporous initial cell), may happen in two ways. First, there may be an identity change or homeosis, where the nucellar cell takes on the identity of a spore. This would be a top-level alteration brought about by identity genes. Alternatively there may be a heterotopic or heterochronic change in which developmental events characteristic of one time or place are shifted in spatial or temporal domain. These alternatives intergrade, and are best viewed as opposite ends of a continuum (Baum & Donoghue, 2002). Nevertheless it is useful to distinguish between the two possibilities in elucidating the biology of apomixis.

To do this, markers of developmental events, such as callose (Tucker et al., 2001) and cytological characteristics (Grimanelli et al., 2003) have been employed.

Most revealing of all, however, has been the use of molecular markers. Tucker et al. transformed sexual and agamospermous *Hieracium* with GUS reporter constructs driven by the promoters of a number of *Arabidopsis* genes that are known to mark key stages in the sexual cycle (Tucker et al., 2003). The genes used were *NZZ*, *SERK1*, *MEA*, *FIS2* and *FIE*.

In sexuals and apomicts *NZZ* promoter-driven GUS is expressed in the megaspore mother cell. However, in apomicts, the *NZZ* promoter-driven GUS expression is absent from the aposporous initial cells, indicating that there is no homeotic transformation of this somatic cell into a megaspore mother cell at the level of the transcriptional regulators of NZZ.

FIS2-associated GUS expression was particularly revealing. In contrast to *Arabidopsis*, in sexual *Hieracium*, the *FIS2* promoter is activated in the three megaspores destined for PCD. The fertile megaspore does not show the *FIS2*-associated GUS until it divides. In agamospermous *Hieracium* all four megaspores show *FIS2*-associated GUS and all four abort. The aposporous initial does not show expression until it divides, like the megaspore in the sexual line. This implies that there is a homeotic transfer of megaspore identity to the aposporous initial, or, at least, a heterotopic change of expression domain of some genes. The later stage expression of the *SERK1* promoter-driven GUS did not change, indicating that the rest of the sexual pathway remains intact despite the upstream changes.

The *FIE/MEA* pathway (with which *FIS2* is associated) is clearly of importance in the establishment of agamospermy, but changes in its expression may be consequences rather than causes of agamospermy. In *Poa pratensis* a *SERK* homologue, Pp*SERK*, has been put forward as the switch specifying the aposporous initials (Albertini et al., 2005), but again changes in SERK expression patterns may be a consequence rather than a cause. The complete elucidation of the molecular mechanism awaits the cloning and characterization of one of the genetically dominant major loci responsible for the inheritance of apospory or diplospory.

5.29 Dormancy and growth of the new sporophyte

In seed plants a period of dormancy may be induced in the embryo that can last years. Dormancy is a very important part of a plant's ecology, and the importance of this trait is a reason for the success of seed plants and the evolution of the seed.

Seeds have evolved many kinds of **dormancy**, some of which are intrinsic in the embryo and require, for instance, cold treatment (**stratification**) for a number of days or weeks before germination (physiological dormancy). Other types of dormancy involve inhibitors in the seed coat as a physical barrier to hydration (physical dormancy). Yet other seeds have a light requirement for germination, or have an immature embryo when the seed is dispersed. Western white pine (*Pinus monticola*) seeds in North America are immature when harvested by Clark's nutcrackers (birds which collect and store pine seeds in caches), but the embryo continues to develop in the seeds after they have been cached.

Abscissic acid (ABA) is the major factor promoting physiological dormancy. White pine seeds exhibit deep dormancy requiring months of moist chilling to germinate. When dry, the seeds have very high levels of ABA, which is catabolized during moist chilling (Feurtado et al., 2004). The importance of ABA is also borne out by the great abundance of ABA response mutants from *Arabidopsis* that have a phenotype of altered germination response (Koornneef, Bentsink & Hilhorst, 2002).

Notable among these ABA response genes are *ABA INSENSITIVE3* (*ABI3*) and its maize homologue *VIVIPAROUS1* (*VP1*). Knockout of *ABI3/VP1* leads to loss of ABA sensitivity and hence complete loss of dormancy and precocious sprouting of the embryo before seed dispersal (**vivipary**). Vivipary in the wild is rare but occurs in the red mangrove (*Rhizophora*; Farnsworth, 2000). *ABI3/VP1* is an important regulator of the ABA signaling pathway that is required for the activation of ABA regulated genes (Suzuki et al., 2003).

A downstream consequence of the ABA response in seeds is the accumulation of the *LATE EMBRYOGENESIS ABUNDANT* (*LEA*) proteins, including the seed-specific group 1 *LEA* genes, during embryo maturation. The precise function of these very hydrophilic proteins is still unclear but they appear to have a role in the **desiccation tolerance**, which is part-and-parcel of seed dormancy. The moss *Physcomitrella patens* has a single group 1 *LEA* homologue, Pp*LEA-1*, which, significantly, is regulated by ABA and osmotic stress providing an intriguing link in mechanisms of desiccation tolerance across land plants (Kamisugi & Cuming, 2005; Noyes, 2000; Oliver, Velten & Mishler, 2005). Desiccation tolerance is a characteristic trait of mosses and other **poikilohydric** organisms, that is those that have little or no endogenous control of water loss. Desiccation tolerance is also likely to have been a quickly evolved trait of early land plant, particularly before complex vascular tissues and closable stomata allow the evolution of **homeohydry**.

The breaking of dormancy starts the growth of the embryo (called **germination** in the context of a seed) and so starts the life of the new sporophyte. The subsequent growth of the sporophyte is a developmental journey to spore formation. In the case of flowering plants it involves the iteration of numerous (perhaps millions) further organs. In the case of the bryophyte it involves merely the enlargement of a unitary and simple sporogonium with foot, seta and capsule but no multiple iteration of those organs (moss sporophytes may occasionally be branched but this is rare and the result of developmental abnormality).

The evolutionary history of the sporangium is a remarkable one, involving a transition from free-sporing plants with dominant gametophytes and sperm swimming in liquid water, to plants that encapsulate the most important transitions in the life cycle, spore, gametophyte and fertilization, all within an integumented megasporangium. By integumenting their life cycle, the seed plants have been able to master the environmental vicissitudes of life on land. It is in the angiosperm seed, with its marvelous capacities for dormancy and dispersal, that the integumented megasporangium reaches the highest level of ecological effectiveness.

CHAPTER 6

Sporophyll to flower

6.1 The origin and history of the sporophyll

In mosses, liverworts and hornworts the sporangium terminates a simple stem-like structure. However, in the extant lycophytes, and many extant euphyllophytes, the sporangia are borne on leaves called **sporophylls**.

In lycophytes the sporangia are borne on the adaxial surface of the microphylls (or at least in the axils). Primitive fossil lycophytes without leaves, such as *Zosterophyllum*†, bear the sporangia as bumps (enations) on the surface of the stems, so it seems that, on the evolution of the **microphyll**, the sporangia were "developmentally captured" by leaves. One theory of the origin of the lycophyte microphyll is that they are sterilized sporangiophores (Kenrick & Crane, 1997). If this is true then a fertile microphyll (sporophyll) of the lycophytes represents a double structure: evolutionary fusion of a sterile and a fertile sporangium (Fig. 6.1).

Lycophytes vary as to whether the sporophylls are strongly differentiated from sterile microphylls. The sporophylls may be in zones on the stem and scarcely distinguishable from the vegetative leaves apart from the presence of sporangia, as in *Huperzia*. Alternatively sporophylls may occur in terminal cones, as in *Diphasiastrum*, in which case the sporophylls are usually clearly differentiated from vegetative leaves. Thus morphological differentiation of the lycophyte sporophyll is linked to determinacy of the fertile shoot. The molecular mechanism for this is not known.

Heterosporous plants produce separate **microsporangia** and **megasporangia**. In some plants, such as the heterosporous fern *Marsilea*, both microsporangia and megasporangia may be produced on the same sporophyll. Alternatively, there may be separate **microsporophylls** and **megasporophylls**, as in angiosperms. Lycophyte leaves have a single sporangium so the leaves are of necessity either megasporophylls or microsporophylls, even though there is usually little discernable difference between the two. When both megasporophylls and microsporophylls occur in the same cone, the megasporangia are usually at the base (**basigynic**). This is true of the fossil lycophyte, *Flemingites scottii*† (*Lepidostrobus veltheimianus*; Scott & Hemsley, 1993), and this arrangement is also found in modern *Selaginella*, including *Selaginella kraussiana*, which has a single megasporophyll at the base of the cone. However, in *Selaginella* species with horizontal cones, the microsporophylls are found on the upper side and the megasporophylls on the lower side (Horner, 1966; Horner & Arnott, 1963). It has been suggested that the differentiation of microsporangia and megasporangia is determined at a late stage in *Selaginella* by the action of ethylene reducing the number of megasporocytes and thus producing megasporangia (Brooks, 1973; Brooks & Tepfer, 1972).

The euphyllophyte leaf, or **megaphyll**, has a separate origin to the leaf in lycophytes and so, therefore, does the euphyllophyte sporophyll. The leaf in euphyllophytes is derived from a branch system. Thus, before there were megaphylls, sporangia were borne at the tips of lateral branch structures and not on leaves. In most groups of euphyllophytes sporangia were incorporated into the evolving leaf structure, giving rise to modern sporophylls. However, some groups of plants apparently retained lateral

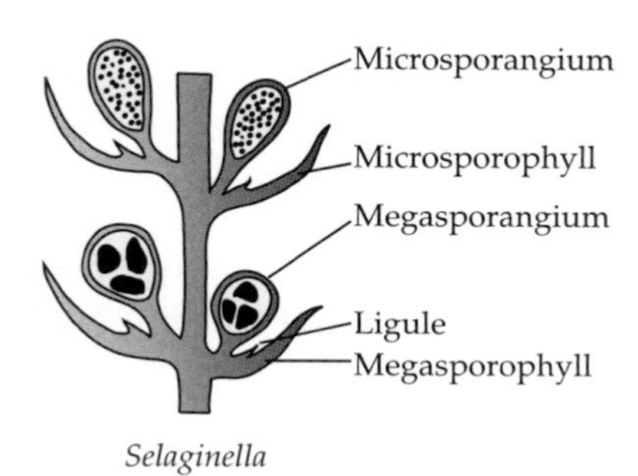

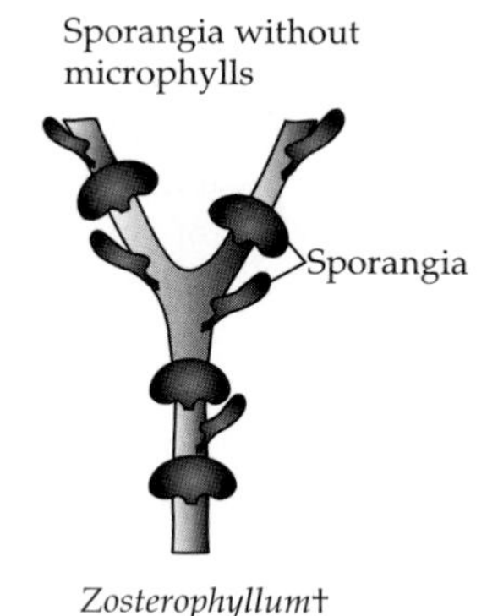

Fig. 6.1. Sporangia of lycophytes: leaf-associated in *Selaginella*, and as simple outgrowths in *Zosterophyllum*.

branch-like structures (**sporoclades**) for their sporangia, even after the evolution of leaves.

6.2 Phyllospory and stachyospory in the euphyllophytes

Plant groups with sporangia terminal on branches are known as **stachysporous**, whereas those with sporangia on leaves are **phyllosporous** (Lam, 1950). Although once erroneously used to make distinctions in the angiosperms (which are all phyllosporous), the terms do capture a fundamental distinction if used for euphyllophytes as a whole. The distinction depends on whether the origin of leaves was a vegetative phenomenon only or a reproductive and vegetative phenomenon. It also suggests a deep phylogenetic split between stachyosporous and phyllosporous groups, before megaphylls were fully elaborated.

In monilophytes (ferns and fern allies), all groups are considered phyllosporous with the exception of horsetails, which are usually interpreted as stachyosporous, with the peltate **sporangiophore** representing a condensed branch system. This is supported by the fossil record, which reveals plants such as *Sphenophyllum*† to have had branched sporangiophores in the axils of leaf-like bracts. They can thus be interpreted as having distinct sporoclades and megaphylls. These organs can be considered to have had a separate evolutionary history going back to leafless equisetophytes, such as *Ibyka*†, which had differentiated sterile and fertile branch systems (Skog & Banks, 1973; Fig. 6.2).

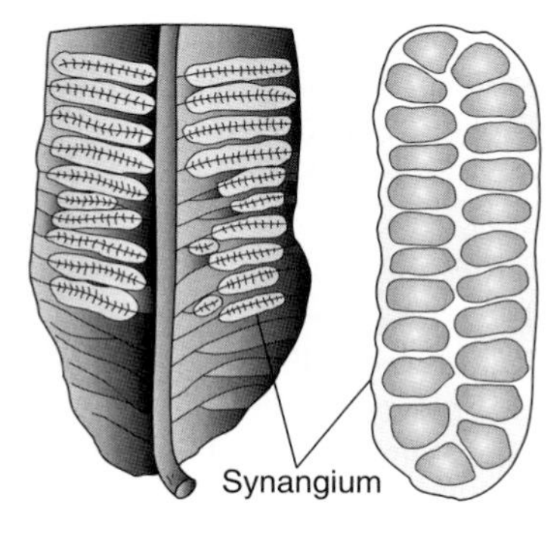

Fig. 6.2. Pinna of leaf of the eusporangiate fern *Danaea* showing the synangia (groups of united sporangia).

The leaf-synangium unit of the Psilotaceae (*Psilotum* and *Tmesipteris*) has also often been considered a condensed branch system axillary to a leaf. However, the recent realization that *Psilotum* is an ophioglossoid (Bateman, 1996; Pryer, Smith & Skog, 1995) has swung the interpretation back in favor of an interpretation of the leaf-synangium unit as a reduced compound leaf (now reduced to a forked flap) with three sporangia (now fused into a synangium). The interpretation of the reproductive unit of *Psilotum* as a reduction was first put forward on the basis of teratology (Sahni, 1923), although the teratologies were then interpreted as indicating a reduction from an axial structure. Bierhorst took an extreme view in homologizing the entire aerial structures of *Psilotum* with single leaves. He explained the fact that these "leaves" are terminal on rhizome axes rather than appendicular by comparing them with known occurrences of nonappendicular leaves in *Stromatopteris* and gleicheniaceous ferns (Bierhorst, 1969, 1977). Whatever the merits of Bierhorst's interpretation, it too is consistent with a phyllosporous Psilotaceae.

In the lignophytes (which include all the extant seed plants) both phyllosporous and stachyosporous groups occur. The angiosperms and the cycads are phyllosporous, while the conifers, ginkgophytes and gnetophytes are generally interpreted as stachyosporous. This division implies very deep phylogenetic divisions within the seed plants, possibly pointing to separate origins of the seed habit from progymnosperm ancestors (Fig. 6.3).

The conifers have complex compound cones with ovules borne on sporoclades in the axils of cone leaves (described below). The reproductive organization of the gnetophytes is problematic to interpret, but the recent realization that the gnetophytes are closely related to the conifers (and possibly nested within them) raises the possibility that the gnetophytes too have reduced compound cones, but one that has undergone reduction and evolutionary shifts. However, in *Ginkgo* the stachyosporous nature of the reproductive organization is easy to see. The male sporoclades are produced in the axils of leaves on the short shoots. They consist on a main axis with lateral branches (**sporangiophores**) each bearing two **microsporangia**. The female sporoclade also arises in the axils of leaves on the short shoots, and likewise consists of a main axis. However, this main axis generally only produces two branches by a terminal bifurcation. Each very short branch terminates in an ovule.

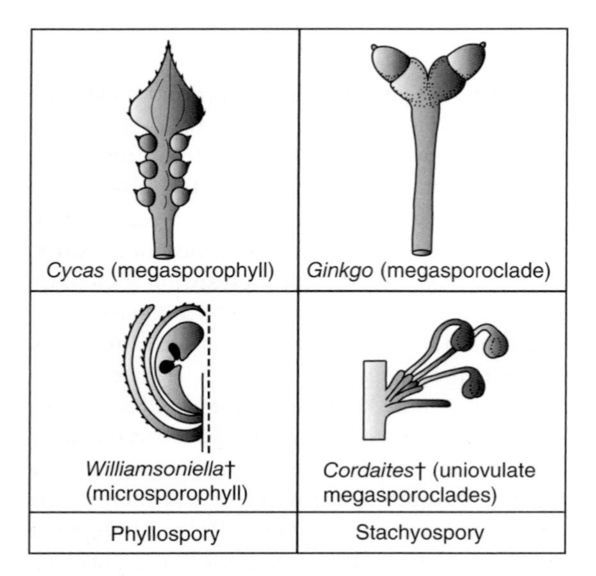

Fig. 6.3. Sporangia borne on axile and leaf-like structures.

Cycads on the other hand are clearly phyllosporous. This is most obvious in species of the genus *Cycas*, which have strikingly leaf-like megasporophylls. In other extant cycads (e.g. *Zamia*, *Encephalartos* and *Dioon*) both the microsporophylls and megasporophylls are borne in cones and are reduced to scale-like organs in comparison to the large compound vegetative leaves. However, there is no convincing evidence that they are anything but leaves.

6.3 Sporophylls of monilophytes

As noted above, all the monilophytes, with the exception of equisetophytes, are generally interpreted as phyllosporous. These comprise the ophioglossoids, the marrattioids and the filicoids (leptosporangiate ferns).

Typical ophioglossoids such as *Ophioglossum* and *Botrychium* have leaves with a sterile blade and a fertile portion apparently arising from the adaxial surface of the leaf. There is good evidence that this fertile portion derives from the union of two basal pinnae of a compound leaf (Deckereisel & Hagemann, 1978; Eames, 1936; St John, 1952; Fig. 6.4).

Perhaps the most familiar type of sporophyll is that found in the leptosporangiate ferns, which often have large compound leaves with patches of sporangia called **sori** either marginal or scattered over the abaxial surface. A sorus may be naked or be protected by a flap of tissue called an **indusium**.

The form and spatial and temporal distribution of sporophylls on the stem are of morphological interest. In some ferns all leaves will bear sporangia at maturity. In others there is an alteration of sterile leaves and sporophylls but there may be little or no morphological differentiation between sterile and fertile leaves. However, in some ferns there is strong differentiation between fertile and sterile leaves. In such cases the sporophyll usually has a highly reduced lamina and the remaining part is covered with sporangia (e.g. *Osmunda, Blechnum*). These sporophylls usually occur in zones on the stem interspersed by the production of sterile leaves.

Production of sterile and fertile leaves may be regulated by seasonal cues. In the Mexican *Acrostichum danaeifolium* sterile leaves are continuously produced, but fertile leaves are season-timed to allow spore release during the rainy season (Mehltreter & Palacios-Rios, 2003).

By analogy it is possible to imagine the mechanism by which **strobili** (**cones**) could evolve from such pulsed production of fertile leaves on an axis. This requires strong differentiation of sterile and fertile leaves and a linkage of sterile–fertile phase change with shoot determinacy at the molecular level, so that a cone of limited growth is produced.

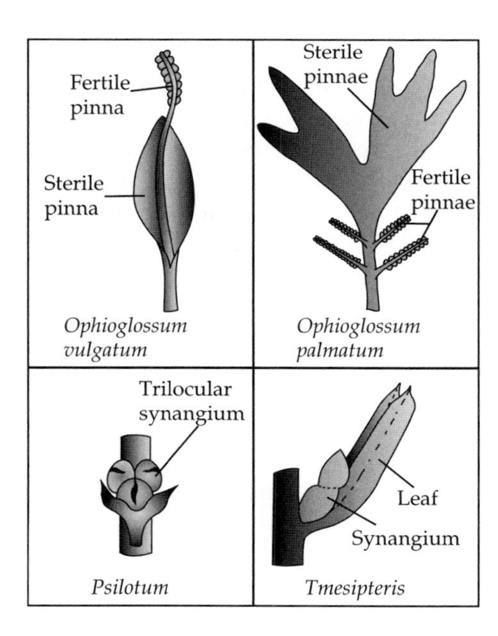

Fig. 6.4. Leaf-synangium structure in ophioglossoids.

An interesting modification of the sporophyll in ferns is the **sporocarp**, exemplified by the water ferns of the family Marsileaceae: *Pilularia, Regnellidium* and *Marsilea*. The sporocarp represents part of a leaf folded up to form a capsule divided into compartments, each segment containing several micro- and megasporangia (Bilderback, 1978).

Sporocarps occur among sterile leaves and function as highly specialized organs of dormancy and dispersal. Not only do they protect the spores in a desiccated state for many years but also by passing undigested through the guts of waterfowl, they effect long-distance dispersal, thus replacing spore dispersal by wind that is characteristic of other ferns. Although the sporocarp clearly represents an evolutionarily specialized innovation, sporocarp-bearing ferns have a surprisingly long fossil history (Lupia et al., 2000).

The sporophyll of *marattioids* is similar to that of the *filicoids*: a large compound leaf with numerous sporangia. However, the sporangia have a eusporangiate development and are gathered together into **synangia**.

6.4 Simple strobili of gymnosperms

A simple **cone** or **strobilus** is a group of **sporophylls** borne terminally on a shoot of determinate growth. They may be at the tip of a shoot that has normal foliage leaves at the base, or (as is usual in the compound cones of conifers) they may represent entire shoots that do not produce any normal foliage leaves. Such simple cones are found in the fossil groups Bennettitales† and Pentoxylales† as well as in extant cycads and angiosperms.

The cones may be **unisexual**, composed only of **microsporophylls** or **megasporophylls**, as is the case in cycads. In contrast, the angiosperm flower represents a bisexual cone in which the megasporophylls are apical (i.e. an **acrogynic** cone). There is only one example of apparently **basigynic** sporophylls in the angiosperms and that is *Lacandonia schismatica* in the Triuridaceae.

Examples of bisexual cones (similar to acrogynic) also occur in fossil Bennettitales†, such as in genus *Cycadoidea*† and sometimes in Williamsonia-type bennettitaleans, in which the bisexual strobili have been called *Amarjolia*† (Bose, Banerji & Pal, 1984). Other Williamsonia-type bennettitaleans have unisexual cones, as in *Weltrichia*† (male), *Williamsonia*† (female; Sahni, 1932).

The bisexual cone of most *Cycadeoidea* species comprises numerous protective hairy bracts at the top of a short peduncle. There is a basal whorl of microsporophylls inside. Each microsporophyll has numerous adaxial projections (**trabeculae**) that bear bean-shaped **synangia** (fused microsporangia). Inside these is a receptacle covered with numerous stalked ovules (each assumed to be a much reduced megasporophyll) intermixed with sterile scales, called **interseminal scales** (assumed to be sterilized megasporophylls). The angiosperm perianth is probably also derived from sterilized sporophylls (i.e. microsporophylls). However, the numerous bracts surrounding some bennettitalean cones could represent either sterilized microsporophylls or modified vegetative leaves.

Another group that has been interpreted as having simple strobili is the Pentoxylales†. In these plants the female reproductive organs are branched axes terminating in small "cones" composed only of densely packed ovules. The male "cones" are made up of a whorl of sporangiophores (each bearing numerous sporangia) on a basal receptacle (Drinnan & Chambers, 1985; Sahni, 1948; Vishnu-Mittre, 1953). Although these reproductive structures have traditionally been called cones because of their whorled or spiral organization (Sahni, 1948), it is unclear whether they are really composed of sporophylls. They could alternatively be interpreted as condensed sporoclades, somewhat reminiscent of those of the Ginkgoales.

In cycads (Cycadales), most species have sporophylls in unisexual cones. An exception is *Cycas*, which has strikingly leaf-like megasporophylls borne not in cones but alternating with the vegetative leaves. The ovules, from two to eight, are borne at the margins. The microsporophylls of *Cycas* are also leaf-like, even though they are borne in cones. The microsporophylls bear numerous microsporangia (up to 1000) on the abaxial surface. Other cycad genera have a strictly strobilar arrangement of their sporophylls.

6.5 The compound strobili of conifers

Strobili may develop as compound structures, by branches originating in the axils of the sterile cone scales. Plants with such **compound strobili** are therefore stachysporous not phyllosporous. Early interpretations of the conifer cone as composed only of sterile bracts and simple sporophylls (ovuliferous scales) proved untenable on examination of the fossil record (Florin, 1951). Fossils of the early conifers *Cordaites*† and *Lebachia*† reveal compound cones made up of a leafy shoot bearing leafless sporoclades in the axils of the cone leaves. Some of the sporoclade branches are sterile and some fertile (terminating in sporangia). Florin showed that there is a series of fossil intermediates connecting these lax compound cones to the cones of modern conifers (Florin, 1951).

Thus, in *Pinus*, the cones are composed of two types of scale: **cone scales** (cone leaves, cone bracts, bract scales) and **ovuliferous scales**. Ovuliferous scales are borne in the axils of the cone scales and are composed of flattened sterile tissue and two ovules. This ovuliferous scale is considered homologous to a sporoclade, with the sterile part composed of united sterile branches of the sporoclade. *Cryptomeria* is interesting in this regard as the ovuliferous scale has five distinct lobes, each of which may represent a sterile branch of the sporoclade.

As the ovuliferous scale is not considered homologous to a leaf but to a reduced stem system that has become flattened, the molecular investigation of how the ovuliferous scale obtained its scale-like (i.e. leaf-like) appearance is an important challenge.

The compound cones of modern conifers (particularly the female cones) have attained enormous diversity through a series of modifications. The main modifications are (1) reduction, (2) fusion of parts and (3) change in physical properties (fleshiness or woodiness). This is seen to the extreme in some Podocarpaceae such as *Dacrycarpus dacrydioides*. Here the cone is reduced to a single ovule, and the sterile cone scale is joined to the ovuliferous scale. The ovuliferous scale (so modified that it is called an **epimatium**) is wrapped around the ovule and fused with the integument. The cone axis below the ovule is swollen into a **receptacle** that promotes bird dispersal.

In some Cupressaceae (*Juniperus*) the cone scales become fleshy and promote bird dispersal. Such a fleshy cone is known as a **galbulus**.

6.6 The angiosperm flower as a simple strobilus

There is no absolute distinction between a simple cone and a flower. However, if a cone has all or most of a series of special features it may be called a flower. These special features include (1) the strong compression of the axis so that the parts are in a tight spiral or in whorls; (2) strong determinacy so that a relatively limited number of parts are produced; (3) production of megasporophylls and microsporophylls in the same bisexual, acrogynous organ and (4) occurrence of basal leaves forming a sterile perianth for the purposes of protection or attraction of pollinators.

These perianth organs may be sterilized microsporophylls or bracts co-opted into the flower. The perianth members are often developed in significant ways, for instance being colored or scented. It is the possession of a perianth that is usually the most obvious feature distinguishing flowers from cones. There are angiosperm flowers that have lost their perianth by reduction, but these are still homologous with true flowers and consequently called flowers.

A group of fossil gymnosperms, the Bennettitales†, have structures similar to a perianth. They also have compressed and determinate cones that are often bisexual. These fully meet any reasonable definition of flowers and there is absolutely no reason not to call them flowers. However, it should be borne in mind that they may be independently derived and so not homologous to angiosperm flowers.

Regarding the angiosperm flower as a simple cone requires that stamens and carpels be homologized as microsporophylls and megasporophylls. This view of the origin of the flower is the classical view and was well articulated by Arber and Parkin, as their "strobilus theory" of the flower (Arber & Parkin, 1907).[1] Arber and Parkin postulated that angiosperms and bennettitaleans evolved from a common ancestor, a hypothetical "amphisporangiate" with both megasporophylls and microsporophylls in a strobilus. In turn they hypothesized that the amphisporangiates and the monosporangiate cycads (with unisexual cones) evolved from a pteridosperm ancestor.

In a subsequent work they developed their theory to interpret the reproductive structures of the Gnetales as simple strobili or "pro-anthostrobili." Thus the Gnetales became, in their view, the closest relatives of the angiosperms (Arber & Parkin, 1908). This part of the strobilar theory proved a major difficulty for many morphologists and subsequently fell into scientific eclipse. It was briefly revived as the "anthophyte hypothesis" when morphological cladistic analyses placed the Gnetales as sister to the angiosperms (Crane, 1985; Doyle & Donoghue, 1986). Subsequently the finding that the Gnetales were related to conifers (Chaw et al., 2000; Wang, 2004; Winter, 1999) meant that the anthophyte hypothesis became untenable (Donoghue & Doyle, 2000; Doyle, 2006). However, there is no evidence that yet rules out a single origin of true simple strobilate groups (e.g. bennettitaleans and angiosperms) from a common pteridosperm ancestor.

The strobilar theory has now received substantial support from molecular-developmental models of the flower. The organs of the flower have been found to be remarkably equivalent at the molecular level, their identities controlled by

[1] This is sometimes called the "euanthial" theory to distinguish it from an alternative, although now unimportant, "pseudanthial" theory.

relatively simple molecular triggers. Homeotic interconversion of all the organs is possible with simple molecular changes. Furthermore, the molecular development of all the organs of the flower has been found to be leaf-like in many details. All this accumulating evidence is completely concordant with the strobilar theory, and it is now hard to see the flower as anything but a simple strobilus.

6.7 Neomorphological interpretations of the flower

As we have seen, there is a convincing case that the angiosperm flower is exactly what it appears to be, a simple strobilus of phyllomes around an unbranched axis. This is the classical view, implicit in the writings of Linnaeus but mainly developed by authors such as Goethe and Candolle (Candolle, 1827; Coen, 2001; Goethe, 1790). Arber and Parkin formalized the classical theory in an evolutionary and paleobotanical context as their "strobilus theory" (Arber & Parkin, 1907). As mentioned above, the classical theory is fully concordant with recent findings of molecular developmental genetics.

However, there are many ingenious alternative interpretations of the flower.[2] Guédès termed these "neomorphologies" to distinguish them from the classical view (Guédès, 1979). These neomorphologies go back a long way and have raised considerable controversy. As Guédès puts it,

> morphology has in fact been harassed by many neomorphologies since the 1830s...That neomorphologies are mostly neglected, it should be granted, is no proof of their falsity. That they are many and none is in agreement with all the others, however, may well hint at the falsity of at least all but one. (Guédès, 1979)

The most notable neomorphologies surviving to the present day are those of Croizat, Meeuse and Melville. Croizat views the carpel as "very variable body of essentially compound nature" (Croizat, 1961). In this view the carpel becomes a leaf–stem composite structure composed of a cauline structure bearing ovules (the placenta), enclosed by a leaf-like structure (the carpel wall). Little evidence has yet accrued for the interpretation of the placenta as an independently derived stem structure. It is true that certain so-called acarpellary flowers exist in which a single basal ovule is attached not to the carpel wall but to the top of the floral axis in an apparently cauline position, as in *Basella* (Sattler & Lacroix, 1988) and some other members of the Centrospermae. However, the phylogenetic evidence points to such flowers being highly derived, and it is easy to see how such flowers could arise by reduction from the carpellary flowers of their close relatives. However, Croizat's hypothesis is testable as it predicts that the placenta is a separate organ to the rest of the carpel, and that the placenta is cauline. We would expect gene expression patterns or homeotic mutations to give an indication of this, but so far there is no evidence forthcoming from molecular studies.

The anthocorm theory of Meeuse may be seen as a modern version of the "pseudanthial theory" of Eichler and others (Eichler, 1876) who viewed the catkin, particularly that of *Casuarina* (Wettstein, 1907), as the primitive form of angiosperm reproductive structure, derived from a gymnospermous cone. The more conventional flower, such as that of *Magnolia*, is in turn derived from the catkin-as-cone by condensation. The theory gained widespread exposure (if not acceptance) from the encyclopedic work of Engler and Prantl, who placed catkin-bearing plants (their group Amentiferae) at the beginning of the Englerian system, although this was merely to indicate their simplicity, not their primitiveness (Engler, 1897). Phylogenetic evidence has ruled out the pseudanthial theory, at least in its original conception, as catkin-bearing plants are all nested within groups characterized by conventional flowers. The catkin is clearly a derived feature representing an inflorescence of reduced flowers, not the forerunner of whole flowers.

In a somewhat similar manner, Meeuse interprets the flower as an ancestral branch system

[2] The following authors are among those who have supported neomorphological interpretations of the flower: Croizat, Emberger, Gregoire, Meeuse, Melville, Nozeran and Plantefol.

bearing simple reproductive structures (the "anthocorm") that has been similarly condensed (Meeuse, 1973, 1975). The flower is thus interpreted as a pluriaxial rather than a uniaxial structure. Other botanists generally received Meeuse's theory unenthusiastically (Corner, 1966), and despite Meeuse's prolific writing on the subject (including numerous books in which he invented an extensive language of anthocormic terminology) few have taken up his ideas. Molecular support for the anthocorm would be forthcoming if homeotic mutations were found that transformed floral parts into anthocormic stem structures or if genes associated with stem development were found to be expressed in floral organs indicating their shoot rather than leaf nature. However, no experimental evidence consistent with the anthocorm theory has yet been forthcoming.

The gonophyll theory of Melville also treats the angiosperm flower as a composite structure. Instead of stamens and carpels as sporophylls, Melville postulated that they were derived from a hypothetical fundamental unit called a "gonophyll." A gonophyll was conceived as consisting of a leaf-bearing epiphyllous, a dichotomously branching stem structure with sporangia (Melville, 1962, 1963). With such a generous starting unit it is not surprising that any type of flower can be derived from the reduction or fusion of multiple gonophylls. Furthermore, by interpreting the reproductive structures of glossopteridalean fossils as gonophylls, the origin of the angiosperm was neatly solved. However, apart from its ability to explain floral structure (albeit less parsimoniously than the classical theory) no clear evidence has accrued for the existence of gonophylls (Corner, 1963). Some molecular-developmental evidence could be expected for the composite stem–leaf nature of stamens and carpels but none has been forthcoming.

On the contrary, the euanthial theory is strongly supported by the ease with which homeotic transitions can be achieved by simple gene mutations, and by the power of the ABC model of floral development (implicit in which is homology between floral organs and leaves). So despite all the neomorphological challenges, the classical theory, now greatly buoyed up by molecular evidence, is still the most widely accepted. As Corner notes, "It is a most elegant theory, comparable with celestial mechanics because it assumes the minimum and derives the most" (Corner, 1966).

6.8 The origin of the carpel from a gymnospermous precursor

A great deal of interest is attached to understanding the ancestry of the angiosperm megasporophyll. The salient features are as follows:

1. Phyllospory.
2. Megasporophyll, a simple leaf. Megasporophylls should be capable of peltation to form an ascidiate carpel.
3. Bitegmic ovules (with the first integument possibly a cupule derived from sterilized sporangial axes (from the stachyosporous phase) and the second integument apparently a leaf- or leaflet-like organ from the phyllosporous phase).

There are numerous candidate gymnosperms that meet these three criteria to a lesser or greater degree. Extant gymnosperms can be ruled out, both on phylogenetic and morphological grounds. First, they form a monophyletic clade sister to the angiosperms (i.e. they are more closely related to each other than to angiosperms). Second, they are all stachyosporous with the exception of cycads, which are not cupulate, so have no candidate structure for the second integument.

Most paleozoic pteridosperms (Permian and earlier) can also be ruled out, as the cupules in these plants are derived from fused sterile axes (telomes) rather than from leaflets (Stewart & Rothwell, 1993). There are a number of more promising groups among the Mesozoic (Triassic and later) fossil gymnosperms.

The Caytoniales (named after Cayton Bay in North Yorkshire) have compound megasporophylls consisting of a stalk-like (apparently abaxialized unifacial) rachis bearing ovule-containing cupular leaflets. The cupular leaflets are excellent candidates for the second integument (Doyle, 2006). There are several ovules per cupule in *Caytonia*† (Caytoniaceae), but in *Umkomasia*† and *Pilophorosperma*† (Corystospermaceae) these have been reduced to one. However, the rachis is less promising as the progenitor of the carpel blade. An adaxial surface would have had to reemerge to form the cross zone necessary for peltate laminar growth, or else angiosperms evolved from an ancestral-type Caytoniales with more blade and less rachis on the megasporophyll. Nevertheless, the Caytoniales provide a plausible path to the angiosperm carpel (Fig).

Another candidate is the group Glossopteridales. These plants have simple leaves. Megasporophylls are similar to foliage leaves but bear what is best interpreted as an ovuliferous petiolulate leaflet.[3] Curiously this leaflet occurs in a basal median position on the adaxial side of the parent leaf. This position for a leaflet is unusual in plants but not unknown. Analogous organization is found in diplophyllous leaves of angiosperms and in the *LAX MIDRIB* (*LXM1-0*) mutant of maize (Schichnes & Freeling, 1993; Schichnes, Schneeberger & Freeling, 1997; Schichnes & Freeling, 1998). In Glossopterids the ovules are borne on what is, topologically, the adaxial surface of the leaflet, facing the adaxial surface of the parent leaf. The leaflet is often called a cupule and is an excellent candidate for the origin of the outer integument, particularly as in some species the cupule inrolls and the ovule number is reduced. If the parent leaf were to become peltate around the cupule, it should not escape our attention that a carpel rather similar to that of *Amborella* would result, with a single ovule in a median position.

The Bennettitales (cycadeoids) have excited some interest as their cones look superficially like angiosperm flowers (and are even amphisporangiate in *Williamsoniella*†). However, it is possible to homologize these cones with a single radialized leaf, rather than an axis (Delevoryas, 1968, 1982). Certainly the naked ovules, inserted directly onto the cone axis and surrounded by **interseminal scales**, provide a particularly poor starting point for the evolution of a carpel with bitegmic ovules.

[3] The morphological interpretation of reproductive structures of the Glossopteridales has been controversial. However, the leaf nature of the ovuliferous structure was clearly established by R. E. Gould and T. Delevoryas in 1977.

6.9 Molecular theories for the origin of the flower

Accepting Arber and Parkin's conception of the flower (Arber & Parkin, 1907) as "an amphisporangiate strobilus of determinate growth with an involucre of modified bracts" requires an explanation for three things (listed here in the possible order in which they arose in evolution):

1. How a cone became **determinate**?
2. How **bisexuality** (i.e. the **amphisporangiate** condition) arose from a unisexual state?
3. How the modified bracts (**perianth**) arose?

To these a further question may be added:

4. How the perianth became differentiated into inner and outer types (**sepals** and **petals**)?

Various attempts have been made to provide models that explain the evolution of the flower at the level of evolving molecular mechanism. At present all such attempts are extremely speculative, but provide hypotheses for testing in the future.

Determinacy and compression, as opposed to a long indeterminate floral axis, is an important aspect of the flower. The current *Arabidopsis*-based model for floral whorl regulation (Jack, 2004; Lohmann et al., 2001; Parcy et al., 1998) supposes that microsporophylls develop within a domain of *UNUSUAL FLORAL ORGANS* (*UFO*) expression. Megasporophylls (carpels), on the other hand, develop within the domain that expresses *WUSCHEL* (*WUS*, which activates C-class genes) but lacks *UFO*. If the axis was elongating, all primordia would soon pass out of the zone of *WUS* expression and these expression zones would not be stable. Therefore, it is clear that the molecular patterning of the modern flower is *dependent* on compression and determinacy.

Determinacy may have come about as a result of the floral meristem coming under the influence of negative regulators (Baum & Hileman, 2006; Lohmann et al., 2001). *APETALA2* regulates stem cell maintenance in the shoot apical meristem (SAM) through the *CLAVATA–WUSCHEL* (*CLV–WUS*) pathway (Wurschum, Gross-Hardt & Laux, 2006), and *AGAMOUS*, the C-class gene, is an important negative regulator of *APETALA2* function, as is the micro-RNA *miR172* (Zhao et al., 2007), a micro-RNA apparently of ancient origin in land plants (Axtell & Bartel, 2005). We may speculate that the establishment of *miR172* expression in floral apices, followed by the evolution of an antagonistic relationship between *AGAMOUS* and *APETALA2* (Mizukami & Ma, 1992), resulted in a determinate cone.

Frohlich first tackled the origin of bisexuality at a molecular level with his **mostly male (MM)** hypothesis (Frohlich, 2003; Frohlich & Parker, 2000). MM assumes a **gynomonoecious** intermediate (with female and **bisexual** strobili) between **monoecy** (separate male and female strobili) and bisexuality of all strobili. It suggests that ovules were **ectopically** added to the tips of male strobili, and when these feminized male strobili evolved into functionally bisexual strobili the female strobili were lost. Consistent with this, the gene *NEEDLY*, a *LEAFY* (*LFY*) homologue that is expressed in female strobili in *Pinus* and other gymnosperms, appears to have been lost in angiosperms. The selective force driving gynomonoecy may have been pollination, as carpels may have produced exudates that attracted pollinating insects. Putting some carpels near the microsporophylls would serve to enhance attraction (Endress, 1996; Frohlich, 2003; Hufford, 1996).

An alternative view to the mostly male hypothesis is the **out of female (OOF)** hypothesis or the **out of male (OOM)** hypothesis (Theissen & Becker, 2004; Theissen et al. 2002). OOM also supposes a gynomonoecious intermediate, but is based on **homeosis** rather than an ectopic mechanism. This homeosis, it is argued, is due to a decline in B-class gene expression toward the cone apex, perhaps due to some biochemical gradient being set up within the cone to which gene expression responds. Baum and Hileman suggest a mechanism based on a *LFY* gradient driving C-class competition (Baum & Hileman, 2006). This hypothesis suggests that increasing levels of *LFY* expression over time within the cone during development will drive increasing B-class and C-class gene expression at the tip. If C-class genes have a higher maximal level of accumulation than B-class, then C-class genes will come to predominate, competitively excluding B-class genes from protein multiplexes (Baum & Hileman, 2006).

However, a simpler explanation has to do with the determinacy of the flower. *APETALA2* promotes B-function, probably through an antagonistic relationship with *AG* (Zhao et al., 2007), but in the center of the flower AP2-function is blocked by *miR172*, to make a determinate flower. This exclusion also has the effect of helping to exclude B-function from the center (Zhao et al., 2007). Thus the evolution of the association between B-function and *AP2* would have the effect of feminizing the AP2 protein-free center of a determinate male cone.

The origin of the perianth is complicated by the two possible origins: **floralized bracts (bracteotepals)** or **sterilized outer stamens (androtepals)**. Baum and Hileman favor an androtepal origin through the co-option of *WUSCHEL*, the meristem gene, as a coregulator of *AGAMOUS*, the C-class gene, (Baum & Hileman, 2006) most likely by the addition of *WUS* binding sites in the C-class gene's *cis*-regulatory regions (Lohmann et al., 2001). As *WUSCHEL* is expressed only in the center of the flower the peripheral regions would be free of C-class expression but have B-class expression.

The origin of the **dimorphic perianth** is an important evolutionary event and one characteristic of the majority of eudicots (Ronse De Craene, 2004). In *Arabidopsis* the B gene, *APETALA3* (*AP3*), is expressed only in regions with both *LFY* and *UFO* expression. As *UFO* is absent from the first whorl of the flower (Parcy et al., 1998), so is *AP3* B-gene expression. Baum and Hileman speculated that coregulation of B-activity by *UFO* may have been the key evolutionary step in the evolution of the eudicot-type of dimorphic perianth (Baum & Hileman, 2006).

6.10 Rules of the flower: patterning of floral organs

There are two very remarkable rules of the flower. The first is the **acrogynous rule**. This rule notes the fact that the gynoecium is always in the center of the flower and that consequently the stamens are inserted below the carpels.

Of some 250,000 angiosperms only one truly breaks this rule. This is *Lacandonia schismatica* an unusual mycoheterotrophic plant in the Triuridaceae from the Lacandon rainforest of Mexico. It came to prominence after the publication of an account of it, and its basigynous flowers, in Flora Neotropica (Maas & Rubsamen, 1986). Similar hermaphroditic flowers are occasionally found in the related dioecious *Triuris brevistylis* (Vergara-Silva et al., 2003).

However, it was pointed out that many members of the order to which *Lacandonia* belongs (the Pandanales) sometimes have reduced flowers (occasionally reduced to single carpels or stamens) clustered into heads or pseudanthia. This opened the possibility that the basigynous flowers of *Lacandonia* were in fact pseudanthial inflorescences (Rudall, 2003). In pseudanthia, as in the Araceae, it is common for the pistils (reduced pistillate flowers) to be at the base of the inflorescence and the stamens (reduced staminate flowers) to be at the top.

The floral development, on the other hand, is consistent with a euanthial model (Ambrose et al., 2006), albeit an unusual one with peripheral carpels. Six common promordia develop. The inner three form three stamens and numerous carpels by polymery. The outer three primordia form numerous carpels by polymery. Thus it seems that Triuridaceae flowers are only inflorescence-like due to carpel multiplication, and flowers in this family may even be derived from an ancestor with only a single free carpel (Rudall & Bateman, 2006). The centrifugal initiation is a strong indication of secondary polymery, as it is in centrifugal stamens as found in Dilleniaceae and Hypericaceae (Corner, 1946).

The molecular change responsible for the inside-out flowers of *Lacandonia* may well involve altered spatial expression (heterotopy) of B-class MADS-box identity genes (Ambrose et al., 2006). However, an upstream regulator of the MADS-box identity genes such as *SUPERMAN* (Bowman et al., 1992; Sakai et al., 2000) may be ultimately responsible (Ambrose et al., 2006).

The second "great rule" of the angiosperm flower is the **four whorl rule**, that there are very commonly four sets of floral organs, corresponding to sepals, petals, stamens and carpels. Even where this is not immediately evident, for instance in secondarily unisexual flowers, on close inspection rudiments of the missing organs can often be found (Parkin, 1957), as **staminodes** or **carpellodes**.

The sterile leaves of the flower comprise the **perianth**. Most angiosperms have a **differentiated perianth**, forming two outer sets of organs, sepals and petals. However, sometimes these undifferentiated morphologically and separable only phyllotactically, in which case they are referred to as **tepals**. It is common for petaloid monocots such as *Tulipa* to have two sets of tepals differentiated phyllotactically as two discrete whorls of three, in which case they are referred to as **inner tepals** and **outer tepals**. In other cases, such as in *Nymphaea* (Nymphaeaceae) *Calycanthus* (Calycanthaceae) and *Camellia* (Theaceae) there is a gradual morphological transition from sepaloid tepals to petaloid tepals, which are sometimes all on the same phyllotactic spiral.

If there is truly a single set of perianth elements it is usually due to the evolutionary loss of one set. It is rare to find a truly **undifferentiated perianth**, in which tepals are completely undifferentiated morphologically and phyllotactically (i.e. by whorl) so no division into two sets can be made. Even in the earliest divergent lineages of angiosperms such as *Amborella* (Buzgo, Soltis & Soltis, 2004) and Austrobaileyales (Endress 2001b) some differentiation occurs and such perianths may be described as **partially differentiated**. In *Amborella* the outer tepals have thin translucent margins like the bracts, whereas the inner tepals have a more stamen-like appearance (Buzgo, Soltis & Soltis, 2004).

Thus in the overwhelming majority of angiosperms a fourfold patterning of floral phyllomes can be seen. This is particularly evident in the flowers of the core eudicots such as the model organisms *Arabidopsis* (Brassicaceae) and *Antirrhinum* (Plantaginaceae). Here, as in most flowers, the four morphologically highly distinct organ sets are obvious and arranged in discrete whorls. The fourfold patterning of the angiosperm flower has its origin in the interaction of three major classes of floral identity genes (see below).

Two other rules of the flower, namely, **acropetal development** and **alternation of whorls**, will be discussed under androecia (below), as it is in androecia (centrifugal stamens and obdiplostemony respectively) that these rules are broken.

6.11 Molecular control of floral organ patterning

Understanding of the molecular control of floral patterning in eudicots began with the detection of floral mutations in *Arabidopsis* and *Antirrhinum* (Coen & Meyerowitz, 1991; Schwarz-Sommer et al., 1990). The mutations could be classified into three general types: **A-class** mutations that affected calyx and corolla; **B-class** mutations that affected the corolla and stamens and **C-class** mutations that affected stamens and carpels (Bowman, Smyth & Meyerowitz, 1989; Carpenter & Coen, 1990).

The genes underlying these phenotypes specify organ identity in an overlapping manner. However, instead of specifying identity in a "one gene/one organ" manner, they interact in overlapping domains (Bowman, Smyth & Meyerowitz, 1991). The fact that B-class function (B-function) overlaps with A and C, but that A-function and C-function are mutually exclusive and antagonistic (Drews, Bowman & Meyerowitz, 1991; Gustafson-Brown, Savidge & Yanofsky, 1994), allows three types of gene to specify four states in an elegant combinatorial code.

Thus A-function on its own specifies **sepals**, A- and B-function together specify **petals**, B- and C-function specify **stamens** and C-function alone specifies **carpels**. This simple scheme of gene interaction, called the **ABC model** (Coen & Meyerowitz, 1991) appears to be fairly well conserved through the angiosperms (Bowman, 1997) and neatly explains the conservation of fourfold patterning of flowers and has served as a "unifying principle" of floral development (Jack, 2004).

The A-function appears to be the least conserved, with the *Arabidopsis* A-function gene *APETALA2* (*AP2*) and the *Antirrhinum* "A-function gene" Am*SQUAMOSA* (Am*SQUA*) functioning as much as floral meristem determining genes as organ identity genes. It is therefore possible to see A-function as being a "ground state" set by floral meristem determining genes and therefore not so dependent on organ identity genes (an idea traceable back to Schwarz-Sommer et al. (1990)). This may be the reason for the greater functional variation in A-function genes. Thus the Am*LIP1* and Am*LIP2* genes (homologues of *AP2*) assist in setting organ identity, but they are not floral meristem determinants (Keck et al., 2003). The other *Antirrhinum* supposed (from orthology) A-function gene, Am*SQUA* is a floral meristem determinant and does not appear to function primarily as an organ identity gene (Huijser et al., 1992).

A-, B- and C-class genes are all MIKC transcription factors (type II MADS-box genes with three extra-conserved domains I, K and C; Becker & Theissen, 2003). Similar genes have been found that control ovule identity, such as *SEEDSTICK* (*STK*; Pinyopich et al., 2003). These genes, originally found in Petunia (Colombo et al., 1995) have been referred to as **D-class** genes in an extension of the ABC model. They include the *AGAMOUS* relatives, *SHATTERPROOF1* and *2* (*SHP1*, *SHP2*) as well as *STK*. All these four function as ovule identity genes (Pinyopich et al., 2003) and all four can form multimeric protein complexes with proteins of the *SEPALLATA* (*SEP*) and B-sister genes (Favaro et al., 2003). B-sister genes are B-class-like genes such as *ABS/TT16*, which take a small but significant role in the normal development of the ovule (Kaufmann et al., 2005).

The proteins produced by the ABC genes function through specific regulatory DNA binding. This is controlled by the MADS-domain (Riechmann, Wang & Meyerowitz, 1996). However, dimerization of the proteins, as mediated by the I- and K-domains (Riechmann, Krizek & Meyerowitz, 1996), is needed for effective DNA binding. The C-domain may mediate formation of even larger protein units. They dimerize not only with each other but with proteins of additional *MIKC* genes of the *SEPALLATA* (*SEP*) group (*SEPALLATA*1–4; Pelaz et al., 2000, 2001). Because the *SEP* genes are necessary cofactors of the ABC gene model, they have been called **E-class** genes, allowing the ABC model to be extended into an

ABCDE model (Theissen, 2001). *SEP* genes are not usually considered to be organ identity genes in their own right, as their expression domains are more general. However, a *SEP* homologue is implicated in determining the identity of the single petaloid calcarate sepal in *Impatiens* (Geuten et al., 2006). *SEP* genes may thus be available to be evolutionarily co-opted as identity specifiers for novel organs.

Generally, it is thought that *SEP* genes operate by forming functionally critical protein dimers with the A-class and C-class proteins, APETALA1 (AP1) and AGAMOUS (AG). Furthermore, there is a strong suspicion that dimers associate into tetramers (assisted by the C-domain) to control organ identity (Kaufmann, Melzer & Theissen, 2005; Melzer, Kaufmann & Theissen, 2006; Theissen & Saedler, 2001). For instance, dimers of APETALA3 (AP3) and PISTILLATA (PI; Winter et al., 2002; Yang, Fanning & Jack, 2003; Yang & Jack, 2004) are thought to associate with AP1/SEP dimers to form an AP3/PI-AP1/SEP tetramer controlling petal identity. This is known as the **quartet model** (Theissen, 2001; Robles & Pelaz, 2005; Fig. 6.5, Table 6.1).

Thus although the ABC model for the fourfold patterning of the flower is simple and elegant, it is deceptively so. Underlying it are extremely complex gene interactions and networks, including miRNA control (Cartolano et al., 2007), which are still being worked out.

Furthermore, the ABC model has best applicability to eudicots with four clearly defined whorls of floral organs. "Basal" angiosperms[4], in contrast, often have floral organs that are helically arranged and that grade into each other, as in the petal/stamen transition in *Calycanthus* (Staedeler, Weston & Endress, 2007). Basal angiosperms have A, B and C genes but their expression is generally more widespread among floral organs and does not show clear boundaries. This particularly applies to the B-class genes (Kim et al., 2005). How then does the ABC model apply to basal angiosperms? As a solution to this problem the "fading boundaries" hypothesis (Soltis et al., 2007) has been put forward, which suggests that organ identity is based on the degree of expression and overlap, and is therefore a more quantitative than qualitative phenomenon. This is a testable hypothesis that forms a promising framework for further investigation. It follows from the "fading boundaries" hypothesis that the critical innovation of the eudicots has been the tight canalization (hard boundaries) of a formerly much looser system. This canalization accompanied the change to a whorled organization and increasing synorganization of the flower.

[4] Early diverging groups of angiosperms is a better term. "Basal" is used here as a shorthand.

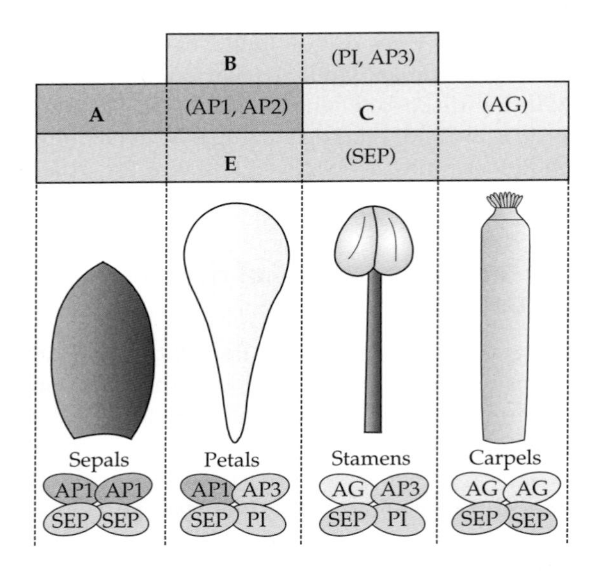

Fig. 6.5. Expression patterns and quartet structure of the MADS-box genes responsible for patterning the flower (after Krizek & Fletcher, 2005).

6.12 The perianth

The **perianth** is the collective name for the sterile basal leaves of the flower. The perianth is usually of two whorls, specified, as noted above, by the A-class and B-class floral organ identity genes. These two whorls are often **heterochlamydeous**, that is, clearly differentiated into an outer **calyx** composed of modified leaves called **sepals** (generally green and leaf-like), and an inner whorl comprising a **corolla** also composed of modified leaves called **petals**. Alternatively,

Table 6.1 A-, B- and C-function MADS-box genes in *Arabidopsis* and *Antirrhinum*.

	Arabidopsis	Antirrhinum
A-function	*APETALA1; APETALA2*	*SQUAMOSA; LIP1/2*
B-function	*APETALA3; PISTILLATA*	*DEFICIENS; GLOBOSA*
C-function	*AGAMOUS; SHATTERPROOF1/2*	*FARINELLI; PLENA*

the two whorls may be **homochlamydeous**, or undifferentiated and morphologically similar, in which case both whorls may be said to comprise a **perigon**, made up of modified leaves called by the neutral term **tepals**. The tepals can take the character of either petals or sepals. The perigon is termed **sepaloid** if the tepals resemble sepals, and it is termed **petaloid** if the tepals resemble petals. Most ornamental monocots (such as daffodils and tulips) have a **petaloid perigon**. Many other monocots, such as Juncaceae, have a **sepaloid perigon**.

The molecular mechanism underlying the petaloid perigon has been examined in *Tulipa* (van Tunen, Eikelboom & Angenent, 1993), *Lilium* (Theissen et al., 2000) and *Agapanthus* (Nakamura et al., 2005). In all cases the results are consistent with an ABC model in which B-function has expanded to encompass both whorls of tepals: in molecular terms they are both "petals." On the other hand, the less convincingly petaloid outer tepals of *Asparagus* do not express B-function genes suggesting other mechanisms are responsible for the patterns (Park et al., 2003, 2004; Park, Kanno & Kameya, 2003).

Occasionally flowers may have only a single perianth whorl, usually by reduction and loss of one whorl. If it is not obvious which whorl has been lost (as is usually the case) the members of the single remaining whorl are usually referred to as tepals. The plant family Aristolochiaceae provides an example. One genus, *Saruma*, has a perianth of two whorls, while other genera, including *Aristolochia*, have a single whorl. Curiously, although *Saruma* shows a pattern of B-class homeotic gene expression consistent with the ABC model, *Aristolochia* shows a pattern that suggests these genes may have been removed from the process of specifying organ identity in these bizarre "Dutchman's pipe" flowers (Jaramillo & Kramer, 2004).

Further reduction may lead to the complete loss of a perianth. Naked flowers of this type are comparatively rare, but do occur, particularly when flowers are much reduced and compacted into large inflorescences, as in the flowers of the Piperaceae.

Rather than being whorled, a perianth may also show helical phyllotaxy. If so, the perianth members may all be similar or submit to slight gradations of form. This single zone of similar perianth members is a perigon of tepals. Alternatively they may exist in two zones with rapid transition of form between the two zones. In the same way as in whorled floral phyllotaxy, the outer zone (generally green and leaf-like) is called the calyx and the inner zone the corolla. The primary organs of the perianth are therefore sepals, petals and tepals. Other organs associated with the perianth, such as the epicalyx (outside the perianth) and nectarial leaves (between the petals and the stamens), will be considered elsewhere (Fig. 6.6, see also Plate 8).

Even when floral phyllotaxis is helical, there is normally an abrupt change in phyllotactic relations in the transition from the foliage leaves to the floral phyllomes. Exceptions are found in some basal angiosperms. In *Amborella* the bracts grade into the tepals both morphologically and phyllotactically (Buzgo, Soltis & Soltis, 2004).

An important question is how the eudicot ABC model applies to these basal angiosperms without the eudicot four-whorled structure.

6.13 Merosity of the perianth

Early in perianth evolution there were multiple changes from numerous and indeterminate numbers of organs to a fixed number usually three (characteristic of monocots) or five (characteristic of eudicots; De Craene, Soltis & Soltis, 2003; Endress & Doyle, 2007; Zanis et al., 2003). This is normally accompanied by a change from spiral phyllotaxy of the perianth members to a whorled phyllotaxy, and undifferentiated to differentiated morphology. In a true whorl the perianth members are inserted at the same level on the floral meristem, have divergence angles summing to 360° and are usually initiated simultaneously. However, some apparent whorls have members that, on close inspection, are initiated successively, possibly as a remnant of spiral phyllotaxy. This is common in perianths that are not strongly differentiated. For instance, in *Rumex* (Polygonaceae) the outer tepals are initiated successively whereas the inner whorl is initiated simultaneously.

Usually there is a single petal whorl, consisting of three or five petals. However, changes of merosity[5], in both number of whorls and number of members within a whorl are quite common. Within Berberidaceae and Annonaceae two trimerous petal whorls are commonly found, and in *Nandina*, five trimerous petal whorls. *Dryas octopetala* has eight petals yet is nested within the actinorhizal Rosaceae clade including *Cercocarpus*, *Chamaebatia* and *Purshia*, all of which have pentamerous flowers. Similarly tetramery (derived from ancestral pentamery) is the norm in the clade comprising the Capparaceae, Cleomaceae and Brassicaceae (the clade that includes *Arabidopsis*).

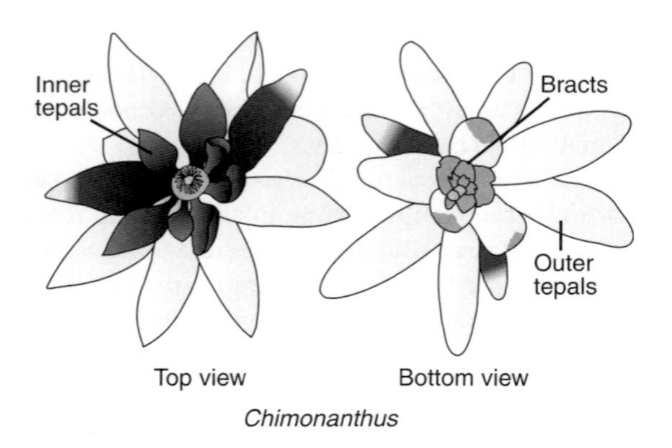

Fig. 6.6. Intergradation of inner and outer tepals (left) and bracts and tepals (right) in *Chimonanthus* (after Arber).

[5] "Merosity" is an awkward derivation from the Greek but distinguishes this term from the literary use of the more correct "merism", i.e. a synecdoche in which a totality is expressed by a listing of parts (e.g. "lock, stock and barrel")

Interestingly the *PERIANTHIA* mutation of *Arabidopsis* is usually pentamerous, implying that *PERIANTHIA* (*PAN*) may be part of the pathway responsible for the innovation of tetramery in the clade. *PAN* encodes a basic region/leucine zipper (bZIP) transcription factor, hypothesized to have been recruited to an existing merosity control pathway (likely acting by primordium inhibition) to modify it from pentamery to tetramery (Chuang et al., 1999). *PAN* thus appears to be responsible for more distant spacing of primordia leading to fewer primordia. At present little is known about *PAN*'s interaction with other genes except that it interacts with the important patterning genes *BLADE-ON-PETIOLE1* and *2* (*BOP1*/*BOP2*; Hepworth et al., 2005). Its action may well prove to be connected to *PETAL LOSS* (*PTL*), a transcription factor of the trihelix family, involved in spacing of sepals by inhibiting growth between sepal primordia (Brewer et al., 2004). As petal primordia subsequently develop in the spaces between the sepal primordia, loss of this gene's function has strong negative effects on petal development.

6.14 Modifications of the perianth members

6.14.1 Fusion

The perianth is enormously variable as it is the chief unit for adaptation to pollination mechanism, and particularly to specific pollinators. The main modifications are connation (ontogenetic fusion of a whorl) and adnation (ontogenetic fusion between whorls) of parts (often to form a tube). Other modifications are proximodistal patterning (differentiation of claw and limb), outgrowths (such as spurs and corona), conversion (for instance into trichomes or lodicules) and secondary polymery.

Congenital fusion of members of perianth whorls to give a **tube** that is **sympetalous**, rather than **polypetalous** whorls of **free** perianth organs, is a very important evolutionary change in angiosperms. It is characteristic of plants with specialized insect pollination as the length of the tube can adapt to the length of a particular proboscis. Generally the perianth organs are only united at the base and the free apical parts are called **lobes**. At the junction of the lobes with the tube (the **throat**) there is usually an inflection, as the lobes are usually reflexed relative to the tube.

Perianth fusion is less often **postgenital** (fusion of parts after formation), although this is the case, for instance, in *Correa* (Rutaceae). In the Apocynaceae, a family with a propensity for postgenital fusion, there may be a mixture of postgenital and congenital fusion. More commonly it is purely **congenital** (failure of separate parts to form). The primordia may start out separate, but soon expand and coalesce eventually forming a torus that elongates as a tube, surmounted by the original free primordia, which develop into the lobes. Congenital fusion is generally due to a failure to maintain boundaries between organs (the boundary hypothesis) rather than to an expansion of meristems caused by an overexpression of meristem determining genes, such as KNOX homeobox genes (the overexpression hypothesis). The boundary hypothesis would result in sympetaly being a recessive trait, whereas the overexpression hypothesis would result in a dominant trait.

Boundary genes are expressed between organs and retard cell growth processes to maintain organ separation (Aida & Tasaka, 2006a,b). Some of these genes are characterized such as the NAC transcription factors *CUP-SHAPED COTYLEDON1–3* (*CUC1–3*) in *Arabidopsis* that affects organ separation on the SAM (Aida, Ishida & Tasaka, 1999). *UFO* as well as genes in the same pathway such as *RABBIT EARS* (*RBE*) and *UFO* modifiers such as *FUSED FLORAL ORGANS1–3* (*FFO1-3*) also appear to play a role in boundary formation (Krizek, Lewis & Fletcher, 2006; Levin et al., 1998; Levin & Meyerowitz, 1995). Any of these genes, and the many boundary control genes that are probably yet to be discovered, are potential candidates for a role in the evolution of sympetaly.

6.14.2 Proximodistal patterning

Perianth members may show strong proximodistal patterning. We have seen how sympetalous perianth whorls may be divided into **tube**, **throat** and **limb**. Whereas the tube is always composed of united perianth parts, the limb may be in the form of free **lobes**. Even in actinomorphic flowers, the relative growth of these units creates a rich array of sympetalous flower shapes, which are very important for pollinator interactions (Table 6.2).

Similarly in polypetalous flowers there may be strong differentiation into a **claw** (basal part) and **limb** (apical part) as is particularly characteristic of the Caryophyllaceae, Brassicaceae and Onagraceae (*Clarkia*). The claw is generally subunifacial and narrow while the limb has more extensive development of the lamina. Thus, in molecular genetic terms, the same mechanisms that prevent lamina outgrowth at the petiole (Chapter 2) may be responsible for the failure of lamina outgrowth in the claw. However, it should be noted that the claw is not necessarily homologous to a petiole.

6.14.3 Outgrowths of the perianth

Perianth members may sport noticeable outgrowths. These outgrowths may be scale-like (a **corona**) or sac-like (a **spur**). The scales (**squamae**) of a corona may be free (as in *Symphytum* or *Silene*) or congenitally fused into a ring (as in *Narcissus*). In *Narcissus*, which has the best-known corona, the six tepals (two whorls of three) are congenitally fused into a tube basally, but free apically. At the throat, a ring of congenitally fused blades produce a prominent raised "trumpet" that functions as a continuation of the tube. Guedes has interpreted the corona of *Narcissus* not as scale-like outgrowths of tepals but as an example of **diplophylly** (Chapter 2) in which the fused lobes of the corona represent the ventral blades (cross leaflets) of the tepals (Guédès, 1966). This makes a simple testable prediction: that the outer surface of the corona is adaxial and the inner abaxial (i.e. reversed from normal perianth orientation). This is testable by

Table 6.2 Type of actinomorphic sympetalous flowers demonstrating the importance of relative growth of tube and limb.

		Relative size of limb		
		Large	**Medium**	**Small**
Relative size of tube	Large	**Hypocrateriform** (salver-shaped, long tube and wide limb: *Syringa*)	**Campanulate** (bell-shaped, wide tube: *Campanula, Atropa*)	**Tubular** (long tube and small limb: *Polygonatum*)
	Medium		**Infundibuliform** (funnel-shaped, tapering tube, narrow at base: *Convolvulus*)	**Urceolate** (tube spherical, limb small: *Erica*)
	Small	**Rotate** (wheel-shaped, limb well-developed, tube small: *Myosotis*)		

examining the identity markers for adaxial and abaxial identity.

Spurs are usually nectarial outgrowths for pollinator attraction and may be sepaloid or petaloid in origin. Spurred (**calcarate**) plants with **sepal spurs** include *Tropaeolum* and *Vochysia*. **Petal spurs** include *Aquilegia, Viola* and *Linaria*. *Delphinium* has both petal and sepal spurs. The Orchidaceae have tepal spurs. *Impatiens* is of particular interest as a single sepal is both spurred and petaloid, whereas the other sepals are typically sepaloid and unspurred. Remarkably, this hetertopic petaloidy is associated with specific expression of a sepallata (E-class) gene (*SEP3*) in the petaloid sepal (Geuten et al., 2006).

Intriguing evince as to how spurs might evolve structurally comes from analysis of dominant mutations of the *Antirrhinum* genes Am*HIRZINA* (*HIRZ*) and Am*INVAGINATA* (INA) which cause ectopic spur-like petal tubes on the *Antirrhinum* corolla (Golz, Keck & Hudson, 2002). *Antirrhinum*, in contrast to the closely related and long-spurred *Linaria*, is unspurred, being shallowly saccate at most. Am*HIRZ* and Am*INA* are both KNOX homeobox genes, which appear to promote cell growth and delayed determinancy in specific cell populations in the corolla tube. This additional growth forms the spur-like petal tubes.

6.14.4 Shape and texture

Another peculiarity of perianth form displayed by *Antirrhinum* is the upfolding that develops in the abaxial corolla, which has the effect of allowing perianth growth to block the mouth of the tube with a hinge mechanism. This type of corolla is termed **personate**, and it is what puts the "snap" in snapdragon. It has the consequence to pollination that only those bees (for instance *Bombus*) with the weight, and strength of thoracic musculature, to open the flower can gain the reward. This is likely to greatly improve the pollinator specificity and the efficiency of pollination as well as reducing the amount of reward-robbing by nonpollinators. Possibly implicated in the development of the fold is the *CINCINNATA* (*CIN*) gene of *Antirrhinum*, as it is expressed in the petal at the fold and acts as a growth-promoting gene, apparently to control aspects of the morphology and micromorphology of individual perianth members (Crawford et al., 2004). It also promotes the formation of conical epidermal cells in petals. Curiously, while *CIN* promotes growth in petals, in leaves it causes growth arrest (see Chapter 2; Fig. 6.7).

Conical cells in the epidermis of petals are enormously important in controlling pollinator-perceived petal texture, presentation, color intensity and brightness (Glover & Martin, 1998; Noda et al., 1994). In *Antirrhinum*, the *MIXTA* gene, an R2R3 MYB transcription factor, is the main determinant of conical epidermal cell shape (Baumann et al., 2007). As related *MIXTA*-like genes control trichome formation in many asterid plants (Jaffe, Tattersall & Glover, 2007; Payne et al., 1999), conical cells can be seen as "failed" (or at least undeveloped) trichomes. It should be noted, however, that the unrelated *MYB* gene, *GLABROUS1*, is required for trichome development in *Arabidopsis*.

Normal *MIXTA* activity produces only conical cells, but in transgenics it can also initiate trichome development, depending on the timing of its expression in development (Payne et al., 1999; Perez-Rodriguez et al., 2005). An R2R3 MYB transcription factor very similar to *MIXTA*, *MYB MIXTA-LIKE1* (AmMYBML1), has multiple functions in the abaxial corolla: in the corolla tube it produces trichomes, on the petal hinge epidermis it produces conical cells and it reinforces the hinge through differential expansion of petal mesophyll cells (Perez-Rodriguez et al., 2005).

Trichomes have many roles in asterid floral function. *Solanum* species have **anther cones** (an adaptation for buzz pollination). In the tomato (*Solanum lycopersicum*) the anthers are stuck together by interlocking trichomes. However, in *Solanum dulcamara* the anther cones are merely stuck together by secretion, and ectopic expression of *MIXTA* in transgenic plants will produce trichomes which break up the anther cones (Glover, Bunnewell & Martin, 2004).

6.14.5 Conversion and polymery

Perianth parts may be converted to strikingly different organs. The grasses have tepal-derived lodicules. These no longer look like tepals, being minute fleshy scales that function to swell at anthesis and open the grass floret. In *Eriophorum* (cotton grass in the Cyperaceae) the tepals have been converted to hairs (cotton) that function in seed dispersal. Similarly, the calyx of Asteraceae and Valerianaceae has been converted to hairs called pappus, which also function in seed dispersal.

Interesting molecular tests have been used to confirm that these "new" organs are indeed perianth-derived. The hypothesis of perianth derivation makes predictions about which MADS-box genes will be expressed (A- and B-function genes). In *Oryza*, B-function genes (*APETALA3* and *PISTILLATA* homologues—*SUPERWOMAN1* (*SPW1*) and Os*MADS1/2*, respectively) are indeed required for lodicule

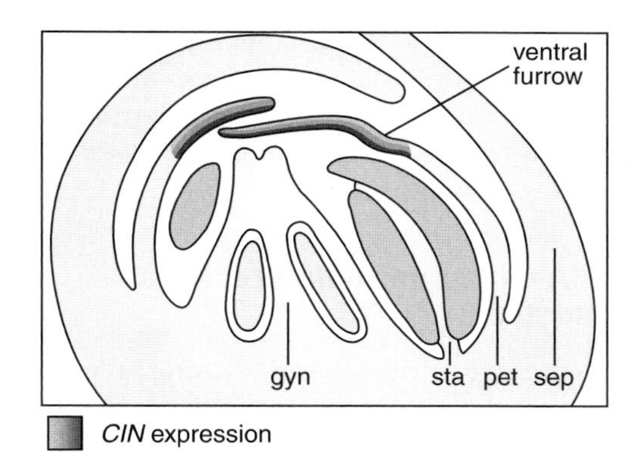

Fig. 6.7. Highly specific expression pattern of cincinnata (*CIN*) in the developing corolla of *Antirrhinum*. This gene has a role in determining the final shape of the corolla (after Crawford et al., 2004).

identity (Kang et al., 1998; Nagasawa et al., 2003). Similarly, in maize (*Zea*) the AP3 homologue *SILKY1* is required (Ambrose et al., 2000).

Similar tests have been performed for the pappus of Asteraceae. In *Gerbera*, patterns of MADS-box gene expression as well as transgenic studies suggest a sepaloid derivation of pappus (Yu et al., 1999).

In many petals the primordia, once formed, can split to produce more than one organ. Such primordia are known as **compound primordia** and the process is known as **secondary polymery**. That such a process occurs is unsurprising as it is possible to induce the formation of two organs from one primordium by surgery (Hicks, 1973). Natural splitting of primordia may lead to mild polymery as in *Ranunculus ficaria*. In *Podophyllum* the upper of the two trimerous petal whorls may produce up to nine petals by division of the primordia. Secondary polymery is important horticulturally as the source of the double carnation, in which fascicles of petals form from the original five petal primordia. Perhaps the best-known case of secondary polymery, centrifugal stamens, will be discussed elsewhere. Secondary polymery may result from the inverse of the mechanism that leads to sympetaly. Instead of a failure of boundary specification, boundaries are specified within a single primordium.

Finally a very important modification of the perianth, one that often affects other floral organs too, is the dorsoventral patterning that alters the symmetry relations of the flower. This is dealt with in the next section.

6.15 A mirror up to nature: floral symmetry

Shakespeare describes the art of acting as "to hold, as 'twere, the mirror up to nature'." However, nature holds the mirror up to itself in many places including the development of the flower.

Symmetry refers to property of an object such that when an operation is performed on that structure, an identical structure is retrieved. Operations include translation (linear movement in space), rotation and reflection. Flowers with parts in whorls are radial structures and remain the same after rotation or reflection. Thus a pentamerous flower may be rotated 72° (or multiples of 72°) and still remain congruent to the original. Similarly there are five planes of reflectional symmetry along which the flower may be reflected without change. Such flowers are therefore known as **polysymmetric** (alternatively **actinomorphic** or **radial**). Symmetry relations thus form a useful descriptive model with which to examine flower morphology (Fig. 6.8; Box 6.1).

Disturbing this symmetry, by changing the morphology on different sides of the flower, obviously will cause the level of symmetry to decline. If the change in morphology is produced equally on both sides of a flower (disturbance symmetrical along one axis), a disymmetric flower results, as in the flower of *Dicentra*. However, if the disturbance is unequal, a **monosymmetric** (**bilaterally symmetric** or **zygomorphic**) flower results, and this is a very common occurrence in floral development (Endress, 1999). This happens when the adaxial part of the flower develops differently to the abaxial part of the flower, as in *Antirrhinum*. The two sides of the flower remain the same so the flower can still be reflected over the adaxial–abaxial axis.

Disturbing the symmetry along two axes results in an **asymmetric flower** with no reflectional symmetry at all. This happens when a flower that is already monosymmetric develops a left–right asymmetry. This may happen due to left–right asymmetric organ abortion as

Box 6.1 Types of perianth floral symmetry.

Symmetry level	Symmetry class	Symmetry type	Example
Asymmetry	Asymmetric	Zygomorphy	*Canna*
Monosymmetry	Bilateral	Zygomorphy	*Antirrhinum*
Disymmetry	—	—	*Dicentra*
Polysymmetric types			
Trisymmetry	Radial	Actinomorphy	*Lilium*
Tetrasymmetry	Radial	Actinomorphy	*Oenothera*
Pentasymmetry	Radial	Actinomorphy	*Linum*
Hexasymmetry	Radial	Actinomorphy	*Lythrum*
Heptasymmetry	Radial	Actinomorphy	*Trientalis*
Octosymmetry	Radial	Actinomorphy	*Dryas*
High polysymmetry	Radial	Actinomorphy	*Sempervivum*

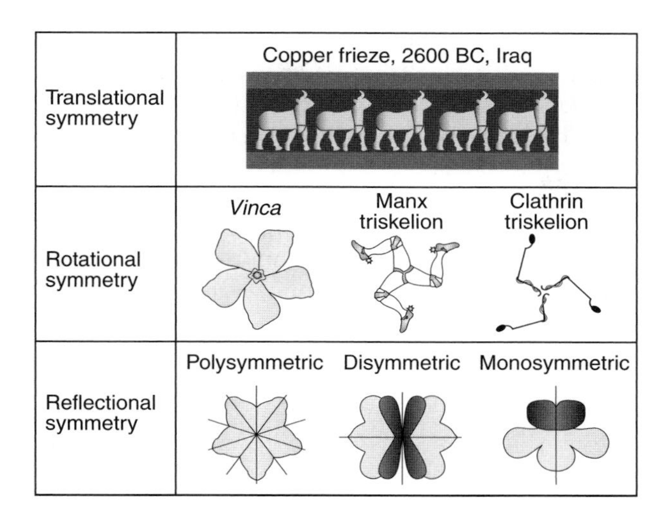

Fig. 6.8. Types of symmetry.

in *Centranthus, Qualea* and *Canna* (Donoghue, Bell & Winkworth, 2003; Kirchoff, 1983; Litt & Stevenson, 2003). Alternatively there may be an asymmetric organ growth as in the keel petals of *Phaseolus*, or the corolla of *Pedicularis*. Alternatively there may be left–right asymmetric organ orientation, as in the **enantiostyly** (lateral style orientation) of plants such as *Saintpaulia*.

A single flower may have one type of symmetry relations in one whorl and a different type in another whorl. Therefore, when describing flower symmetry relations it is very important to specify symmetry whorl by whorl or of the flower as a whole. The most notable expression of symmetry relations is in the corolla and this is usually what is meant when a whole flower is ascribed to a particular symmetry. However, it should be remembered that symmetry often differs in interesting ways whorl by whorl.

The monosymmetric flower has developed numerous times in flowering plants (Cubas, 2004; Donoghue, Ree & Baum, 1998), but this is particularly characteristic of the asterid clade, legumes and orchids. It is a powerful mechanism for promoting pollinator specificity in flowers and thus for increasing the efficiency of pollen use (i.e. for improving male function; Cronk & Moller, 1997). There is a functional and a developmental reason for why monosymmetry almost always develops in the adaxial–adaxial plane. First, when a flower is formed laterally on an inflorescence axis (as in a raceme) the axis itself forms a powerful referent to orient the flower. Diffusion of signals from the axis will create a biochemical asymmetry even where there is no morphological asymmetry. Any plant that links morphological developmental genes to this underlying adaxial–abaxial asymmetry can become monosymmetric (Clark & Coen, 2002; Cubas, Coen & Zapater, 2001). This underlying asymmetry is called a **prepattern** (Almeida, Rocheta & Galego, 1997; Clark & Coen, 2002). Furthermore, monosymmetry in this plane makes functional sense as, if the inflorescence axis is vertical, floral symmetry will be in the same plane as the dorsoventral symmetry of a flying pollinator, thus allowing for the evolution of close adaptation between floral morphology and pollinator morphology and **animal behavior**.

A potential problem arises if the inflorescence is not vertical. Usually this is compensated for by **gravitropic** orientation of the flowers by means of a twisting of the pedicel. Monosymmetric flowers are generally much more strongly responsive to gravity than actinomorphic flowers, for obvious reasons: they depend on orientation to interact correctly with approaching pollinators. Pollinators generally do not like flying upside down. In monosymmetric flowers with hanging racemes, such as *Laburnum*, the flowers are **resupinate**, twisting themselves 180° to come to the correct orientation.

Another potential problem arises with cymose rather than racemose inflorescences. In dichasial cymes, the axis terminates with a flower, and produces branches symmetrically beneath symmetrically bearing more orders of terminal flowers. In such a system there is no axis to produce a positional referent for the adaxial–abaxial polarity in the flower. This explains why monosymmetric flowers are very rare in plants with dichasial cymes. It also explains why in racemose monosymmetric plants, when a terminal flower is produced as an aberration, that terminal flower is polysymmetric. Being terminal it has no abaxial–adaxial axis along which to be monosymmetric.

Such aberrations have been studied at the molecular level in *Antirrhinum*. The racemose inflorescence of *Antirrhinum* is maintained in an indeterminate condition by the gene *CENTRORADIALIS* (*CEN*), a homologue of the *TERMINAL FLOWER1* (*TFL1*) gene in *Arabidopsis*. When *CEN* is not functional the inflorescence axis becomes determinate and produces a terminal flower that is polysymmetric, not as a direct result of the *cen* knockout, but as a pleiotropic effect (normal developmental positional signals are not available to this flower; Bradley et al., 1996, 1997). The production of polysymmetric flowers in normally monosymmetric species is called **peloria**, after a name used in a study by Linnaeus in *Linaria* (Cubas, Vincent & Coen, 1999; Linnaeus & Rudberg, 1749; Rudall & Bateman, 2003). Another example of **terminal peloria**, like the *CEN* mutant of *Antirrhinum*, is found in naturally occurring individuals of *Digitalis* (foxglove). Other examples of peloria affect all the flowers on a plant, as they result from mutations in the genes directly responsible for monosymmetry (next section).

Given the above, it may seem difficult to explain those cases of monosymmetric flowers with cymose inflorescences. Indeed, plants with single terminal flowers, or terminal regular cymes, are never monosymmetric. However, where cymes are small and lateral, adaxial–abaxial positional information can be provided by the main stem, and subsequently from intrinsic developmental asymmetries in the cyme itself (monochasial cymes are intrinsically asymmetrical; Clark & Coen, 2002). Thus developmental asymmetries laid down early in development (when the whole inflorescence is in close proximity to a main axis) can be propagated through the cyme as an "asymmetry cascade" (Box 6.1)

6.16 The molecular mechanism underlying monosymmetry

Using a transposon mutagenesis approach, the genes underlying the peloric (radially symmetric) mutant of the monosymmetric *Antirrhinum* have been cloned (Luo et al., 1996). Darwin himself studied the peloric mutant of *Antirrhinum* (Darwin, 1868) and scientific studies on the peloric mutant of the closely related *Linaria* go back even further (Linnaeus & Rudberg, 1749). The cloning of the genes involved was thus of considerable interest, and a considerable achievement.

In *Antirrhinum* two closely related loci, *CYCLOIDEA* (*CYC*) and *DICHOTOMA* (*DICH*), were shown to encode transcription factors of the TCP class (Cubas et al., 1999). *CYC* and *DICH* show partial redundancy, but the *cyc/dich* double knockout mutant is fully radial. Interestingly, the radial phenotype is that of the abaxial part of the wild-type flower implying that the abaxial part is the default and it is the adaxial part that has been altered by *CYC/DICH* to create a monosymmetric flower. This was confirmed by expression studies that showed *CYC* and *DICH* to be restricted in expression to the adaxial part of the flower (the top two corolla lobes). The mechanism of monosymmetry seemed elegantly clear. By responding to an adaxial signal, *CYC/DICH* is expressed adaxially and triggers morphological changes at the top of the flower only, so creating a monosymmetric flower.

However, this does not explain how the lateral petals also came to be somewhat different to the abaxial (default) petals. It seems that *CYC* acts at a distance to alter the lateral petals. This was explained by the analysis of another peloria-type

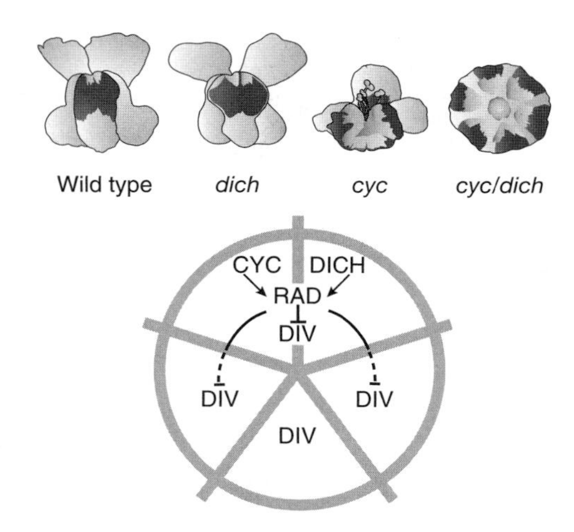

Fig. 6.9. Gene interaction in the symmetry patterning of *Antirrhinum* (after Corley et al., 2005).

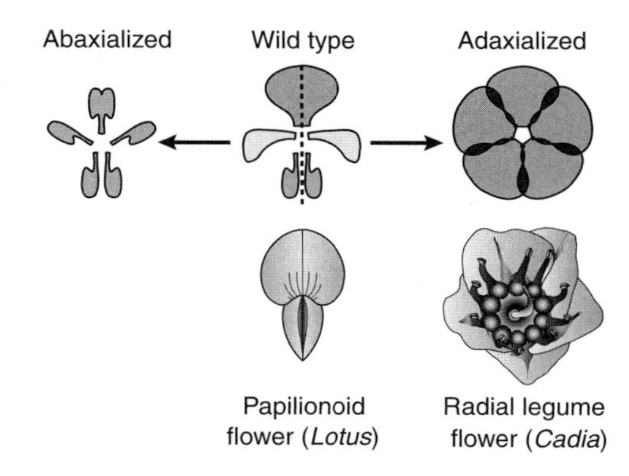

Fig. 6.10. Diagram showing the evolution of a radial flower phenotype (*Cadia*) in legumes by adaxialization of petal identity (after Cronk 2006).

mutant, *RADIALIS* (*RAD*). *RAD* encodes a MYB transcription factor that acts downstream in the CYC/DICH pathway (Corley et al., 2005; Costa et al., 2005). Its expression is activated by CYC, so it has a similar expression pattern. However, the protein is very small (93 amino acids) and is possibly able to move to adjacent parts of the floral meristem, thus affecting the lateral petals (Corley et al., 2005). If protein movement is the correct explanation for the noncell-autonomous action of the *CYC* and *RAD* pathway, it is reminiscent of the movement of CAPRICE (CPC) protein in root hair development (Fig. 6.9, see also Plate 9).

Although *CYC* has some *RAD*-independent effects, much of the morphological influence of *CYC* is mediated by *RAD*, which probably functions by being a binding-site competitor with other MYB proteins (Corley et al., 2005; Costa et al., 2005).

A third gene in this interacting network is the gene responsible for maintaining abaxial identity, *DIVARICATA* (*DIV*; Almeida, Rocheta, Galego, 1997; Galego & Almeida, 2002; Perez-Rodriguez et al., 2005). *DIV* is another MYB transcription factor and the DIV protein is antagonized by RAD, which keeps *DIV* function (but not *DIV* expression) out of the adaxial parts of the flower (Corley et al., 2005). Thus each part of the flower has a unique gene function set: the adaxial part has *RAD/CYC/DICH* function. The lateral part has *RAD* function (and probably some *DIV* function) and the abaxial part has *DIV* function only. Much of *RAD*'s phenotypic effect can be explained by its role in eliminating DIV function. However, *RAD* has some *DIV* independent effects, and so may antagonize other, as yet unknown, MYB proteins.

The *CYC/DICH* pathway has whorl-specific effects. In the corolla of *Antirrhinum* it triggers downstream genes to enlarge the adaxial lobes, which are showy. On the other hand, in the androecium it causes the adaxial stamen to be reduced to a staminode. Changes in both the expression pattern of *CYC* and in the whorl-specific effect of the *CYC* pathway appear to be the mechanisms underlying changes in the details of monosymmetry from species to species. In *Mohavea* (a relative of *Antirrhinum*) expansion of the CYC expression domain is correlated with the abortion of the lateral stamens as well as the adaxial stamens (Hileman, Kramer & Baum, 2003). In *Cadia* (Fabaceae) even greater expansion of the expression domain of a *CYC* homologue is correlated with almost complete adaxialization of the floral phenotype and the creation of a radial flower with an adaxial phenotype (in contrast with peloric flowers which are radial but abaxialized; Citerne, Pennington & Cronk, 2006). The mode of CYC control of monosymmetry in *Iberis* (Brassicaceae) is interesting. Here, early expression is symmetric, but when petal growth is at its maximum CYC expression becomes asymmetric and suppresses growth of the inner, shorter petals (Busch & Zachgo, 2007; Fig. 6.10, see also Plate 10).

Mention has been made of the partial redundancy of the close paralogues *CYC* and *DICH*. *DICH* has a more restricted expression domain and a lesser phenotypic effect. It appears to be intrafunctionalized relative to *CYC*, to which it bears a helper relationship. It is also subject to less purifying selection (Hileman & Baum, 2003). In particular it increases localized morphological differentiation within the adaxial corolla lobes, so conferring some internal asymmetry. In *Mohavea* the expression domain of *DICH* has been reduced and there is an associated increase in the internal symmetry of the adaxial corolla lobes (Hileman, Kramer & Baum, 2003).

At first view it seems remarkable that although legumes (Fabaceae) and *Antirrhinum* (Plantaginaceae) certainly gained their characteristic forms of floral asymmetry independently, they have recruited the same genes to produce it. Legumes have the morphology of adaxial part of the flower set by the expression of *CYC* homologues (Citerne, Pennington & Cronk, 2006; Feng et al., 2006). However, this is less surprising when it is realized that *TCP* genes respond to adaxial signals even in plants, like *Arabidopsis,* with no monosymmetry (Cubas, Coen & Zapater, 2001). Thus the evolution of monosymmetry results from downstream floral developmental genes coming to respond to *CYC*. *CYC* already has, in preexistence, the necessary functionality for monosymmetry. Thus, monosymmetry throughout the eudicots is likely to involve *CYC* homologues. The situation in monocots is less clear, and the mechanism of monosymmetry in the most diverse monosymmetric group of all, the orchids (Orchidaceae), at the time of writing remains unknown.

6.17 Outside the flower: the epicalyx, calycle and petaloid bracts

The flower is a strongly modified shoot, but only occasionally is the distinction between the vegetative shoot and the flower a sharp transition. More commonly the developmental mechanisms producing flowers have strong effects on adjacent leaves, which commonly take the form of bracts and bracteoles, specialized leaves of the inflorescence. The whole flower-bearing shoot is thus affected by the reproductive transition, not just the flower itself. In the early divergent angiosperm, *Amborella*, there is no clear distinction between bracts and outer tepals, which grade into each other. In the eudicots on the other hand, there is usually a clear distinction, in morphology and phyllotaxy, between the outer perianth (sepals) and the extrafloral leaves (bracts and bracteoles).

However, there are some cases in which the extrafloral leaves have become **floralized**, and apparently absorbed into some floral developmental programs that have expanded to include them. Floralization is therefore of considerable interest in understanding how developmental pathways are spatially regulated. Examples of floralization include the trimerous **epicalyx** of the Malvaceae (in the broad sense), in which bracts have taken on a whorled phyllotaxy characteristic of the floral organs, as well as a position adjacent to the perianth, so forming a fifth whorl. In *Goethea strictiflora* the epicalyx (and sepals) are even petaloid. The epicalyx is thought to have evolved from a basic unit of the malvalean inflorescence, the so-called bicolor unit after *Theobroma bicolor*, bearing three bracts and several flowers. It is possible that on reduction of these bicolor units to a single flower the three bracts were retained, now in intimate association with the remaining flower (Bayer, 1999). An epicalyx is also characteristic of the condensed inflorescences of the Dipsacales, where it is formed of bracts and bracteoles (Roels & Smets, 1994).

Another example is the adoption of petal-like characters by bracts. These **petaloid bracts** are usually associated with the **transference of function** of pollinator attraction from the usually small and insignificant flower to the showy bracts. The bracts may be arranged around a compact inflorescence in a manner resembling perianth parts. The whole therefore forms a **pseudanthium**, or false flower. Examples of petaloid bracts are

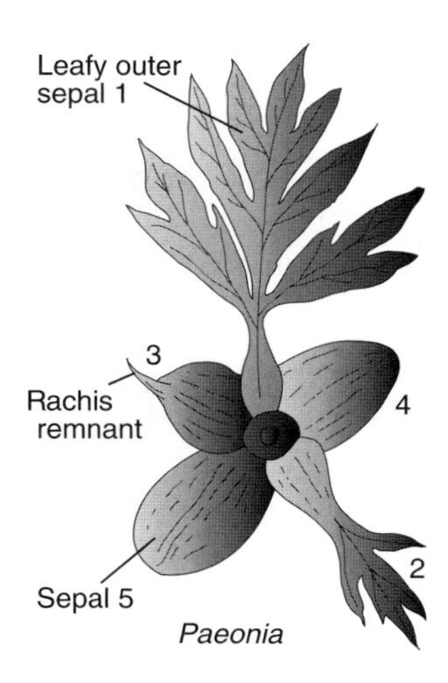

Fig. 6.11. Intergradation between leaves and sepals sometimes seen in *Paeonia* (after Arber).

found in some species of *Cornus* (Cornaceae), and in *Euphorbia* (Euphorbiaceae), such as the familiar poinsettia of commerce (Fig. 6.11).

One apparent instance of a fifth whorl of sepaloid organs resembling an epicalyx is commonly found in the Rosaceae (including *Fragaria, Potentilla* and *Hagenia*). However, this is not an epicalyx but a **calycle**, formed from the stipules of the sepals and not from extrafloral bracts.

Other specialized bracts, intimately associated with flowers or **pseudanthia**, such as **involucral bracts**, **spathe**, **spathilla**, **utricle** and **glumes** will be treated in the section on inflorescence leaves (Section 6.37).

6.18 The floral nectary

Most floral organs have highly conserved spatial relations with other organs. In contrast, **nectaries** are the vagrants of plant morphology. They generally consist of discrete patches of small, highly proteinaceous cells that secrete a liquid containing sugars and other substances, including amino acids (Carter et al., 2006). They are a highly characteristic and important part of the animal pollinated angiosperm flower. However, the wide range of nectarial structure indicates that they are unlikely all to be homologous. Some nectaries even consist of secretory hairs, or even modified stomata on the labellum of orchids through which nectar is secreted (Davies, Stpiczynska & Gregg, 2005).

Nectaries are often composite organs as they frequently depend for their function on nectarial modifications of the organ that bears them, which may be **receptacle**, **sepal**, **petal**, **filament** or **carpel**, or even organs outside the flower in the case of **extrafloral nectaries**.

In many monocots, adjacent carpel walls (that would normally be joined as septa of the ovary) bear nectaries, which may have an opening via slits at the septal radii of the ovary. These are called **septal nectaries** (Rudall, 2002). In the Iridaceae, septal nectaries are probably the original nectary type, but members of the family are found with nectaries on the perianth, the perianth tube, the filaments and the style base. This diversity is conceivably due to heterochronic changes in the timing of nectary development to coincide with the development of other organs (Rudall, Manning & Goldblatt, 2003).

Ring-shaped nectaries around the flower axis are called **discs**, or more specifically **nectarial discs**. The discs of the Tamaricaceae and Rhizophoraceae may represent stamen outgrowths, but most are probably receptacular in origin, being thickenings of the floral axis. In the Brassicaceae the nectaries form on the floral receptacle but in close association with floral organs (typically at the base of stamens).

Nectaries are often associated with perianth members, and with modifications of perianth parts as in **sepal spurs** or **petal spurs**. In *Ranunculus* (Ranunculaceae) the petals are **diplophyllous** structures with a small flap of tissue on the adaxial side caused by **epipeltation**. This forms a small cup for the nectary. In *Helleborus* and *Nigella* (also Ranunculaceae) the **peltation** is more extreme and the inner petals take the form of **ascidiate** cups filled with nectar.

Epipeltation is caused by the spread of abaxial identity at the expense of adaxial identity in floral phyllomes. Members of the YABBY gene family are involved in specifying abaxial identity. It is therefore interesting, although probably coincidental, that a YABBY gene *CRABS*

CLAW (*CRC*) is involved in nectary development in *Arabidopsis*, even though the *Arabidopsis* nectary is receptacular in origin.

CRC is necessary but not sufficient for nectary formation, as *crc* knockout mutants have no nectary but ectopic expression of *CRC* does not cause ectopic nectaries (Bowman & Smyth, 1999). In *Arabidopsis* the nectaries are associated with the third whorl, even in some homeotic mutants that change third whorl identity from stamens to other organs (Baum, Eshed & Bowman, 2001). However, the ABC floral homeotic genes also influence nectary formation (Lee et al., 2005a). This implies a complex regulatory network in nectary development involving *CRC* as well as floral organ identity (*FOI*) genes and floral meristem identity (*FMI*) genes such as *LFY* and *UFO* (Lee et al., 2005a).

CRC has a taxonomically wide role in nectary development in rosids and asterids (Lee et al., 2005b). However, *CRC* homologues do not appear to be associated with nectary development in basal eudicots such as *Aquilegia* (Ranunculaceae) in which *CRC* is restricted to carpel development. This may be due to the tendency of basal eudicots to have perianth-associated nectaries whereas the core eudicots tend to have nectaries more centrally in the flower. *CRC* functions in the radial development of carpels and it has been suggested that the movement of nectaries inward in the flower may be linked to nectary development coming under the control of a gene expressed centrally in the flower (Lee et al., 2005b).

6.19 The androecium

Stamens together are called an **androecium**. In eudicots the **androecium** is commonly composed of one or two whorls of five stamens (three stamens in monocots). However, there may be many whorls, or the androecium may be helically arranged rather than whorled. In such circumstances there may be many stamens. By convention if there are more than 20 stamens the flower is said to have an **indefinite number of stamens** and to be **polyandrous**. If there is only one whorl of stamens, they are known as **haplostemonous**, and if two **diplostemonous** (Fig. 6.12).

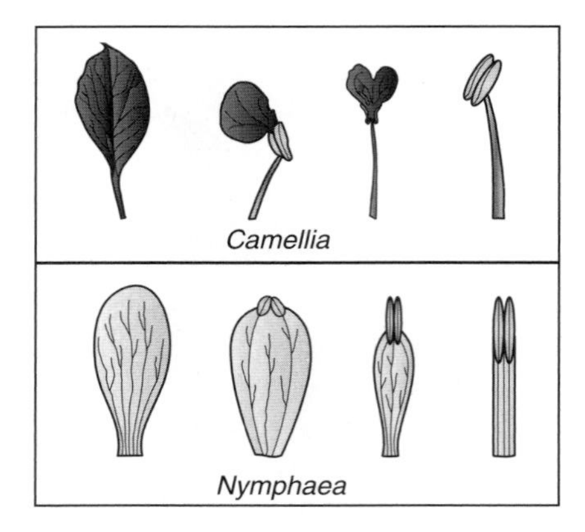

Fig. 6.12. Different patterns of intergradation between petals and stamens seen in *Camellia* (eudicot) and *Nymphaea* (basal angiosperm).

A fairly common but anomalous pattern is that there are two whorls of stamens and the outer whorl is opposite the petals (**antipetalous**) instead of alternating with the petals as would be expected. In this state the flower is called **obdiplostemonous**. Obdiplostemonous androecia break a cardinal rule of the flower, that of the **alternation of successive whorls**. When organs of successive whorls lie against each other as in antipetalous stamens, it is usually taken as an indication that an intervening whorl has been lost.

Stamens have a great evolutionary propensity for **congenital fusion**, either with themselves or with other organs. The filament bases in an androecium may be fused into an **androecial tube**. If all the filaments are united into one tube the stamens are said to be **monadelphous** (e.g. *Althaea officinalis* and *Laburnum anagyroides*). If there are two groups they are said to be **diadelphous**. This is a very common pattern in Fabaceae, for instance in *Trifolium* and *Lotus*, in which there is a bundle of nine stamens and one free stamen (the 9+1 pattern).

The filaments may also, but rarely, be united with the style to form a compound organ called a **gynostemium**. Such flowers are said to be **gynandrous** (a term now rarely used). In Asteraceae it is of common occurrence for the filaments to be free but the anthers to be united into a tube. This condition is known as **syngenesious**. It is also found in species of *Viola*, *Solanum* and *Gentiana*.

The propensity for congenital fusion often makes it appear that the androecium is inserted on another whorl. The most common form of displaced insertion is when there is a **corolla tube** and the stamens are inserted on the tube. This is very common in the Lamiales, Boraginaceae and Asteraceae. Other forms of **displaced insertion** occur in **epigyny** and **perigyny**.

Occasionally plants are met with that have numerous **bunches** (**fascicles**) of stamens and these are called **polyadelphous** (e.g. *Ricinus communis*, *Citrus aurantium* and *Hypericum* spp.). The explanation for these fascicles is generally that the stamen primordia are **compound primordia** and divide to produce more stamens in a **basipetal** (**centrifugal**) direction. Such **centrifugal stamens** break another cardinal rule, the rule of the flower that is of **acropetal organ initiation**. Floral primordia are (almost) always **acropetal** (**centripetal**) in developmental succession, with the youngest organs at the tip, just as are primordia in vegetative shoots.

6.20 The form of the stamen

The **stamen** is the **microsporophyll** of the angiosperms, bearing two **bilocular synangia**. A synangium is a concrescence of sporangia, and thus the angiosperm stamen is generally tetrasporangiate, as each synangium is composed of two **sporangia**. The four sporangia together comprise the **anther**. The sporangia of the anther are dealt with in Chapter 5. This account will focus on the form of the stamen as a whole.

The stamens of some plants in early divergent clades of angiosperms are leaf-like with a pronounced blade notably *Degeneria* (Degeneriaceae) and *Austrobaileya* (Austrobaileyaceae; Bailey & Smith, 1942; Bailey & Swamy, 1949). The characters possessed by the *Austrobaileya* stamen are good candidates for primitive ones. These include a pronounced dorsiventral blade and anthers superficial (i.e. not marginal) on the adaxial side. The stamens of *Degeneria* in the Magnoliales are more problematic as they bear anthers in an abaxial position, which is anomalous in basal angiosperms.

Further evidence for the leaf-like nature of stamens comes from species with a transition series from **petals** or **tepals** to stamens. Such plants have leaf-like (petaloid) outer stamens even though the inner stamens are not leaf-like. A well-known example is the water lily, *Nymphaea*, which produces organs intermediate between petals and stamens in the transition zone between the two organ types. *Nymphaea* has a floral phyllotaxy of multiple whorls so transitional positions occur (Endress, 2001b).

In plants with definite **whorls** the organ identities are usually very constant, as the transition zones found in helical phyllotaxy are absent. Nevertheless, **homeotic transformations** between stamens and petals occur. One that is very common in horticulture is a form of **double flower**, in which stamens are converted to petals.

Usually stamens do not resemble leaves, as the **blade** of the microsporophyll is reduced to a simple **stalk** bearing the anther. This stalk is the **filament** and is often thin, cylindrical and colorless. The filament is an abaxialized **unifacial** organ. For this reason the eudicot stamen as a whole has been interpreted, with good evidence from transitional forms, as a **peltate** (actually **diplophyllous**) structure, with the anther connective to the remains of the peltate blade (Baum & Leinfellner, 1953). Under this interpretation the stamen can be considered a "rolled up" version of a leaf-like sporophyll. The molecular basis of peltation in leaves is given in Chapter 2, but little is yet known about the molecular origin of peltation in stamens.

Filament length is frequently of biological interest. In *Cassia*, for example, the abaxial stamens are long and the adaxial stamens are short. This occurs very frequently in four-stamened Lamiales. Here there are two long and two short stamens, and either the adaxial or the abaxial may be the longer, depending on the example. This case is common enough to have received a special name, and they are referred to as **didynamous**. Alternatively different whorls of stamens can be of different lengths as is the case in **tristylous** plants (e.g. *Lythrum*) with three style lengths and three stamen lengths, variously displayed in different individuals as part of a **heteromorphic self-incompatibility** system. Another case of stamens in different whorls being of different lengths occurs in the Brassicaceae where there are four long stamens and two short. This pattern is distinctive enough to have received the name **tetradynamous**.

The attachment of the anther to the filament is variable: it may be innate, adnate or versatile. **Innate anthers** are attached to the filament by their base, as for instance in *Ranunculus* spp., and commonly in many dicots. **Adnate anthers** are attached to the filament by their whole length, as in *Magnolia*. **Versatile anthers** are attached to the filament by a small and flexible connection positioned in the middle of the anther. This allows the anther to swing about, which in turn promotes pollen dispersal. Perhaps for that reason versatile anthers are often found in wind pollinated plants, notably the grasses (Poaceae). However, versatile anthers are also found in many insect pollinated plants including lilies (*Lilium*).

Anther dehiscence is another variable character. Two apertures must form, one for each lobe (theca). Dehiscence may be **introrse**, **extrorse**, **marginal**, **porate** or **valvate**. In introrse dehiscence the aperture forms as longitudinal slits on the **adaxial** (inner, facing the carpels) side of the anther (*Viola*; Asteraceae). Extrorse is the same but on the **abaxial** (outer, facing the perianth) side, as in *Iris pseudacorus*. If the slits are on the lateral margins, as in *Begonia* and *Fritillaria imperialis*, the dehiscence is termed **lateral**. In porate dehiscence a pore-like aperture forms, generally at the top of each anther lobe as in *Solanum* and *Rhododendron*. In valvate dehiscence pores form by the uplift of a flap as in *Berberis*.

6.21 The staminode

Loss of stamens occurs frequently. This may take the form of the loss of individual stamens in a whorl or the suppression of whole whorls. Where such loss has occurred, a **vestigial stamen** is left called a **staminode** (De Craene & Smets, 2001). It is rare for stamens to be lost without trace.

A staminode is a sterile stamen, and although usually reduced and vestigial, they can be large and even conspicuous. Far from being vestigial, these staminodes, although sterile, have been converted to alternative functions in floral biology (Walker-Larsen & Harder, 2000). In *Parnassia palustris* the staminodes form conspicuous false and true **nectaries** to attract insects (Sandvik & Totland, 2003). The conspicuous hairy staminodes of *Penstemon* function mechanically in enhancing pollination (Walker-Larsen & Harder, 2001).

The showy **petaloid labellum** in the Zingiberaceae has a staminodial origin (Kirchoff, 1991). It is therefore possible that organs that look like petals have a staminodial origin. This **staminodial origin of petals** may be important in angiosperm evolution. Thus the petals of *Ranunculus* and *Helleborus* in the Ranunculaceae have been suggested to have a staminodial origin (Erbar, Kusma & Leins, 1998), along with other groups in the Ranunculales and Caryophyllales (Aizoaceae). It appears that the angiosperm petal has two origins: as a **bracteopetal** (derived from outer floral members) or as a **staminopetal** (derived from inner floral members; Albert, Gustafsson & Di Laurenzio, 1998; Ronse De Craene, 2003; Ronse De Craene & Smets, 2001).

6.22 The carpel

The **carpel** is the **megasporophyll** of the flowering plant. It bears the **integumented megasporangia** called **ovules**. Ovules have been discussed in Chapter 5. Here the general features of their enclosing organs will be dealt with.

In wild normal flowers, although intermediates between petals and stamens are very frequently found, intermediates between stamens and carpels are rarely found. However, they are frequent enough in some species, such as *Sempervivum tectorum* to indicate clear homology between stamens and carpels at the level of sporophylls (Guédès, 1972). Interesting stamen–carpel intermediates are also found as **terata**, such as in *Nicotiana* cybrids (Fitter et al., 2005).

There has been much debate as to the primitive form of the carpel, hindered by the absence of credible phylogenetic data. Now that the main features of angiosperm phylogeny are well established it is possible to reconstruct the likely ancestral state from the pattern of carpel morphologies in the early divergent clades of angiosperms. From an examination of the carpels of basal angiosperms, it seems that the ancestral carpel was much as in *Amborella*. Primitive carpels were several, free (**apocarpous**), helically arranged, shortly stalked (**stipitate**), with **marginal ovules**, sac-like (**ascidiate**) by **peltation**, congenitally fused below and postgenitally fused above (Doyle & Endress, 2000; Endress, 2001b).

Ascidiate carpels are usual in basal angiosperms such as *Amborella* and the Nymphaeales (Buzgo, Soltis & Soltis, 2004; Endress, 2005), and it is a common part of the development of carpels in other species too (Guédès, 1971; Guédès & Schmid, 1978). In *Amborella* (Buzgo, Soltis & Soltis, 2004) there is a single ovule that develops from the **cross meristem** of the ascidiate (peltate) carpel. Thus it is a marginal ovule, but median rather than lateral. Lateral placement of the marginal ovules is usually the case when there are multiple ovules as in some Austrobaileyales, and it is strikingly seen in the follicle of the basal eudicot *Caltha* (Ranunculaceae).

Despite the frequent occurrence of the ascidiate carpel in the Austrobaileyales (Endress, 2001b; Tucker & Bourland, 1994), *Illicium* (Williams, Sage & Thien, 1993) departs from this pattern by having folded or **plicate carpels**, implying that peltation has been overridden in that genus. It was formerly argued (Bailey & Swamy, 1951) that the primitive carpel was a folded (plicate) type, characteristic of *Drimys* and other Winteraceae, called **conduplicate** (flat-folded with the ovules on the sides rather than the margins of the walls). These carpels, it was thought, evolved into the **induplicate** carpels (inrolled-folded with ovules on the margins) such as the follicle of *Caltha* (Fig. 6.13).

It now appears that both types of plicate carpel are derived. However, the carpel of *Drimys* presents a possible intermediate between the ascidiate carpel and the strongly plicate carpel of *Caltha*. The *Drimys* carpel is unifacially stipitate, and ascidiate at the base. Above that, the walls, with their hypertrophied stigmatic crest, are conduplicate, having outgrown the basal peltation.

If the basic ascidiate nature of the carpel is accepted, then peltation has clearly played a major role in carpel evolution. Peltation is due to the spread of the abaxial domain in what is otherwise a dorsoventral structure with both adaxial and abaxial surface. As the abaxial domain comes to surround the adaxial, meristematic adaxial–abaxial boundary growth will occur all round the adaxial surface, which soon comes to be enclosed in a sac. Thus carpels have adaxial surface on the inside and abaxial surface on the outside. When carpels are stalked (**stipitate**), as they often are, the surface of the stipe is completely abaxial, as in the petiole of a peltate leaf (e.g. *Tropaeolum*).

YABBY genes are abaxial identity genes, and a particular YABBY gene, namely, *CRABS CLAW*

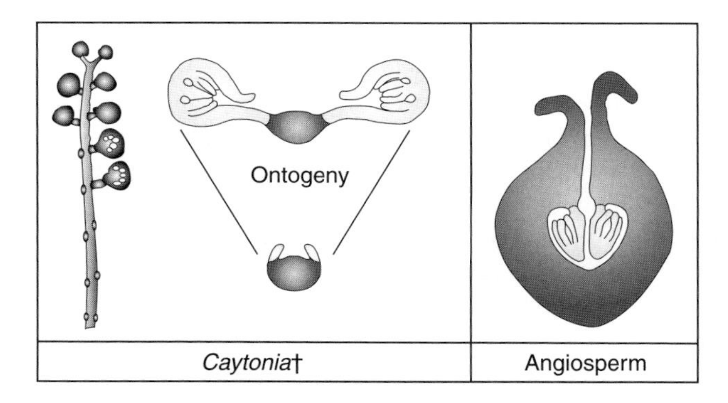

Fig. 6.13. Scheme of possible homology of structures between the fossil *Caytonia* and angiosperms, showing the difficulties. If the cupules of *Caytonia* are equivalent to ovules (gray) then the leaf part of the carpel is equivalent to the nonleaf-like cupule stalk in *Caytonia* (black) (after Doyle, 2006).

(*CRC*) is implicated in carpel development of *Arabidopsis*. Loss-of-function *crc* mutants have abnormal gynoecia, with a failure of carpel fusion at the tip (Bowman & Smyth, 1999; Eshed, Baum & Bowman, 1999). A putative orthologue of *CRC* has been found in *Amborella* raising the possibility that *CRC* has been central to carpel development throughout the history of the angiosperms. This is supported by the finding that the expression pattern of *CRC* is highly conserved between *Amborella* and *Arabidopsis*, being characteristic of the abaxial domain of the carpel wall in both (Fourquin et al., 2005, 2007; Scutt et al., 2006). This similarity suggests that there is conservatism of **radial patterning** in angiosperm carpels.

Against this, however, is the apparent *CRC* orthologue in rice (*Oryza*) called *DROOPING LEAF* (*DL*). *DL* has a different expression pattern in the carpel and is expressed in the leaf. Loss-of-function of *dl* mutants produce stamens in place of carpels (Yamaguchi et al., 2004), suggesting that *DL* may function in rice as a carpel identity gene rather than a carpel development gene. *DL* and *CRC* clearly do not have a conserved function.

One striking difference in *Amborella* compared to *Arabidopsis* is the expression of *AmbCRC* in the stamen filaments (Fourquin et al., 2005). Stamens of *Amborella* superficially look remarkably similar to carpels (Buzgo, Soltis & Soltis, 2004), and the presence of *CRC* expression in stamens could possibly indicate an ancestral role for *CRC* in the peltation of all sporophylls, a role that has been restricted to carpels in eudicots. A greater understanding of the complete network of genes involved in carpel development is greatly needed.

Another gene of interest in carpel development is *TOUSLED* (*TSL*). *TSL* is protein kinase that is involved in the proximodistal patterning of the gynoecium in *Arabidopsis* as it shows a peak of expression in the carpel apex and is required for correct development of the apex (Roe, Nemhauser & Zambryski, 1997; Roe et al., 1993). A putative *TSL* orthologue has been detected in the basal angiosperm *Cabomba* (Cabombaceae). This too is expressed at a high level in the apex of the carpel (Fourquin et al., 2005), raising the possibility that the genetic mechanism of **proximodistal patterning** in the carpel may be widely conserved through the angiosperms.

6.23 The gynoecium

The female part of the flower is made up of one or more carpels and collectively called the **gynoecium**. The gynoecium may be composed of a single carpel (a single **ovary**), many free carpels (many ovaries), or many carpels congenitally or postgenitally fused into a single **compound ovary**. This is a **pluricarpellate** female organ. As expected, **tricarpellate** ovaries are common in monocots and **pentacarpellate** ovaries are common in eudicots.

A **pistil** is the term given to an **ovary** and a **style** together. Like ovaries pistils may be designated simple (of one carpel) or compound. In **epigynous** and **perigynous** flowers the ovary may be surrounded by receptacle tissue.

When the gynoecium is composed of one (Fabaceae) or more (Ranunculaceae) **free carpels** the gynoecium is termed **apocarpous**. On the other hand, if the gynoecium is a compound ovary of fused carpels it is termed **syncarpous**. If the carpels are free, an **abaxial line** and an **adaxial suture** can usually be discerned. The adaxial (ventral) suture corresponds to the meeting place of the two margins of the megasporophyll and the ovules are usually borne here making them, morphologically, **marginal ovules**. The abaxial (dorsal) line corresponds to the midrib of the megasporophyll.

The compound ovary contains the ovules in one or more chambers, called **loculi**. It consists of an outer wall and **inner walls** (**dissepiments**). The inner walls typically correspond to the united walls of adjacent carpels, and if so are called **septa**. In this case each locule will represent one carpel. However, the dissepiments may be formed as outgrowths of the placenta, as in the **false septum** or **replum** of the Cruciferae, or as outgrowths of the midribs of the carpels as in the false septa of flax (*Linum usitatissimum*).

The gynoecium is occasionally stalked and this stalk is termed a **gynophore** (e.g. Capparaceae,

Leguminosae). The gynophore of the peanut, *Arachis*, may be up to 20 cm long, whereas that of *Trifolium* is very short. This stalk may be a receptacular feature from elongation of the receptacle, or a carpellary one by elongation of the carpel base. If carpellary, it is tempting to suggest that it is homologous with the petiole of the megasporophyll, but there is no evidence for this.

It has long been known that even though it is required for geotropism, auxin has an inhibitory effect on the gynophore elongation of *Arachis* (Jacobs, 1951). Therefore it is of little surprise that mutations in auxin response factor genes, such as *ETTIN* (*ETT*) or *MONOPTEROS* (*MP*), cause a significant elongation of the gynophore in *Arabidopsis* (Nishimura et al., 2005). In wild-type *Arabidopsis*, the gynophore is minute and scarcely noticeable. The auxin sensitivity is the reason why gynophores are an early developmental feature of gynoecia, as developing fruits are a major source of auxin.

Finally, in unisexual flowers, males may have vestigial sterile carpels. These are called **carpellodes** in counterpoise to **staminodes**.

6.24 Placentation

The ovary contains regions of tissue to which the ovules are attached, called **placentae**. The distribution of the placentae is variable, and particular types are often characteristic of plant families, so **placentation** has been well studied as a taxonomic character.

The main types of placentation are parietal, axile and central. In **parietal placentation** (e.g. *Viola*) the placentae are on the outer wall of the ovary, while in **axile placentation** (e.g. *Tulipa*) the placentae are in the middle of a **plurilocular ovary**. In **central placentation** (e.g. *Primula*) the placentae are on a **central column** of a **unilocular ovary**, which may extend all the way from top to bottom of the ovary or only part of the way up (**free central placentation**).

Placentae appear as outgrowths of the carpel margins, consistent with the notion that ovules are fundamentally marginal on the carpel. Sometimes the placentae can be very substantial as in the bicarpellate gynoecia of Brassicaceae. Here they arise at the carpel margins and form a plate right across the gynoecium: the **false septum** (**replum**) characteristic of the family.

The placentae are cushions of physiologically active tissue. It is probable that the evolution of a well-developed placenta is an adaptation to buffer the ovules from stress. The placenta is the site of phloem unloading for sugars *en route* to the ovule. Furthermore, the placenta is physiologically responsive to plant water stress that might damage the ovules. It has been shown that stress gene activation (Yu & Setter, 2003) and abscissic acid (ABA) levels (Wang, Mambelli & Setter, 2002) in the placenta of maize (*Zea*) increase dramatically on plant droughting.

Basal angiosperms with ascidiate, and often uniovulate, carpels, such as *Amborella*, do not have prominent placentae. Therefore a strong development of placental tissue must be considered to be an evolutionary innovation: an innovation characteristic of many but not all angiosperms, probably in response to spread into more xeric habitats. Likewise the absence of strongly developed placentae in some basal angiosperms is consistent with an origin for angiosperms in wet habitats.

Two evolutionary processes, **congenital fusion** and **reduction** of the inner carpel walls (septa), are largely responsible for the different placentation types. When many carpels are brought together in a simple plurilocular ovary, the adaxial carpel margins bearing placentae come to be in the center of the gynoecium (axile placentation). Congenital fusion means that each placenta is derived from the two margins of a single carpel (and consequently often has two rows of ovules). A plurilocular ovary with axile placentation is the most basic type of syncarpous ovary.

If the carpel walls are reduced, and develop only partially, and there is no division between the carpels (locules), then the ovary is said to be unilocular. The wall reduction can take two forms. If the abaxial parts of the carpel wall are the only parts left, then the placentae are adjacent to the outer wall of the ovary and placentation is parietal. Each placenta in parietal placentation is derived from the margins of adjacent but different carpels.

However, if only the adaxial part of the carpel wall develops strongly, and is ovuliferous, then a central column may form (central placentation). The central placenta is derived from the adaxial part of all the carpel walls, congenitally fused in the center of the ovary.

6.25 Style and stigma

The apical part of carpels or compound gynoecia forms the style and stigma. The **style** is the organ through which the pollen tube must grow for fertilization. In the typical primitive carpel the style is formed from the mouth of the sac-like carpel grown up into a tubular beak, which is then plugged with mucilage through which the pollen tubes grow. The carpel is technically open and each carpel has its own style.

A syncarpous ovary may have several styles, and if so the number usually equals the number of carpels comprising the ovary. Alternatively, there may be a single style, indicating that the styles corresponding to each carpel have been evolutionarily united into a single column. The union of styles may not be complete but may take place only apically or basally. Nevertheless, any form of stylar union represents a very important evolutionary event as it gives any germinating pollen grain the potential to fertilize any ovule in the flower, much more efficient than the separate pollinations required for each style if they are separate. A point of stylar union, allowing such crossing over of pollen tubes, is called a **compitum**. The presence of a compitum is a good example of the evolutionary advantages to be had from synorganization of the gynoecium. Running through the center of the style, there may be a **stylar canal** of soft **conducting tissue**, or even **mucilage**, providing easy passage for pollen tubes.

The **stigma** is the pollen-receptive end portion of the style. It may be sticky and **papillate** making for easy pollen attachment. The "wet stigma" is correlated with gametophytic incompatibility systems. However, many angiosperms (including *Amborella*) possess a **dry stigma**—a trait associated with sporophytic incompatibility (Heslop-Harrison & Shivanna, 1977).

If the style derives from more than one carpel it may branch into several stigmatic arms at the apex. The number of stigmas will generally equal the number of carpels and they are normally positioned in line with the center of the locules. In certain plants such as poppies (*Papaver*) the style is reduced so the stigmatic surface is sessile on the ovary. In the case of poppies, these stigmas represent the marginal stigmas of two carpels, and they are therefore positioned over the septa. These are called **commissural stigmas**.

At the junction of the style and ovary there may be a swelling of the style called a **stylopodium**, which is often nectar-bearing. It is a particularly common feature in the Apiaceae.

6.26 The sinking seed: epigyny and the receptacle

The axis of the flower on which the floral phyllomes are borne is called the receptacle. The receptacle at growth arrest is often dome-shaped. This is to be expected, as this is the normal shape of a shoot apex. However, in some species, such as *Magnolia* (Magnoliaceae) and *Myosurus* (Ranunculaceae) the receptacle is notably elongated. This is associated with the delayed determinacy involved in producing large numbers of helically arranged carpels. Alternatively, if apical growth arrests before that of the flanks, a flattened receptacle can be formed as in many Crassulaceae (*Sedum, Aichryson*). If the axis flanks continue to grow (or secondarily start into growth) after the arrest of the axis tip, a bowl-shaped receptacle (**floral cup**) can be formed with the carpels in the center sunken below the level of the stamens and petals at the edge of the bowl (e.g. *Exochorda*). Accentuation of this process has led to the complete concealment of the gynoecium within a deep floral cup.

The perianth and androecium are conventionally inserted below the level of the carpels, and they are therefore said to be **hypogynous** (as in *Ranunculus, Geranium, Papaver*). However, when these outer whorls are inserted above the carpels on the rim of a floral cup, they are said to be **perigynous** (as in *Calycanthus, Prunus, Daphne*). If the ovary wall is fused to the floral cup, the perianth and androecium appear to be inserted right on top of the ovary and they are called **epigynous** (as in *Fuchsia,* Apiaceae, Asteraceae). In this instance, the flower is said to have an **inferior ovary**, in contrast to the **superior ovary** of hypogynous and perigynous flowers. In some families, such as the Asteraceae, it is characteristic of the entire family. In others, such as the Rosaceae and Saxifragaceae (Soltis & Hufford, 2002), it is a variable character. In the Rosaceae, for instance, there are genera with perianths that are variously hypogynous (*Fragaria*), perigynous (*Rosa*) and epigynous (*Pyrus*).

Such relative sinking of the gynoecium has taken place independently in many groups of plants and is a major trend in angiosperm evolution. The most likely suggestion is that sinking the gynoecium removes the danger of damage to the carpels by pollinators (Grant, 1950a,b) such as pollen feeding beetles (which may graze on carpels if they are readily available) or birds (which may damage carpels accidentally while probing for nectar). In addition, the provision of two walls (the ovary wall and the wall of the floral cup) may provide extra protection against ovule and seed feeding insects.

In the above it has been assumed that the floral cup is an axial structure formed from outgrowths of the receptacle. However, it is also possible to imagine a floral cup formed of congenitally fused bases of the perianth and androecium (called a **hypanthium**). In this case it is a phyllomic rather than an axial structure but would have the same effect of bringing the apparent insertion of the outer whorls above that of the carpels. There are cases in which a **hypanthial tube** extends a long way above the ovary as in *Fuchsia* and some species of *Passiflora,* in which the tube is part of a hummingbird pollination floral syndrome.

In the early years of the last century a debate raged as to which of these explanations of the floral cup was correct in various species (Douglas, 1944, 1957). If the ovary wall is an early developmental feature this is a profitless quest, as the growth of the base of the flower occurs before there is much histogenesis characteristic of axis or phyllomes. It is best simply to conclude that the ovary wall is formed from the base of the flower and to leave the homology with late developmental features of the flower moot. However, an axial component can clearly be demonstrated if the receptacular growth that forms the floral cup commences late in development after the receptacle has been vascularized and is deep-seated enough to carry that vascularization up into the floral cup as bulges or loops (Puri, 1952). Such is the case in *Darbya* (Santalaceae) in which the vascular bundles rise up the floral cup to the perianth and stamens and then descend all the way down to supply the gynoecium at the bottom of the floral cup (Smith & Smith, 1942). The floral cup of *Darbya* is thus unequivocally wholly axial. Other floral cups are more likely to be phyllomic and it is possible that some are of mixed origin.

Box 6.2 Major trends in floral evolution.

The following trends can be seen as parts of a single trend for greater **synorganization** within the flower, from indeterminate aggregations of individual organs to the flower behaving, developmentally and functionally, as a single unit. The shift from indeterminate to determinate numbers of floral organs is the essential first step for synorganization to proceed (Endress, 2001a).

1. Helical phyllotaxy of floral phyllomes to whorled phyllotaxy (Section 6.12)
2. Indeterminate to determinate numbers of floral phyllomes (Section 6.13)
3. Free perianth parts to perianth parts united into tubes (Section 6.14)
4. Apocarpy (free carpels) to syncarpy (fused carpels; Section 6.22)
5. Superior to inferior ovaries (Section 6.26)

Actinomorphy (polysymmetry) to zygomorphy (monosymmetry or asymmetry; Section 6.15)

In principle, it should be possible (Gustafsson & Albert, 1999) to distinguish axial and and phyllomic components of floral cups by means of molecular developmental tests. For instance, if the tissues of the floral cup express, during development, identity genes specifying the outer floral whorls, it would strongly argue for a phyllomic origin (Box 6.2).

6.27 The fruit

A **fruit** is an **ovary** when fully developed after pollination (together with any accessory parts). In structure it is therefore similar to the ovary, but it is developmentally interesting as it shows further differentiation between parts and new features such as color and fleshiness. In plants with an **apocarpous** gynoecium as in *Ranunculus*, there are as many fruits (**achenes**) that develop from each flower, as there are carpels. Each carpel constitutes a separate ovary and therefore a separate fruit. However, in **syncarpous** flowers there is only one ovary and therefore only one fruit.

The simple fruit is composed of **pericarp** (the tissue surrounding the seeds) and the seeds themselves. Following the terminology introduced for the ovary, the pericarp can be divided into the **wall** (the outer wall), the **dissepiments** (walls between chambers) and the **placentae** (tissue that bears the seeds). The wall is often usefully divided into **exocarp**, **mesocarp** and **endocarp**. For instance in a peach, plum or cherry (*Prunus*), the exocarp is represented by the leathery skin, the mesocarp by the flesh and the endocarp by the hard stone. There is a single seed within the endocarp, and the seed is protected by the endocarp. In this genus the endocarp is composed of extremely hard woody sclerenchyma tissue.

In addition to the ovary, certain other parts may be added to the fruit as **accessory organs**. These are often important to the fruit for protection, ripening or dispersal. In **epigynous** flowers the ovary is invested in receptacle-derived tissue and this becomes part of the fruit if there is a single ovary as in an apple (*Malus*). If there are multiple ovaries, as in fruit developed from a **perigynous** flower, such as rose hips (*Rosa*), the receptacle is an accessory organ containing many small **simple fruits** (fruitlets). A similar situation exists in the strawberry (*Fragaria*) where the numerous simple fruits are embedded on the red, juicy accessory organ, the "strawberry" derived from a single fleshy receptacle. Both rose hips and strawberries are therefore a type of **false fruit**, derived from many simple fruitlets of a single flower. This false fruit is called an **anthocarp**. Another type of false fruit is derived from an inflorescence, either because the fruits of many different flowers have grown together (*Morus*) or because the main functional part of the fruit is formed by growth of the peduncle of the inflorescence (*Ficus*). This type of false fruit is called a **syncarp**.

Other possible accessory organs include the style and stylopodium, which may be persistent. The floral receptacle may be fleshy and an important accessory part of the fruit. Bracts may be important as in pineapple (*Ananas sativus*). A pineapple is a syncarp, representing a spike-like inflorescence in which fleshy bracts and simple fruits have grown together in one mass. In the Fagaceae and Nothofagaceae an important accessory part of the fruit is the cupule. The cupule is a **coenosome** derived from a cymose ramification system (Fey & Endress, 1983).

The calyx may be persistent and an important accessory organ for fruit protection or fruit dispersal. The "Chinese lanterns" or inflated calyx tubes in *Physalis* (Solanaceae) provide an example of fruit protection. The calyx, initially normal-sized, continues growing after flowering (i.e. it is **accrescent**). The calyx comes to surround completely the mature fruit (a berry). Accrescence of the calyx has been shown to involve a MADS-box transcription factor, *MPF2*, which positively regulates cell proliferation in the sepals (He & Saedler, 2005; He et al., 2007). The potato orthologue of this gene (*StMADS16*) is expressed in vegetative organs not flowers and potato does not have an accrescent phenotype. It therefore seems that heterotopic expression of this gene in the evolution of the *Physalis* lineage has brought about the accrescent phenotype. Interestingly the phenotype is linked to fertilization and fruit formation so the expression of *MPF2* is not, by itself, sufficient. The action of the gene requires the plant hormones cytokinin and gibberellin, which act as cofactors in the accrescence (He & Saedler, 2007).

The calyx modified for dispersal is exemplified by the scaly, bristly or hairy "pappus" of the Asteraceae. The sepaloid nature of the pappus in the Asteraceae has been investigated at the molecular level (Yu et al., 1999), using *Gerbera* as a model system, and the results of MADS-box gene expression studies are quite consistent with the morphological nature of pappus bristles as modified sepals, as expected.

6.28 The classification of fruits

The fruit, as the mature ovary of the plant, is of exceeding ecological importance. In consequence it is evolutionarily highly labile. While genera typically have similar fruit types there is great variation within families. Thus the Rubiacaceae, Solanaceae and Rosaceae all contain both dry and fleshy fruits. Two important characteristics are generally used as the basis for classifying fruits: *fleshy versus dry* and *dehiscent (splitting open) versus indehiscent*. Both these characters are of considerable ecological and functional significance.

Fleshy fruits are animal dispersed and almost always indehiscent for the simple reasons that drying of the ovary wall is generally required for dehiscence, to set up the tensions required to rip through the pericarp. There are some exceptions to this rule and a few fleshy fruits, such as *Myristica*, can split (in these instances they tend to be leathery rather than truly fleshy). Furthermore, in the fleshy fruit, individual seeds are separately dispersed through the destructive or digestive activity of animals so there is no need for a dehiscence mechanism.

Dehiscence is a means of individually dispersing seeds from many seeded fruits without relying on animals. Although dehiscence does not generally allow for the long-distance dispersal of seed that is provided by birds and mammals, there are exceptions, such as when seeds have seed hairs for dispersal as in *Asclepias*, *Populus* and *Aeschynanthus*. Dehiscence confers reliability (providing dry conditions occur, it is a difficult strategy in the everwet tropics) and dispersal in particular environmental conditions and in some cases over a long period of time. An example is provided by the "censer mechanism" of capsules with porate dehiscence like the poppy (*Papaver*). In strong winds a few seeds will shake through the pores and it many take weeks or months for the capsule to empty. There is a potential conflict between favorable conditions for dehiscence being dry and favorable conditions for growth being wet. This has been solved elegantly by the twisted-fruit Gesneriaceae such as *Streptocarpus*. The fruit splits open on drying but on further drying twists shut. High atmospheric humidity untwists the fruits and lets seed out of the valves, until dry conditions cause it to twist shut again. Again, complete emptying of the capsule can take months.

Attempts to classify fruit are fraught with difficulty as form of fruit is multivariate rather than hierarchical. Thus any attempt at hierarchical classification is bound to contain anomalies. The value of such an attempt is didactic rather than scientific, and in this spirit a traditional treatment is presented in Table 6.3.

6.29 Distribution of sex in flowers and inflorescences

We have seen that the organs of sexual function, stamens and carpels, have a regulated distribution within the **bisexual** (**hermaphrodite** or **perfect**) flower. However, as we have also seen, not all flowers are bisexual, and so there is a possibility of regulated distribution of stamens and carpels between and within individuals quite separate from patterning within the flower.

If a plant species contains more than one gender, then these genders are usually determined in individuals by segregating genetic factors. These may be whole chromosomes, as in *Marchantia*, *Ginkgo*, *Silene* and *Rumex*, or one or more segregating loci on autosomal chromosomes (Dellaporta & Calderon-Urrea, 1993; Ming et al., 2007; Tanurdzic & Banks, 2004; Vyskot & Hobza, 2004).

Table 6.3 Synopsis of some common temperate fruit types.

Dry dehiscent	**Follicle** (fruit of a single carpel)	**True follicle** (dehiscence along the ventral suture: *Paeonia, Delphinium, Caltha*) **Legume** (dehiscence along the ventral suture and dorsal line: Fabaceae)
	Capsule (fruit, formed of a syncarpous ovary, releasing seeds through apertures. Capsules can be further divided according to the nature of these apertures: next column)	**Loculicidal valvate** (aperture corresponds to almost the entire carpel midrib, the commonest type: *Syringa, Lilium*) **Septicidal valvate** (aperture corresponds to the line of union between two adjacent carpels: *Gentiana, Colchicum*) **Toothed** (separation only at the top: *Lychnis, Cerastium*) **Cleft** (as valvate dehiscence but splitting only partial: Orchidaceae, *Oxalis*) **Porate** (dehiscence by small apertures usually at the top of the fruit: *Papaver, Antirrhinum, Campanula*) **Circumscissile** (the top of the capsule is thrown off as a circular lid or operculum: *Hyoscyamus, Plantago, Anagallis*—this type of fruit is sometimes called a **pyxis**) **Silique/silicle** (fruit of two carpels joined at the margin only, but separated by a false septum (**replum**) of outgrown placentae; dehiscence is by two valves that remain attached to the replum at their tops: Brassicaceae—a silicle is simply a broad version of a silique)
	Schizocarp (fruit splits into one-seeded portions—mericarps)	**Diachenial to polyachenial schizocarps** (fruit splits into two or more mericarps, or "nutlets," which remain closed: *Galium* [2 mericarps]; *Tropaeolum* [3]; Lamiaceae, Boraginaceae [4]; Malvaceae [many]) **Diachenial samara** (fruit splits into two mericarps each of which is winged: *Acer*) **Dicoccous to pentacoccous schizocarp** (fruit splits into two or more opening mericarps: *Mecurialis* [2 mericarps]; *Euphorbia* [3]; *Geranium* [5])
Dry indehiscent (single-seeded)	**Caryopsis** (grain; the pericarp and the seed-coat cohere: Poaceae)	
	Achene (pericarp leathery, no coherence between pericarp and seed coat)	**True achene** (*Ranunculus, Rosa, Fragaria*) **Cypsela** (formed from an inferior ovary: Asteraceae, Valerianaceae)
	Nut (pericarp brittle or woody; no coherence between pericarp and seed coat)	**True nut** (usually woody: *Quercus, Fagus, Corylus*) **Samara or key** (winged nut: *Fraxinus, Ulmus*)
Fleshy indehiscent	**Berry** (one/many seeds free within a fleshy fruit)	**True berry** (flesh from fruit wall: *Solanum, Musa, Vitis, Cucumis*) **Hesperidium** (flesh formed from hairs on inner surface of fruit wall: *Citrus*)
	Drupe (stone fruit, seeds enclosed within a stony endocarp or pyrene)	**Single-stoned** (monopyrenous) drupe (*Prunus, Olea*, and individual fruits within the aggregate fruit of *Rubus*) **Multistoned** (dipyrenous to polypyrenous) drupe: *Cornus* [2]; *Sambucus* [3]; *Ilex* [4]; *Sararanga* [many] **Pome** (drupaceous fruit with leathery rather than stony endocarp: *Malus*)

Not treated here are aggregate fruits (**etaerio**) as in raspberry (*Rubus*), which is an aggregate of drupes; **anthocarps** (**accessory fruits**) such as strawberry (*Fragaria*), which is a collection of achenes on a fleshy receptacle, or **syncarps** (multiflower fruit) such as fig (*Ficus*). Furthermore, much tropical fruit diversity is hard to place in such a scheme. For instance, *Litchi chinensis* (lychee) is a nut by virtue of its tough pericarp, yet fleshy because of the aril on the seed; *Myristica* (nutmeg) is a berry (fleshy fruit), yet dehiscent by two **valves** (flaps of pericarp left after the splitting of the fruit).

In plants where the gametophyte generation is dominant, the segregating sex factor will have its action in the gametophyte, and the sporophyte is sexless. This is not the case in plants where the sporophyte is dominant. In seed plants the sex of the gametophyte is not determined by the genetic sex factor it carries. Instead it is determined epigenetically by the sporophyte. The sexual identity of the gametophyte is thus at the "whim" of the sporophyte, gametophytes can be thought of as having "epigender" only. At the genetic level the gender of gametophytes is marked only by the methylation, or lack of it, of certain genes (Spielman et al., 2001). As E.J.H. Corner puts it in *The Life of Plants*: "By transferring maleness and femaleness, that is sexuality, from the gametophyte to the sporophyte, the gametophyte part of the life cycle is eliminated as a free-living state of the plant." As Corner also notes, this process of "transference of sexuality" from gametophyte to sporophyte starts, in a small way, with heterospory (homosporous sporophytes being quite sexless).

In plants with a single gender the distribution of male, female and hermaphrodite flowers on the plant is regulated by hormonal signaling, the details of which vary between plant species but frequently involve ethylene and/or gibberellin. This has been studied in maize, a monoecious grass. The ancestors of maize were andromonoecious, with male and bisexual florets. The first step in the evolution of sexuality in maize is therefore the elimination of pistils from some flowers, which in the panicoid grasses is effected by targeting of the cells of the gynoecium for apoptotic cell death very early in development (Malcomber & Kellogg, 2006). *TASSELSEED2* (*TS2*) is the key pistil-destroying gene and it has been implicated as a likely sex determination gene, working in the same pathway as other *TASSELSEED* genes (*TS1* and *TS5*) and the *SILKLESS1* (*SK1*) gene. All these genes specifically affect gynoecial development but *TS2* appears to be upstream of the others. *TS2* encodes a short-chain dehydrogenase/reductase (SDR) thought to be involved in hormone metabolism (Delong, Calderon-Urrea & Dellaporta, 1993). Further elucidation of this mechanism requires the characterization of *SK1* with which *TS2* has an important interaction. *SK1* is a pistil-promoting gene, and it may be that *TS2* acts by keeping *SK1* out of male flowers (Veit et al., 1993) or alternatively that *SK1* protects bisexual and female flowers from the masculinizing effect of *TS2* (Calderon-Urrea & Dellaporta, 1999).

The masculinizing effect of *TS2* and its partners is sufficient for the evolution of andromonoecy, but maize has gone further and become fully monoecious by anther abortion in the ear florets. This abortion is mediated by gibberellin (GA). Mutations that negatively affect GA synthesis can restore anthers to the ear. One such mutant is *ANTHER EAR1* (*AN1*), which is is involved in the synthesis of ent-kaurene, the first tetracyclic intermediate in the GA biosynthetic pathway (Bensen et al., 1995).

Cucumber (*Cucumis sativa*) is another plant well studied for sexual development. Wild-type cucumber is trimonoecious, producing first female flowers then bisexual flowers then male flowers in orderly sequence (Shifriss, 1961). Unisexual flowers are produced by the developmental suppression of stamens or gynoecia, in a whorl-specific manner (Kater et al., 2001). Several loci have been found that affect sex expression, either pushing it in a female or male direction. Of particular interest is the *FEMALE* (*F*) locus. The dominant *F* mutation has the effect of completely feminizing the plant. The basis of the *F* mutation has been shown to be a novel duplicate copy of the gene encoding 1-aminocyclopropane-1-carboxylate synthase (ACS), the rate-limiting enzyme in the ethylene biosynthesis pathway (Mibus & Tatlioglu, 2004; Trebitsh, Staub & O'Neill, 1997). Not only do *F* plants have an extra copy of this gene (*CsACS1*) but the copy (*CsACS1g*) has some different putative *cis*-regulatory elements. It is these that may explain the different responsiveness of *CsACS1g* to developmental and hormonal factors (Knopf & Trebitsh, 2006).

Recently some progress has been made in the determination of sex in dioecious and trioecious species at the genomic level (Ming & Moore, 2007). *Silene latifolia* has a relatively recently evolved sex chromosome (XY) system, and the Y chromosome

Table 6.4 A table of the eight genders of plants (top), and the nine ways in which these genders are associated in plant species (left).

	Bisexual flowers	Male flowers	Female flowers	Male flowers and female flowers	Male flowers and bisexual flowers	Female flowers and bisexual flowers	Female flowers, male flowers and bisexual flowers
Hermaphroditism	+						
Monoecy				+			
Andromonoecy					+		
Gynomonoecy						+	
Trimonoecy							+
Dioecy		+	+				
Androdioecy	+	+					
Gynodioecy	+		+				
Trioecy	+	+	+				

shows three "evolutionary strata" (Bergero et al., 2007) of degeneration (degeneration is inevitable after recombination stops). These strata show how a sex locus (a recombination dead spot) on an autosome expanded until eventually recombination ceased over the whole chromosome and a full Y chromosome was born.

Carica papaya (pawpaw) is a dioecious or trioecious species, which has recently been shown to have a large male-specific region associated with one of the autosomes, which may be an incipient Y chromosome (Liu et al., 2004). The male-specific region shows a high degree of recombination suppression and occupies about 10% of the chromosome. Now that the genomic regions responsible for sex determination have been identified in these systems, there is the potential for full elucidation of the determining genes and the genetic networks they entrain (Table 6.4).

6.30 Floral reduction and pseudanthy

The primitive flower consists of an acrogynous strobilus with an indefinite number of sporophylls possibly helically arranged and surrounded by bracts and bracteotepals. However, early in angiosperm evolution multiorgan flowers such as this began a process of progressive reduction in several lineages. Reduction was probably driven by (1) evolution of wind rather than insect pollination, (2) reduction in size of the vegetative plant leading to smaller meristems with less available space for formation of large organs and (3) lower investment in reproduction relative to vegetative growth as ecological competition in the vegetative phase becomes more important with the rise of angiosperm-dominated plant communities.

Reduction involves both the reduction in size of organs and, more strikingly, the suppression of primordia so that organs are lost altogether. The extreme end point of such reduction is a male flower consisting of a single stamen and a female flower consisting of a single carpel, and this extreme situation is found in the basal angiosperm family Hydatellaceae (Saarela et al., 2007), where it is associated with extreme reduction in vegetative size and wind pollination.

The family Chloranthaceae, although not in the basal ANA grade (i.e. *Amborella*, Nympheales, Austrobaileyales), is a family that probably diverged just above the ANA grade. In this family of tropical understory shrubs, considerable floral reduction has also taken place. Flowers are generally perianthless (although the female flowers of *Hedyosmum* have three tepals) with a small number of stamens and a single carpel and are often unisexual, as in *Ascarina* (Doyle, Eklund & Herendeen, 2003).

In the monocots and eudicots too, extreme floral reduction has occurred. In *Lemna* (the dwarf floating aquatic aroid) both the flowers and the inflorescence have been severely reduced. There is a tiny **spathe** containing the evolutionary remains of a **spadix**: a single pistil representing a female flower and two stamens, each representing a male flower). In *Euphorbia*, the female flowers have been reduced to a single trilocular ovary and the male flowers to a single stamen. Revealingly, the anther stalk in *Euphorbia* bears a **joint** (seen as a slight ring-like constriction) indicating that the anther stalk consists of both a filament (above the joint) and a floral axis (below the joint).

A rather different form of reduction occurs in many genera of tropical legumes, of which *Bauhinia* is a good example. Ten stamens are expected (as is usual in legumes). However, in *Bauhinia* there is a **reduction series**, and species occur with ten, five or three stamens, and there are several species with only a single stamen. Some of the species with a single stamen have a reduced number of petals also, down to a single petal in some cases (Wunderlin, 1983). These species have large showy flowers and the reduction in organ number is not due to dwarfism, or wind pollination, but rather due to complex coevolution with animal pollinators. The ginger family Zingiberaceae has similarly evolved single stamen flowers, in this case with the stamen intimately associated with the style. This is also a mechanism for highly precise animal pollination.

As organ loss is highly positionally specific, it is likely to involve targeted active suppression rather than loss of organogenetic potential. Suppression mutations would obviously be dominant, whereas loss of organogenetic potential would be recessive. In *Lepidium* (Brassicaceae), species with four-stamen flowers occur by loss of the two lateral stamens of the flower. Genetic analysis shows this to be a dominant mutation. Intriguingly, it has also been argued that speciation by allopolyploidy will spread this trait. If one parent has the dominant suppression gene then this trait will be incorporated stably in the derivative allopolyploid genome (Lee, Mummenhoff & Bowman, 2002).

In theory at least, loss of suppression should enable evolution to reverse and the missing organs to reappear. However, in lineages where reduction is deep-seated, such reversal is exceedingly rare; it seems that after a certain amount of evolutionary time, the ability to produce the missing organs deteriorates. It is apparent that there is often selective pressure for reversal to large flowers but this is instead accomplished by a developmental trick. Reduced flowers may be (and frequently are) aggregated into **pseudanthia** (literally "false flowers"). A familiar type of pseudanthium is the capitulum of the Asteraceae that brings together numerous small flowers into what is functionally a single unit. However, the small flowers (florets) of the Asteraceae are perfect. They are small by virtue of small size and not organ loss, and the pseudanthium is therefore obvious to interpret. Things are more difficult when flowers have been reduced to single stamens or pistils. The **cyathium** of *Euphorbia* with a single ovary at the center (female flower) surrounded by stamens (male flowers) looks very much like a single flower (and was interpreted as one by Linnaeus). However, there are clues that this is a pseudanthium, notably the cymose arrangement of the stamens and the joints on the anther stalks.

Other instances give rise to real (and often unresolved) debate. The flower of *Triglochin* has been interpreted both as pseudanthial (Charlton, 1981; Eames, 1961; Rudall, 2003) and euanthial (Lieu, 1979). Similarly the "inside-out" (**basigynal**) flowers of *Lacandonia* have a possible pseudanthial interpretation (Rudall, 2003) but also admit of a euanthial one (Ambrose et al., 2006; Rudall & Bateman, 2006). Of relevance in *Lacandonia*, many pseudanthia, such as the spadices of Araceae, are basigynal, as opposed to the almost universally **acrogynal** nature of the flower.

The question may be raised as to whether molecular tests will help resolve the euanthial–pseudanthial debate. This may be difficult. When flowers are reduced to single stamens and carpels, their molecular signature may be similarly reduced. Bringing such flowers together in a single apex might have the effect of reassembling the molecular signature of single flower. Alternatively the existence of a flower-like pseudanthium might result from the continued existence of molecular-developmental pathways capable of developing floral organization, under the control of which single organ flowers might fall to be organized. Such floral control pathways might include the floral meristem identity genes, such as *LFY* and *APETALA1/CAULIFLOWER* (*AP1/CAL*; Lamb et al., 2002; Sablowski, 2007), or *UFO* (Hepworth, Klenz & Haughn, 2006).

From either way of looking, that is, molecular-developmental reassembly of floral developmental patterns or a florally organized inflorescence, one could argue that there is partial homology, at the molecular level, established between pseudanthia and flowers, and that these pseudanthia are in a **refloral state**. It has also been suggested that some monocot pseudanthia represent not a reassembly from reduced flowers but instead a breakdown of the inflorescence/flower identity boundary, this identity boundary perhaps not having been established very strongly anyway, in contrast to eudicots (Rudall, 2003). The fact that some floral meristem identity genes, such as *LFY*, also function as inflorescence identity genes gives a possible mechanism by which the boundary might be dissolved. Such flowers would be in a **quasifloral state**. Further understanding awaits detailed molecular studies of pseudanthia.

6.31 The inflorescence: phase change in the SAM

Plants, particularly trees, may have a long juvenile phase and only flower when a certain level of maturity is reached. This may mean only on branches at the top of tall trees or in the case of rosette plants, only when the rosette reaches a certain size and age. A phase change happens which either converts preexisting SAMs to reproductive mode or allows new lateral SAMs to be produced that are reproductive rather than vegetative. The "green fuse" that drives the flower is the **florigen**, the protein of *FLOWERING LOCUS T* (*FT*) that is transported from leaves, through the phloem, to the shoot apex (Corbesier et al., 2007). Once at the shoot apex, FT stimulates floral meristem genes such as *APETALA1* in an effect mediated by bZIP protein FD (Wigge et al., 2005).

The conversion of a vegetative SAM to reproductive mode, as terminal inflorescences or flowers, generally terminates vegetative growth. Mostly, the inflorescence terminates the shoot either by a terminal flower being produced or by abortion of the SAM after a certain number of lateral flowers or inflorescences have been produced. However, in a few cases there is grow-through of a terminal inflorescence or even grow-through of the terminal flower. Examples of inflorescence grow-through are provided by many species of Myrtaceae including the genus *Callistemon*. Bands of flower production may be seen on a single branch. Another example is pineapple (*Ananas*): in this the inflorescence continues to produce foliage leaves after flower production has ceased. This can be seen as the leafy top on pineapples. That the SAM is still fully functional is evidenced by the fact that pineapples can be propagated from these tops.

Grow-through of flowers is rare but the terminal flower of *Impatiens* can be manipulated through environmental changes to revert from a floral to a vegetative SAM (Battey & Lyndon, 1984). Another example is Goethe's proliferous rose, which had

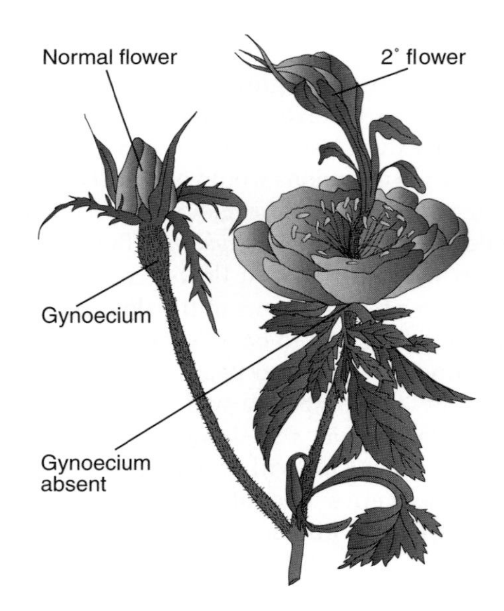

Fig. 6.14. Goethe's proliferous rose (after Candolle, 1827).

a stem bearing another flower issuing from the center of a lower flower (Fig. 6.14). As Goethe himself describes it: "Calyx and corolla are arranged and developed around the axis; however, the seed container in the center is not contracted nor are the male and female sexual parts arranged on and around it in ordered sequence; instead, the stalk shoots upward again" (Goethe, 1790). The phenomenon of reversion has been studied extensively in *Impatiens*, where there seems to be an innate susceptibility to reversion, probably resulting from the developmental control of flower formation in this species. It has also been shown that *Impatiens*, unusually, requires a continuous leaf-produced signal to complete flowering (Tooke & Battey, 2000). The molecular basis of the "lack of commitment" in the *Impatiens* floral meristem is still obscure (Ordidge et al., 2005; Tooke et al., 2005; Fig. 15).

6.32 Inflorescences

The distribution of flowers (or other strobili) on the plant, and relative to each other, shows great evolutionary variation and is morphologically interesting. Flowers are often produced in groups called inflorescences. There is a fundamental morphological difference between two main types of inflorescence, cymose and racemose. In racemose inflorescences flowers are produced laterally on a main stem, which continues growing. A raceme is therefore monopodial and indeterminate. It should be noted, however, that although a raceme is in theory indeterminate it is, in practice, determinate by abortion of the SAM after a certain number of flowers are produced. Furthermore there are occasional examples of a racemose inflorescence terminating in a flower as, for example, in *Berberis*, *Rubus* and *Prunus laurocerasus* (Fig. 6.16).

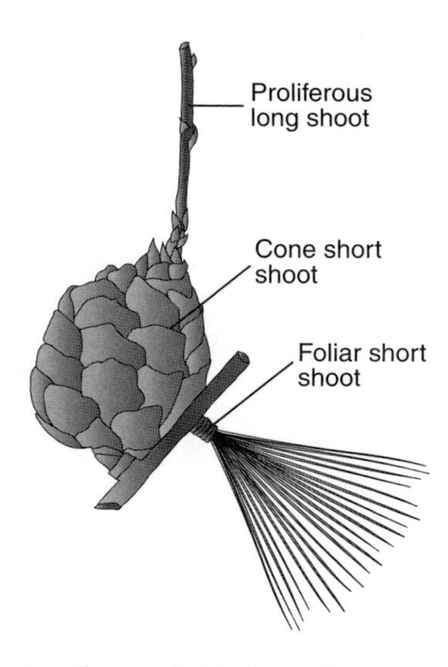

Fig. 6.15. A conifer equivalent to the proliferous rose: a proliferous cone in larch (*Larix*) (after Candolle, 1827).

A cyme on the other hand is an inflorescence in which early in development the main axis terminates by producing a single flower. Further flower production can only occur by branching beneath this flower. These branches terminate in a flower and further branches may again be produced below, and so on. This is therefore a sympodial structure. In racemes the flowers at the bottom open first followed successively by flowers higher up the stem. This development is therefore acropetal. In cymes, in contrast, the central flower opens first followed by lateral flowers in the successive orders of branching. Flower opening therefore often appears to be centrifugal.

6.33 Molecular basis of determinacy in the inflorescence

Some plants, such as *Magnolia*, bulbous monocots such as tulips (*Tulipa*) and some eudicots such as

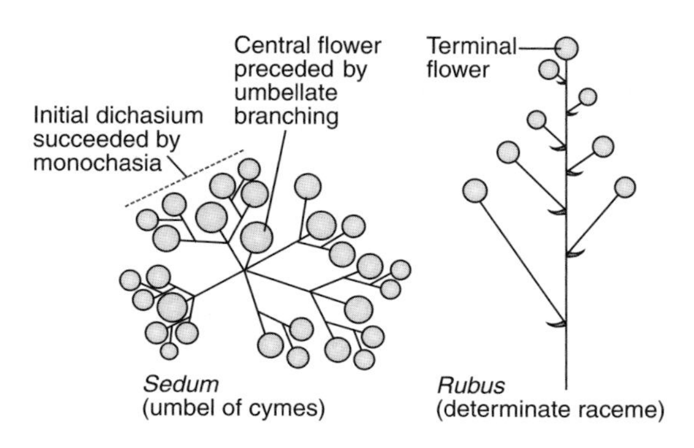

Fig. 6.16. Types of inflorescences. *Rubus* departs from typical racemes by having a terminal flower. *Sedum* shows a mixed (thyrsoid) inflorescence (after Rickett, 1955).

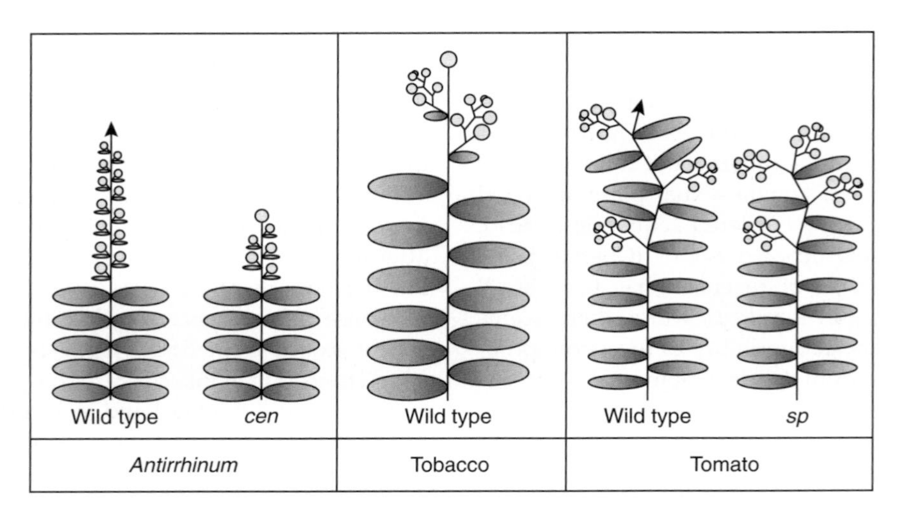

Fig. 6.17. Changes in inflorescence architecture due to loss-of-function mutation in *CENTRORADIALIS* or its homologues. The mutant phenotype is more determinate. The wild-type tobaccosomewhat resembles the *cen* mutation of *Antirrhinum* (after Amaya et al., 1999).

Papaver somniferum, terminate their vegetative shoots with a single **terminal flower**. This can be seen as a poor use of a SAM. **Terminal inflorescences** are common: in these, although the vegetative shoot is used up, some indeterminacy is maintained so that numerous lateral flowers can be thrown off. Examples are *Aloe, Agave, Digitalis, Helianthus, Fragaria, Rubus* and the Brassicaceae, all **racemose inflorescences**. If determinacy is weaker still, the plant can "grow-through" the inflorescence and reestablish vegetative shoot identity, as in *Callistemon*. Finally, some plants can enter the reproductive phase with no determinacy of the vegetative shoot, merely throwing off lateral flowers while maintaining vegetative shoot identity, as in *Galanthus*. Lateral flowers and inflorescences will be discussed below.

These mechanisms can be seen as a continuum of varying determinacy in flowering, from absolute (converting a leaf-producing vegetative shoot into a single flower) to none (maintaining vegetative shoot identity while producing lateral flowers). All plants reside somewhere along this continuum. This continuum has been called "vegetativeness" (veg), and associated with two meristem genes: *TERMINAL FLOWER 1* (*TFL1*), which increases veg, and *LFY*, which reduces veg (Prusinkiewicz et al., 2007). Using a model assuming different timing and magnitude of expression, a variety of inflorescence types, mapping to real world types, can be generated (Prusinkiewicz et al., 2007; Fig. 6.17).

The relationship between *TFL1* (*CENTRORADIALIS* in *Antirrhinum*) and *LFY* (*FLORICAULA* in *Antirrhinum*) is antagonistic (Liljegren et al., 1999; Ratcliffe, Bradley & Coen, 1999). This antagonism functions to exclude the activity of floral identity genes from the inflorescence shoot apex and to exclude indeterminacy from floral apices (Amaya, Ratcliffe & Bradley, 1999; Ratcliffe, Bradley & Coen, 1999). This is critical in setting a clear distinction between flower and inflorescence. A failure of this mechanism, leading to strong coexpression of both *LFY* and *TFL1* might lead to an indeterminate organ somewhere between a flower and an inflorescence. Such structures are not found in nature, except perhaps for the flower-like terminal structures (FLTS) in the racemes of the Piperales and some other plants (Sokoloff, Rudall & Remizowa, 2006).

Despite the antagonism, there appears to be some important cross talk between LFY and TFL1 in patterning the inflorescence. The TFL1 protein is mobile and spreads throughout the inflorescence shoot meristem, even when the

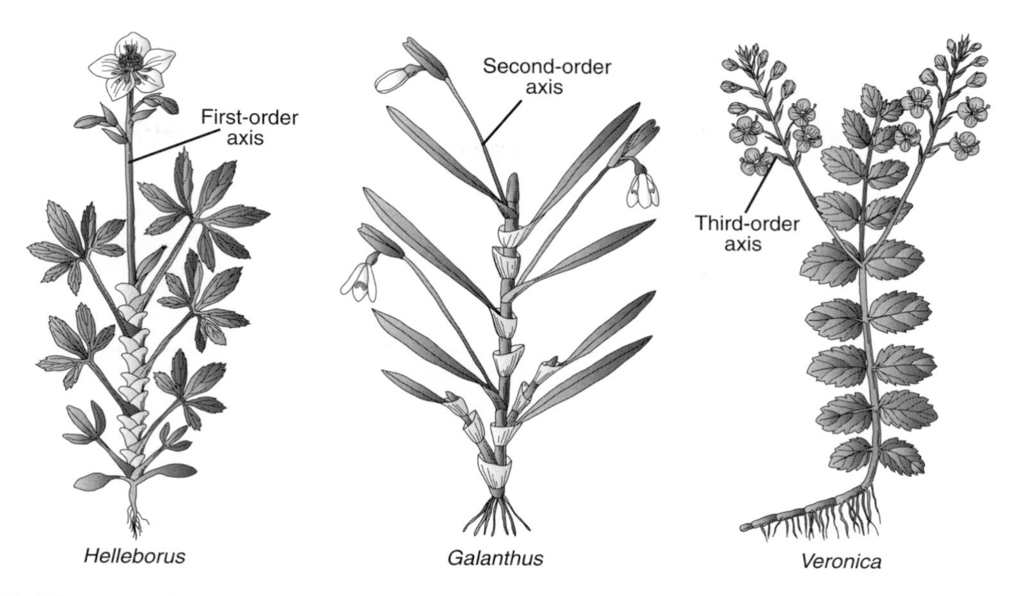

Fig. 6.18. Inflorescences relative to orders of branching in the plant (after Moll).

gene is expressed only in the center. Surprisingly this mobility is stopped in *lfy* mutants, suggesting that LFY expressing floral meristems signal to the vegetative inflorescence meristem (Conti & Bradley, 2007).

6.34 Terminal and lateral inflorescences

A single terminal flower is the ultimate in determinacy. However, there is a further twist for those plants producing a single terminal flower: another way of escaping from absolute determinacy. They may create a "compound flower" by branching from the prophylls below the terminal flower. This is a **cymose inflorescence**, and a good way of drawing out more flowers from a terminal flower system (Fig. 6.18). The various different ways of compounding the terminal flower, leading to many different types of **cyme**, will be discussed in another section.

In plants with terminal inflorescences, vegetative growth ends when flowering occurs. Further growth can only occur sympodially, by the growth of axillary buds below the inflorescence. If this does not occur, growth is monocarpic, and the plants die after flowering. This occurs in many annuals and in biennials and hapaxanthic plants, like the century plant (*Agave*). An alternative to terminal inflorescences is to mask the main shoots from the reproductive transition and produce lateral inflorescences. This masking is carried out through the *action of genes* such as *AtTERMINAL FLOWER 2* (*TFL2*). *TFL2* is the *Arabidopsis* homologue of the animal/yeast chromodomain protein gene *HETEROCHROMATIN PROTEIN1/SWITCHING GENE6* (*HP1/SWI6*). Curiously, in *Arabidopsis*, it targets euchromatic genes and is capable of silencing B-class, C-class and E-class MADS-box floral identity genes (Nakahigashi et al., 2005). It is a powerful antiflower gene.

The advantage of the **lateral inflorescences** is that the main axis can continue growing uninterruptedly, so growth can be monopodial. Lateral inflorescences are very common, as for instance in the Leguminosae: *Vicia*, *Trifolium*, *Lotus*, *Medicago*, and so on. Some plants have **scattered lateral flowers**, in which flowers are produced in the axils of foliage leaves and not in any specific region of the stem that can be identified as an inflorescence. These **solitary** flowers may be considered developmentally separate from the lateral flowers that form in an inflorescence in the axils of bracts. An extreme case of lateral flowering

Table 6.5 Shape-based classification of racemose inflorescences.

	Flat-topped	Elongate or pyramidal	Narrow
Shape determined by differential growth of pedicels and internodes	**Corymb** (flat-topped by differential growth growth); **Umbel** (as corymb but branching whorled)	**Raceme** (pedicels moderately short, flowers well-spaced by internodes, giving cylindrical inflorescences); **Panicle** (pyramidal inflorescence)	**Spike** (pedicels and internodes usually short, giving a compact, narrowly cylindrical inflorescence)
Shape determined by swelling of the peduncle	**Capitulum** (of Asteraceae); **Coenanthium** (e.g. *Dorstenia*, flowers set on a flat or dish-shaped swollen peduncle)		**Spadix** (flowers set in a swollen club-shaped peduncle (Araceae))

Note: the inflorescence of *Dorstenia* is, strictly speaking, not racemose but thyrsoid.

is provided by the phenomenon of cauliflory. In this, flowers or inflorescences are produced from slow developing or dormant reproductive axillary buds in old stems or even in the trunks of trees. Examples include *Theobroma* (cocoa) and *Cercis siliquastrum* (Judas tree).

Many families contain both lateral and terminal flowering species. In the Arecaceae, the date palm (*Phoenix*) is lateral flowering, while palms like *Corypha umbraculifera* are monocarpic, dying after flowering as the single apical meristem is entirely used up in inflorescence production.[6] In the Orchidaceae many orchids such as *Vanilla* and *Epipactis* are terminal flowering while others like *Phaius* are lateral. In the Musaceae, the banana (*Musa*) is terminal and the banana stem dies after flowering. However, in the banana relative, the traveller's palm (*Ravenala*), the inflorescences are lateral allowing the main stem to continue growing monopodially to produce a tall trunk.

6.35 Types of racemose inflorescence

Racemose inflorescences may be simple or compound. **Simple racemes** usually take the form of a loose elongated inflorescence with well-developed pedicels on the flowers. However, if the **pedicels** or **internodes** are exceptionally developed or undeveloped in different parts of the inflorescences, different shapes are produced which are distinguished by a variety of names (Table 6.5). Some insight into how raceme shape may evolve is gained from mutants in *Arabidopsis*. Wild-type *Arabidopsis* inflorescences are conspicuously not flat-topped (**corymbose**) as, first, early elongation of the pedicels puts the young flowers above the central buds, and, second, elongation of the older inflorescence internodes elevates the flowers and buds above the level of the older flowers and developing fruit. Mutations in the *ERECTA* (*ER*) gene give a flat-topped tendency as *er* plants are negatively affected in pedicel growth, so preventing the flowers from rising above the buds. Alternatively, mutations in the *CORYMBOSA2* (*CRM2*) gene negatively affect internode growth and so prevent the young flowers and buds from overtopping the spent ones (Suzuki, Takahashi & Komeda, 2002). It is easy to see how genes that affect pedicel and internode length in inflorescences (such as *ER* and *CRM2* and probably others) could together produce an array of raceme shapes, including naturally corymbose inflorescences such as those of *Iberis*. Another important factor is inflorescence branching, which is, for instance, important in the development of the corymbose compound raceme of the elderberry, *Sambucus nigra*.

Raceme shapes are of considerable ecological importance. Corymbose inflorescences provide a flat landing pad for insects such as flies and butterflies that like to forage vertically into flowers. Insects such as bees that commonly have a

[6] *Corypha umbraculifera* (Talipot palm) has the largest inflorescence of any plant. Observations made by H. F. Macmillan in Peradeniya in 1899 estimated the number of flowers in a single large inflorescence to reach 60 million.

horizontal approach into flowers are suited by laterally placed flowers on an elongated raceme, such as those of *Antirrhinum*.

Many grasses have spike-like inflorescences, but these are compound inflorescences (spikes of spikes) as the spikelet is itself a miniature spike. The term **catkin** is often used for the characteristic (usually unisexual and pendulous) spike-like inflorescences of monoecious and dioecious trees. However, when these are examined closely, they are sometimes found to be derived from compound inflorescences, as in the Betulaceae.

Heads of flowers (**capitula**) are usually very compact, as in the Asteraceae, Campanulaceae and Dipsacaceae. The shape of a capitulum is strongly influenced by the shape of the peduncle (the **inflorescence receptacle**). The inflorescence of the Asteraceae is particularly interesting for the division of labor that frequently occurs. Many Asteraceae have evolved dimorphic flowers, placing attractive male-sterile, zygomorphic flowers (**ligulate florets** or **ray florets**), with an attractive petaloid outgrowth on the abaxial side, at the margin of the capitulum. Smaller less showy, actinomorphic, hermaphrodite flowers (**disc florets**) occupy the center.

Where the peduncle itself is the dominant organ in the form of an extensive plate or cup (*Dorsteinia*, *Antiaris*) the term **coenanthium** is used. The extreme of this is reached with the fig-inflorescence (**syconium**) of *Ficus*. The term "head" is also used for the **loose head** (of *Trifolium*, for instance, or the heads of male strobili in *Cephalotaxus fortunei*), for which the term **glomerulus** is better.

Compound racemes produce lateral racemes (**racemules**), of the same type as the main raceme. There may be one or more iterations of such branching. Compound spikes (*Lolium*, *Gunnera*), compound umbels (*Daucus*) or compound heads (*Echinops*). Alternatively there may be a change of raceme type thus *Hedera* has racemes of umbels. *Avena* has a compound raceme of spikelets.

It should be noted that it is sometimes difficult to draw the line between inflorescences and inflorescence branches in compound racemose

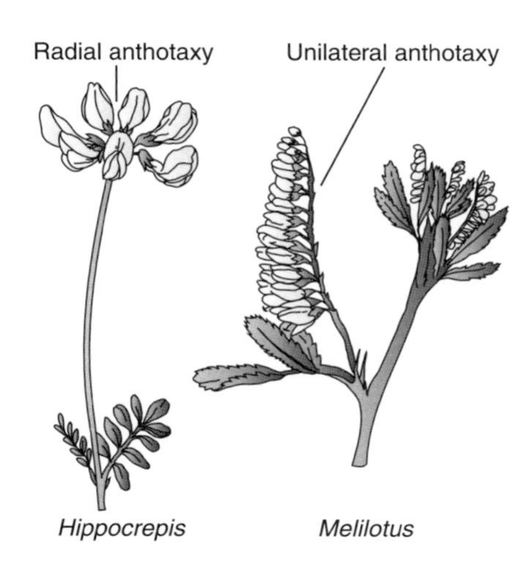

Fig. 6.19. Different positioning of the flowers (anthotaxy) leading to functionally different racemose inflorescence shapes in two leguminous examples.

inflorescences, particularly those that are loosely organized. The basal inflorescence branches may be large and they may reiterate the main inflorescence axis (and are often subtended by foliage-like leaves too). Although they are associated with the main inflorescence they have many properties of independent inflorescences rather than inflorescence branches. The term **coflorescence** has been usefully used for these. In *Arabidopsis* the inflorescence gives rise to two types of lateral axis: (1) flowers (lateral axes that terminate in a single flower) that are not subtended by a bract and (2) coflorescences (lateral axes which recapitulate the main inflorescence) that are subtended by a bract (Fig. 6.19).

Racemes often do not end in a terminal flower, instead the inflorescence meristem just peters out. However, in many cases they do end in a terminal flower, as in *Rubus*. In actinomorphic species the terminal flower is normal (or slightly enlarged) but in zygomorphic species the terminal flower is actinomorphic (for the reasons discussed in the section on floral symmetry). In many cases terminal flowers are produced as terata (Rudall & Bateman, 2003), as in *Digitalis*, *Campanula* and *Antirrhinum*. In *Antirrhinum* the terminal flower terata have been shown to result from *cen* mutations (Bradley et al., 1996).

6.36 Types of cymes

Cymes are determinate inflorescences that branch below the terminal flower, leading to more flowers, and so on. The amount of branching below the terminal flower follows the number of bracteoles (prophylls). A single bracteole below flowers leads to a single branch at any one branch-point (**monochasium**). Alternatively, there may be two bracteoles and two branches at any branch-point (**dichasium**) or even three (**trichasium**) as in thyrsoid inflorescences such as *Eriogonum*.

The monochasium, in particular, is the source of fascinating variation depending on the pattern of branching (Fig. 6.20) created as it interacts with the axes of polarity within the inflorescence and the plant (Buys & Hilger, 2003). Branches may be produced abaxially (**sickle** or **drepanium**), as in the Juncaceae (or as an analogue in the segmentation of the *Amorphophallus* leaf). Alternatively, branches may be produced adaxially (**rhipidium**), or laterally (**screw** or **bostryx**), either laterally left (**sinistral screw**) or right (**dextral screw**). Branches may also be produced alternatively left and right (**cincinnus**). The cincinnus is a cyme familiar to all, being the inflorescence predominant in the Boraginaceae (e.g. *Heliotropium*, *Cordia*, *Symphytum*). The cincinnus may look superficially like a raceme but it can be recognized by usually being rolled up (scorpioid), at least at the top, and the flowers being in two rows (as a result of the alternate left and right branching).

Two special cymes should be mentioned as they are enormously important in two major families. The first is the **verticillaster**, common in the Lamiaceae. In this, two dichasia are produced in the axils of opposite leaves that then form an encircling ring of flowers around the

Types of monochasial cyme			
name	diagram	branching type	example
sickle or drepanium	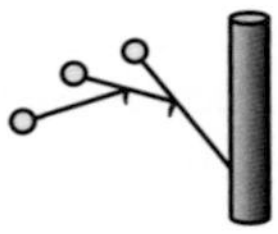	abaxial	*Juncus*
fan or rhipidium	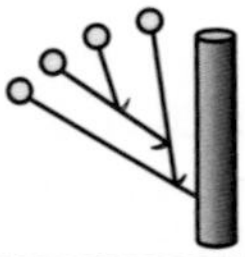	adaxial	*Iris*
screw or bostryx		dextral	*Hemerocallis*
	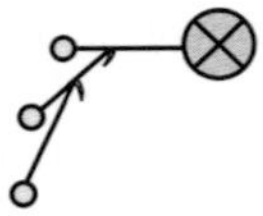	sinistral	
cincinnus		alternating	Boraginaceae

Fig. 6.20. Types of monochasial cyme. It should be noted that the drepanium has additional bracts and is best considered a cyme-like monochasium rather than a cyme in the strict sense.

stem, superficially resembling a **whorl** (**verticil**) of flowers. The other is the **cyathium**, a very interesting thyrsoid inflorescence characteristic of the genus *Euphorbia*. It has a central female flower surrounded by five bracts forming an **involucre**, in the axils of which are cincinni of male flowers.

In addition, it should be noted that inflorescences are frequently of a mixed type, even mixtures of cymes and racemes. In such **mixed inflorescences** the branching pattern of the primary inflorescence unit is different from that of the secondary inflorescence unit. A raceme of cymes, one of the commonest of the mixed inflorescence patterns, is given its own name, a **thyrse** (Table 6.6).

6.37 The leaves of inflorescences

Although the peduncles (inflorescence stalks) and pedicels (flower stalks) of some inflorescences have leaves resembling foliage leaves this is not usual. More often the leaves of inflorescences are reduced to bract-like organs, which may be **bracts** (subtending inflorescence branches) and **bracteoles** (small bracts subtending or preceding flowers). Furthermore, there may be **sterile bracts** (subtending nothing) on the peduncle and pedicel respectively. The leaf or leaves at the first node of a branch are often referred to as **prophylls**, especially if they have a distinctive form or position. In species where there is one leaf on the pedicel (which is common) this will be a bracteole (as it precedes the flower) but it will also be a prophyll (as it is at the first node on a branch).

The reduction of normal foliage leaves to bracts in an inflorescence is a consequence of, and a marker of, the major shoot identity shift (from vegetative to reproductive) on transition to flowering. Generally, leaves of some sort are essential on an inflorescence as all inflorescence branches spring from the axils of these leaves. It is likely that the presence of bracts in inflorescences is primitive. Many basal angiosperms, such as *Amborella* (Endress & Igersheim, 2000), are well endowed with bracts, which probably perform a protective function.

Some of the most distinctive inflorescence leaves (particularly in the monocots) are those that are specialized for an obvious protective function. These leaves have their own names (see Table 6.7). It should be noted that the epicalyx, a protective ring of bracteoles subtending the flower, and particularly common in the Malvaceae, indicates a multiflowered inflorescence unit reduced to a single flower, as detailed in Section 6.17.

Table 6.6 Mixed inflorescences.

Primary unit	Secondary unit				
	Heads	Umbels	Dichasia	Screws	Cincinni
Raceme of	*Solidago*				*Aesculus*
Spike of			*Betula, Alnus*		
Head of				*Haemanthus*	*Armeria*
Screw of	*Cichorium*	*Caucalis, Chelidonium*			
Sickle of	*Juncus*				

Table 6.7 Protective leaves in the inflorescence.

Part enclosed	Single leaf	Multiple leaves
Flower	Utricle (Cyperaceae)	Epicalyx or involucel (Malvaceae)
Partial inflorescence	Spathilla (Arecaceae)	Glumes (Poaceae)
Whole inflorescence	Spathe (Araceae)	Involucre (Asteraceae, Dipsacaceae)

However, there is a negative side to having bracts, particularly in compact or evolutionarily reduced inflorescences. The bract primordia can crowd the developing perianth primordia causing developmental abnormalities including chimaeric organs (Masiero et al., 2004). Probably for this reason some plants suppress all or some of these leaves and the inflorescence may even appear to be naked. *Arabidopsis* has lost bracts in the inflorescence, and something is known of the molecular events leading to bract suppression in this plant.

Rather than loss by a loss-of-function (recessive) mutation, the loss of *Arabidopsis* bracts is a case of active suppression as a result of a gain of function (dominant) mutation in a suppressor gene. Despite the absence of bracts in A*rabidopsis*, there is vestigial expression of bract-specific genes where bracts would be expected. These vestiges have been called "cryptic bracts." The leaf developmental gene *JAGGED* is required to form bracts (Dinneny et al., 2004) and this is excluded from the inflorescence of *Arabidopsis* by the action of the *BLADE ON PETIOLE* (*BOP*) genes acting in concert with the developmental gene *LFY* (Norberg, Holmlund & Nilsson, 2005). With this system it can be seen that bract can easily return, through loss-of-function mutations in the suppression gene network (desuppression). As the suppressed gene, *JAGGED*, has an important role in leaf development elsewhere in the plant, there is no likelihood that it will degenerate, making the reoccurrence of bracts impossible.

A rather different mechanism appears to have been recruited for the suppression of bracteoles in *Antirrhinum*. Here the MADS-box gene *INCOMPOSITA* (*INCO*) acts to suppress bracts. The loss-of-function mutation of *INCO* shows the "reemergence" of bracteoles (Masiero et al., 2004).

It might be asked why in both these cases bracts are lost by active suppression (reversible by loss of suppression) rather than loss of gene function (probably irreversible). The most likely explanation is that, as bracts are leaves, any loss of function will lead to undesirable effects on other leaves, whereas targeted suppression does not. A consequence of the inherent reversibility of bract loss means that this character can potentially "flicker" on and off in evolutionary time (Endress, 2001a). Loss and gain of bracts are of frequent occurrence in nature. The effect is particularly noticeable in the Apiaceae: parsnip, fennel and dill (*Pastinaca, Foeniculum, Anethum*) have no significant bracts, yet in carrot, hemlock and celery (*Daucus, Conium, Apium*) they are obvious.

6.38 Leaves of grass partial inflorescences

Grasses (Poaceae) have numerous specialized leaves on their partial inflorescences. These partial inflorescences are called **spikelets**, and in common parlance the bract-like leaves they bear are called **chaff**. There are two leaves called **glumes** at the base of the peduncle of the spikelet. These are both usually referred to as bract but more specifically one of these (the lower) is likely the bract of the spikelet and the other the first prophyll. The **palea** and **lemma** both appear to subtend the flower and have generally therefore been considered bracteoles. Under this interpretation, the perianth is reduced to lodicules (small lumps of tissue that swell to push back the lemma and palea and expose the sexual organs at anthesis).

The interpretation of lodicules as inner perianth is elegantly supported by molecular studies.

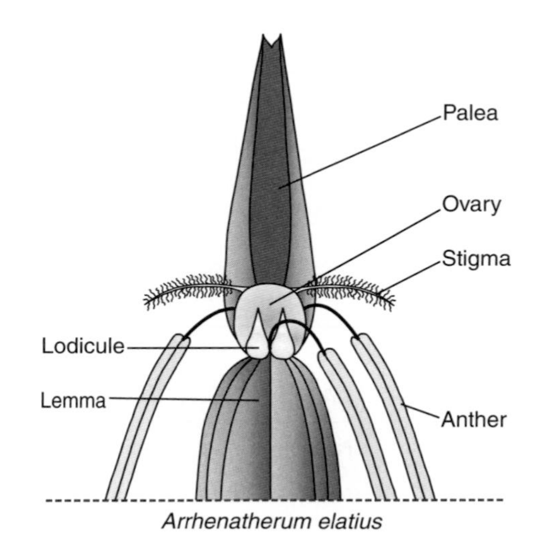

Fig. 6.21. Grass floret.

The lodicules are marked by B-class gene expression that is characteristic of the stamens and inner tepal whorl of monocots (Whipple et al., 2007), whether it be the petaloid inner tepal whorl of petaloid monocots, the bracteoid inner tepal whorl of the basal grass *Streptochaeta* (that does not have lodicules) or lodicules themselves (the inner tepal whorl of the majority of grasses; Fig. 6.21).

What then of the outer tepal whorl? Has it simply been lost? A recent interpretation suggests that the traditional designation of the lemma and palea as bracteole-derived organs may be incorrect. There is now some molecular support for the idea that the lemma and palea are derived from the "missing" outer tepal whorl (Preston & Kellogg, 2007).

It has been proposed (Preston & Kellogg, 2007) that the monocot A-class gene *FRUITFULL2* (*FUL2*) interacts with the *SEP*-like gene, *LEAFY HULL STERILE1* (*LHS1*) to specify the identity of the outermost floral organs, and that subsequent evolutionary change in *FUL2* action has led to the modification of sepals (i.e. outer tepals) to form the bract-like lemma and palea. In grasses, *FUL1* and *FUL2* are the duplicated paralogues of the *Arabidopsis AP1/FUL* genes, and expression studies (Preston & Kellogg, 2007) indicate that whereas *FUL1* has a general role in floral organ identity, *FUL2* has a more specific role in floral outer whorl identity. If *FUL2* expression is taken to reveal the outer tepal nature of palea and lemma, then it is reasonable to suggest that the palea is actually derived from the connation of the two adaxial tepals of the outer whorl and, further, that the lemma may represent the abaxial tepal.

This book started with minute Silurian land plants, it ends here at the adaxial tepal of a grass. In between, sketched so briefly and superficially, lies the 400 million year upward struggle of the embryophytes. This struggle that has made and remade the terrestrial ecosphere so many times. It is also the story of genetic inheritance that is constantly being duplicated, co-opted, changed and, has created the expanding morphological possibilities. This energetic brew of genes has propelled some disc of green cells barely covering a penny, such as is today found in *Coleochaete*, to the giant fig that sheltered Alexander's army of 7,000 under limbs that stretch not only to the sky, but deep into human culture.[7]

[7] Alexander the Great, in his conquest of the Punjab in 326 BCE, is said to have found this huge tree, probably *Ficus bengalensis*, covering a huge area by means of prop roots. The notes made by his scientists were used by Theophrastus, who discussed the morphological nature of the prop roots.

References

Abraham, A. & Mathew, P. M. (1963). Cytology of coconut endosperm. *Annals of Botany* **27**, 505–512.

Adams, K. L. & Wendel, J. F. (2005). Polyploidy and genome evolution in plants. *Current Opinion in Plant Biology* **8**, 135–141.

Ahlers, F., Bubert, H., Steuernagel, S. & Wiermann, R. (2000). The nature of oxygen in sporopollenin from the pollen of *Typha angustifolia* L. *Zeitschrift Fur Naturforschung C-A Journal of Biosciences* **55**, 129–136.

Ahlers, F., Thom, I., Lambert, J., Kuckuk, R. & Wiermann, R. (1999). H-1 NMR analysis of sporopollenin from *Typha angustifolia*. *Phytochemistry* **50**, 1095–1098.

Aida, M., Beis, D., Heidstra, R., Willemsen, V., Blilou, I., Galinha, C. et al. (2004). The *PLETHORA* genes mediate patterning of the *Arabidopsis* root stem cell niche. *Cell* **119**, 109–120.

Aida, M., Beis, D. & Scheres, B. (2004). PLETHORA1 and PLETHORA2 are involved in the formation and maintenance of *Arabidopsis* root stem cells. *Plant and Cell Physiology* **45**, S16.

Aida, M., Ishida, T. & Tasaka, M. (1999). Shoot apical meristem and cotyledon formation during *Arabidopsis* embryogenesis: interaction among the *CUP-SHAPED COTYLEDON* and *SHOOT MERISTEMLESS* genes. *Development* **126**, 1563–1570.

Aida, M. & Tasaka, M. (2006a). Genetic control of shoot organ boundaries. *Current Opinion in Plant Biology* **9**, 72–77.

Aida, M. & Tasaka, M. (2006b). Morphogenesis and patterning at the organ boundaries in the higher plant shoot apex. *Plant Molecular Biology* **60**, 915–928.

Albert, V. A., Gustafsson, M. H. G. & Di Laurenzio, L. (1998). Ontogenetic systematics, molecular developmental genetics, and the angiosperm petal. In *Molecular Systematics of Plants. II. DNA Sequencing* (eds D. E. Soltis, P. S. Soltis and J. J. Doyle), pp. 349–374. Kluwer Academic Publishers, Boston.

Albertini, E., Marconi, G., Reale, L., Barcaccia, G., Porceddu, A., Ferranti, F. et al. (2005). *SERK* and *APOSTART*. Candidate genes for apomixis in *Poa pratensis*. *Plant Physiology* **138**, 2185–2199.

Albrecht, C., Russinova, E., Hecht, V., Baaijens, E. & de Vries, S. (2005). The *Arabidopsis thaliana* SOMATIC EMBRYOGENESIS RECEPTOR-LIKE KINASES1 and 2 control male sporogenesis. *Plant Cell* **17**, 3337–3349.

Albrecht, H., Yoder, J. I. & Phillips, D. A. (1999). Flavonoids promote haustoria formation in the root parasite *Triphysaria versicolor*. *Plant Physiology* **119**, 585–591.

Allen, J. & Martin, W. (2007). Out of thin air. *Nature* **445**, 610–612.

Almeida, J., Rocheta, M. & Galego, L. (1997). Genetic control of flower shape in *Antirrhinum majus*. *Development* **124**, 1387–1392.

Altesor, A., Silva, C. & Ezcurra, E. (1994). Allometric neoteny and the evolution of succulence in cacti. *Botanical Journal of the Linnean Society* **114**, 283–292.

Amaya, I., Ratcliffe, O. J. & Bradley, D. J. (1999). Expression of *CENTRORADIALIS* (*CEN*) and *CEN*-like genes in tobacco reveals a conserved mechanism controlling phase change in diverse species. *Plant Cell* **11**, 1405–1417.

Ambrose, B. A., Espinosa-Matias, S., Vazquez-Santana, S., Vergara-Silva, F., Martinez, E., Marquez-Guzman, J. et al. (2006). Comparative developmental series of the Mexican triurids support a euanthial interpretation for the unusual reproductive axes of *Lacandonia schismatica* (Triuridaceae). *American Journal of Botany* **93**, 15–35.

Ambrose, B. A., Lerner, D. R., Ciceri, P., Padilla, C. M., Yanofsky, M. F. & Schmidt, R. J. (2000). Molecular and genetic analyses of the *silky1* gene reveal conservation in floral organ specification between eudicots and monocots. *Molecular Cell* **5**, 569–579.

Amor, B. B., Shaw, S. L., Oldroyd, G. E., Maillet, F., Penmetsa, R. V., Cook, D. et al. (2003). The NFP locus of *Medicago truncatula* controls an early step of Nod factor signal transduction upstream of a rapid calcium flux and root hair deformation. *Plant Journal* **34**, 495–506.

An, H., Roussot, C., Suárez-López, P., Corbesier, L., Vincent, C., Piñeiro, M. et al. (2004). CONSTANS acts in the phloem to regulate a systemic signal that induces photoperiodic flowering of *Arabidopsis*. *Development* **131**, 3615–3626.

Anastasiou, E., Kenz, S., Gerstung, M., MacLean, D., Timmer, J., Fleck, C. et al. (2007). Control of plant organ size by KLUH/CYP78A5-dependent intercellular signaling. *Developmental Cell* **13**, 843–856.

Ane, J. M., Kiss, G. B., Riely, B. K., Penmetsa, R. V., Oldroyd, G. E., Ayax, C. et al. (2004). *Medicago truncatula* DMI1 required for bacterial and fungal symbioses in legumes. *Science* **303**, 1364–1367.

Arber, A. (1918). The phyllode theory of the monocotyledonous leaf, with special reference to anatomical evidence. *Annals of Botany*, **32**, 465–501.

Arber, A. (1922). On the nature of the "blade" in certain monocotyledonous leaves. *Annals of Botany*, **36**, 329–351

Arber, A. (1925). *Monocotyledons*. Cambridge University Press, Cambridge. (Reprint, 1961; J. Cramer, Weinheim.)

Arber, A. (1931). Studies in floral morphology. II. Some normal and abnormal crucifers: with a discussion on teratology and atavism. *New Phytologist* **30**, 172–203.

Arber, A. (1950). *The Natural Philosophy of Plant Form*. Cambridge University Press, Cambridge.

Arber, E. A. N. & Parkin, J. (1907). On the origin of angiosperms. *Journal of the Linnean Society* **38**, 29–80.

Arber, E. A. N. & Parkin, J. (1908). The relationship of the angiosperms to the Gnetales. *Annals of Botany* **22**, 489–515.

Atsatt, P. R. (1973). Parasitic flowering plants: how did they evolve? *American Naturalist* **107**, 502–510.

Avery, G. S. (1933). Structure and development of tobacco leaves. *American Journal of Botany* **20**, 565–592.

Axtell, M. J. & Bartel, D. P. (2005). Antiquity of microRNAs and their targets in land plants. *Plant Cell* **17**, 1658–1673.

Bago, B., Pfeffer, P. E. & Shachar-Hill, Y. (2000). Carbon metabolism and transport in arbuscular mycorrhizas. *Plant Physiology* **124**, 949–958.

Bailey, I. W. & Smith, A. C. (1942). Degeneriaceae, a new family of flowering plants from Fiji. *Journal of the Arnold Arboretum* **30**, 64–70.

Bailey, I. W. & Swamy, B. G. L. (1949). The morphology and relationships of *Austrobaileya*. *Journal of the Arnold Arboretum* **30**, 211–226.

Bailey, I. W. & Swamy, B. G. L. (1951). The conduplicate carpel of dicotyledons and its initial trends of specialization. *American Journal of Botany* **38**, 373–379.

Balasubramanian, S. & Schneitz, K. (2002). *NOZZLE* links proximal–distal and adaxial–abaxial pattern formation during ovule development in *Arabidopsis thaliana*. *Development* **129**, 4291–4300.

Barcaccia, G., Arzenton, F., Sharbel, T. F., Varotto, S., Parrini, P. & Lucchin, M. (2006). Genetic diversity and reproductive biology in ecotypes of the facultative apomict *Hypericum perforatum* L. *Heredity* **96**, 322–334.

Barlow, P. W. (1995). Structure and function at the root apex—phylogenetic and ontogenetic perspectives on apical cells and quiescent centre. In *Structure and Function of Roots* (eds F. Baluska, M. Ciamporora, O. Gasparikova and P. W. Barlow), pp. 3–18. Kluwer, Dordrecht, The Netherlands.

Bateman, R. M. (1994). Evolutionary-developmental change in the growth architecture of fossil rhizomorphic lycopsids—scenarios constructed on cladistic foundations. *Biological Reviews of the Cambridge Philosophical Society* **69**, 527–597.

Bateman, R. M. (1996). Two steps forward, one step back: non-floral iteration and paedomorphosis in living and fossil land-plants. In *Homoplasy and the Evolutionary Process* (eds M. J. Sanderson and L. Hufford). Academic Press, London.

Bateman, R. M., Crane, P. R., DiMichele, W. A., Kenrick, P. R., Rowe, N. P., Speck, T. et al. (1998). Early evolution of land plants: phylogeny, physiology, and ecology of the primary terrestrial radiation. *Annual Reviews of Ecology and Systematics* **29**, 263–292.

Battey, N. H. & Lyndon, R. F. (1984). Changes in apical growth and phyllotaxis on flowering and reversion in *Impatiens balsamina* L. *Annals of Botany* **54**, 553–567.

Battjes, J., Chambers, K. L. & Bachmann, K. (1994). Evolution of microsporangium numbers in *Microseris* (Asteraceae, Lactuceae). *American Journal of Botany* **81**, 641–647.

Baum, D. A. & Donoghue, M. J. (2002). Transference of function, heterotopy, and the evolution of plant development. In *Developmental Genetics and Plant Evolution* (eds Q. C. B. Cronk, R. M. Bateman and J. Hawkins), pp. 52–69. Taylor & Francis, London.

Baum, D. A. & Hileman, L. C. (2006). A developmental genetic model for the origin of the flower. In *Flowering and its Manipulation* (ed. C. Ainsworth), pp. 3–27. Blackwell Publishing, Sheffield, UK.

Baum, H. (1952a). Die doppelspreitigen Petalen von Ranunculus auricomus und neapolitanus. *Österreichischen botanischen Zeitschrift* **99**, 65–77.

Baum, H. (1952b). Normale und inverse Unifazialität an den laubblättern von Codiaeum variegatum. *Österreichischen botanischen Zeitschrift* **99**, 421–451.

Baum, H. & Leinfellner, W. (1953). Die ontogenetischen Abänderungen des diplophyllen Grundbaues der Staubblätter. *Oesterreichische Botanische Zeitschrift* **100**, 91–135.

Baum, S. F., Dubrovsky, J. G. & Rost, T. L. (2002). Apical organization and maturation of the cortex and vascular cylinder in *Arabidopsis thaliana* (Brassicaceae) roots. *American Journal of Botany* **89**, 908–920.

Baum, S. F., Eshed, Y. & Bowman, J. L. (2001). The *Arabidopsis* nectary is an ABC-independent floral structure. *Development* **128**, 4657–4667.

Baumann, K., Perez-Rodriguez, M., Bradley, D., Venail, J., Bailey, P., Jin, H. L. et al. (2007). Control of cell and petal morphogenesis by R2R3 MYB transcription factors. *Development* **134**, 1691–1701.

Bäurle, I. & Laux, T. (2003). Apical meristems: the plant's fountain of youth. *BioEssays* **25**, 961–970.

Bayer, C. (1999). The bicolor unit—homology and transformation of an inflorescence structure unique to core Malvales. *Plant Systematics and Evolution* **214**, 187–198.

Beck, C. B., Schmid, R. & Rothwell, G. W. (1982). Stelar morphology and the primary vascular system of seed plants. *Botanical Review* **48**, 691–815.

Becker, A. & Theissen, G. (2003). The major clades of MADS-box genes and their role in the development and evolution of flowering plants. *Molecular Phylogenetics and Evolution* **29**, 464–489.

Becraft, P. W. & Asuncion-Crabb, Y. (2000). Positional cues specify and maintain aleurone cell fate in maize endosperm development. *Development* **127**, 4039–4048.

Bell, P. R. (1989). The alternation of generations. *Advances in Botanical Research Incorporating Advances in Plant Pathology* **16**, 55–93.

Bell, P. R. (1996). Megaspore abortion: a consequence of selective apoptosis. *International Journal of Plant Sciences* **157**, 1–7.

Benfey, P. N. (2005). Developmental networks. *Plant Physiology* **138**, 548–549.

Bensen, R. J., Johal, G. S., Crane, V. C., Tossberg, J. T., Schnable, P. S., Meeley, R. et al. (1995). Cloning and characterization of the maize *An1* gene. *Plant Cell* **7**, 75–84.

Benson, M. (1904). *Telangium scotti*, a new Species of *Telangium* (Calymmatotheca) showing structure. *Annals of Botany*, Old Series18, 161–177.

Benzing, D. H., Friedman, W. E., Peterson, G. & Renfrow, A. (1983). Shootlessness, velamentous roots, and the pre-eminence of Orchidaceae in the epiphytic biotope. *American Journal of Botany* **70**, 121–133.

Berger, F. (2003). Endosperm: the crossroad of seed development. *Current Opinion in Plant Biology* **6**, 42–50.

Berger, F., Haseloff, J., Schiefelbein, J. & Dolan, L. (1998). Positional information in root epidermis is defined during embryogenesis and acts in domains with strict boundaries. *Current Biology* **8**, 421–430.

Bergero, R., Forrest, A., Kamau, E. & Charlesworth, D. (2007). Evolutionary strata on the X chromosomes of the dioecious plant *Silene latifolia*: evidence from new sex-linked genes. *Genetics* **175**, 1945–1954.

Berking, S., Hesse, M. & Herrmann, K. (2002). A shoot meristem-like organ in animals; monopodial and sympodial growth in Hydrozoa. *International Journal of Developmental Biology* **46**, 301–308.

Berleth, T., Scarpella, E. & Prusinkiewicz, P. (2007) Towards the systems biology of auxin-transport-mediated patterning. *Trends in Plant Science* **12**, 151–159.

Bernhardt, C., Lee, M. M., Gonzalez, A., Zhang, F., Lloyd, A. & Schiefelbein, J. (2003). The *bHLH* genes *GLABRA3* (*GL3*) and *ENHANCER OF GLABRA3* (*EGL3*) specify epidermal cell fate in the *Arabidopsis* root. *Development* **130**, 6431–6439.

Bernhardt, C., Zhao, M. Z., Gonzalez, A., Lloyd, A. & Schiefelbein, J. (2005). The *bHLH* genes *GL3* and *EGL3* participate in an intercellular regulatory circuit that controls cell patterning in the *Arabidopsis* root epidermis. *Development* **132**, 291–298.

Bessey, C. E. (1897). The phylogeny and taxonomy of the angiosperms. *Botanical Gazette* **24**, 145–178.

Bharathan, G., Goliber, T. E., Moore, C., Kessler, S., Pham, T. & Sinha, N. R. (2002). Homologies in leaf form inferred from *KNOXI* gene expression during development. *Science* **296**, 1858–1860.

Bicknell, R. A., Borst, N. K. & Koltunow, A. M. (2000). Monogenic inheritance of apomixis in two *Hieracium* species with distinct developmental mechanisms. *Heredity* **84**, 228–237.

Bicknell, R. A. & Koltunow, A. M. (2004). Understanding apomixis: recent advances and remaining conundrums. *Plant Cell* **16**, S228–S245.

Bidartondo, M. I. (2005). The evolutionary ecology of myco-heterotrophy. *New Phytologist* **167**, 335–352.

Bierhorst, D. W. (1969). On stromatopteris and its ill-defined organs. *American Journal of Botany* **56**, 160–174.

Bierhorst, D. W. (1977). Systematic position of *Psilotum* and *Tmesipteris*. *Brittonia* **29**, 3–13.

Bilderback, D. E. (1978). Development of sporocarp of *Marsilea vestita*. *American Journal of Botany* **65**, 629–637.

Birnbaum, K. & Benfey, P. N. (2004). Network building: transcriptional circuits in the root. *Current Opinion in Plant Biology* **7**, 582–588.

Blackwell, W. H. (2003). Two theories of origin of the land–plant sporophyte: which is left standing? *Botanical Review* **69**, 125–148.

Blanc, G. & Wolfe, K. H. (2004). Widespread paleopolyploidy in model plant species inferred from age distributions of duplicate genes. *Plant Cell* **16**, 1667–1678.

Blancaflor, E. B. & Masson, P. H. (2003). Plant gravitropism. Unraveling the ups and downs of a complex process. *Plant Physiology* **133**, 1677–1690.

Blazquez, M. A., Soowal, L. N., Lee, I. & Weigel, D. (1997). *LEAFY* expression and flower initiation in *Arabidopsis*. *Development* **124**, 3835–3844.

Blilou, I., Xu, J., Wildwater, M., Willemsen, V., Paponov, I., Friml, J. et al. (2005). The PIN auxin efflux facilitator network controls growth and patterning in *Arabidopsis* roots. *Nature* **433**, 39–44.

Bonilla, J. R. & Quarin, C. L. (1997). Diplosporous and aposporous apomixis in a pentaploid race of *Paspalum minus*. *Plant Science* **127**, 97–104.

Bonke, M., Thitamadee, S., Mahonen, A. P., Hauser, M. T. & Helariutta, Y. (2003). APL regulates vascular tissue identity in *Arabidopsis*. *Nature* **426**, 181–186.

Bose, M. N., Banerji, J. & Pal, P. K. (1984). *Amarjolia dactylota* (Bose) comb. nov., a bennettitalean bisexual flower from the Rajmahal Hills, India. *Palaeobotanist* **32**, 217–229.

Boss, P. K., Sreekantan, L. & Thomas, M. R. (2006). A grapevine TFL1 homologue can delay flowering and alter oral development when overexpressed in heterologous species. *Functional Plant Biology* **33**, 31–41.

Boss, P. K. & Thomas, M. R. (2002). Association of dwarfism and floral induction with a grape 'green revolution' mutation. *Nature* **416**, 847–850.

Bowe, L. M., Coat, G. & dePamphilis, C. W. (2000). Phylogeny of seed plants based on all three genomic compartments: extant gymnosperms are monophyletic and Gnetales' closest relatives are conifers. *Proceedings of the National Academy of Sciences of the United States of America* **97**, 4092–4097.

Bower, F. O. (1908). *The Origin of a Land Flora*. Macmillan & Co., London.

Bower, F. O. (1935). *Primitive Land Plants*. Macmillan & Co., London.

Bowman, J. L. (1997). Evolutionary conservation of angiosperm flower development at the molecular and genetic levels. *Journal of Biosciences* **22**, 515–527.

Bowman, J. L., Sakai, H., Jack, T., Weigel, D., Mayer, U. & Meyerowitz, E. M. (1992). Superman, a regulator of floral homeotic genes in *Arabidopsis*. *Development* **114**, 599–615.

Bowman, J. L. & Smyth, D. R. (1999). *CRABS CLAW*, a gene that regulates carpel and nectary development in *Arabidopsis*, encodes a novel protein with zinc finger and helix-loop-helix domains. *Development* **126**, 2387–2396.

Bowman, J. L., Smyth, D. R. & Meyerowitz, E. M. (1989). Genes directing flower development in *Arabidopsis*. *Plant Cell* **1**, 37–52.

Bowman, J. L., Smyth, D. R. & Meyerowitz, E. M. (1991). Genetic interactions among floral homeotic genes of *Arabidopsis*. *Development* **112**, 1–20.

Bradley, D., Carpenter, R., Copsey, L., Vincent, C., Rothstein, S. & Coen, E. (1996). Control of inflorescence architecture in *Antirrhinum*. *Nature* **379**, 791–797.

Bradley, D., Ratcliffe, O., Vincent, C., Carpenter, R. & Coen, E. (1997). Inflorescence commitment and architecture in *Arabidopsis*. *Science* **275**, 80–83.

Brand, U., Fletcher, J. C., Hobe, M., Meyerowitz, E. M. & Simon, R. (2000). Dependence of stem cell fate in *Arabidopsis* on a feedback loop regulated by CLV3 activity. *Science* **289**, 617–619.

Bremer, B., Bremer, K., Chase, M. W., Reveal, J. L., Soltis, D. E., Soltis, P. S. et al. (2003). An update of the Angiosperm Phylogeny Group classification for the orders and families of flowering plants: APG II. *Botanical Journal of the Linnean Society* **141**, 399–436.

Bremer, K., Chase, M. W., Stevens, P. F., Anderberg, A. A., Backlund, A., Bremer, B. et al. (1998). An ordinal classification for the families of flowering plants. *Annals of the Missouri Botanical Garden* **85**, 531–553.

Brewer, P. B., Howles, P. A., Dorian, K., Griffith, M. E., Ishida, T., Kaplan-Levy, R. N. et al. (2004). *PETAL LOSS*, a trihelix transcription factor gene, regulates perianth architecture in the *Arabidopsis* flower. *Development* **131**, 4035–4045.

Brook, W. J., DiazBenjumea, F. J. & Cohen, S. M. (1996). Organizing spatial pattern in limb development. *Annual Review of Cell and Developmental Biology* **12**, 161–180.

Brooks, K. E. (1973). Reproductive biology of *Selaginella*. 1. Determination of megasporangia by 2-chloroethylphosphonic acid, an ethylene-releasing compound. *Plant Physiology* **51**, 718–722.

Brooks, K. E. & Tepfer, S. S. (1972). Reproductive biology of *Selaginella*—experimental control of sex expression. *American Journal of Botany* **59**, 675 [abstract].

Brownlee, C. (1994). Tansley review No-70—signal-transduction during fertilization in algae and vascular plants. *New Phytologist* **127**, 399–423.

Brundrett, M. C. (2002). Coevolution of roots and mycorrhizas of land plants. *New Phytologist* **154**, 275–304.

Buchholz, J. T. (1920). Embryo development and polyembryony in relation to the phylogeny of conifers. *American Journal of Botany* **7**, 125–145.

Busch, A. & Zachgo, S. (2007). Control of corolla monosymmetry in the Brassicaceae *Iberis amara*. *Proceedings of the National Academy of Sciences of the United States of America* **104**, 16714–16719.

Buvat, R. (1952). Structure, évolution et fonctionnement du méristème apical de quelques dicotylédones. *Annales des Sciences Naturelles* (*Series 11, Botany*) **8**, 199–300.

Buys, M. H. & Hilger, H. H. (2003). Boraginaceae cymes are exclusively scorpioid not helicoid. *Taxon* **52**, 719–724.

Buzgo, M., Soltis, P. S. & Soltis, D. E. (2004). Floral developmental morphology of *Amborella trichopoda* (Amborellaceae). *International Journal of Plant Sciences* **165**, 925–947.

Byrne, M. E., Barley, R., Curtis, M., Arroyo, J. M., Dunham, M., Hudson, A. et al. (2000). Asymmetric leaves1 mediates leaf patterning and stem cell function in *Arabidopsis*. *Nature* **408**, 967–971.

Calderon-Urrea, A. & Dellaporta, S. L. (1999). Cell death and cell protection genes determine the fate of pistils in maize. *Development* **126**, 435–441.

Calladine, A. & Pate, J. S. (2000). Haustorial structure and functioning of the root hemiparasitic tree *Nuytsia floribunda* (Labill.) R.Br. and water relationships with its hosts. *Annals of Botany* **85**, 723–731.

Calladine, A., Pate, J. S. & Dixon, K. W. (2000). Haustorial development and growth benefit to seedlings of the root hemiparasitic tree *Nuytsia floribunda* (Labill.) R.Br. in association with various hosts. *Annals of Botany* **85**, 733–740.

Calonje, M., Cubas, P., Martinez-Zapater, J. M. & Carmona, M. J. (2004). Floral meristem identity genes are expressed during tendril development in grapevine. *Plant Physiology* **135**, 1491–1501.

Camp, W. H. & Hubbard, M. M. (1963). On the origins of the ovule and cupule in lyginopterid pteridosperms. *American Journal of Botany* **50**, 235–243.

Canales, C., Bhatt, A. M., Scott, R. & Dickinson, H. (2002). EXS, a putative LRR receptor kinase, regulates male germline cell number and tapetal identity and promotes seed development in *Arabidopsis*. *Current Biology* **12**, 1718–1727.

Candolle, A. P. D. (1827). *Organographie végétale*. Déterville, Paris.

Cano-Delgado, A., Yin, Y. H., Yu, C., Vafeados, D., Mora-Garcia, S., Cheng, J. C. et al. (2004). BRL1 and BRL3 are novel brassinosteroid receptors that function in vascular differentiation in *Arabidopsis*. *Development* **131**, 5341–5351.

Capoen, W., Goormachtig, S., De Rycke, R., Schroeyers, K. & Holsters, M. (2005). SrSymRK, a plant receptor essential for symbiosome formation. *Proceedings of the National Academy of Sciences of the United States of America* **102**, 10369–10374.

Carafa, A., Duckett, J. G., Knox, J. P. & Ligrone, R. (2005). Distribution of cell-wall xylans in bryophytes and tracheophytes: new insights into basal interrelationships of land plants. *New Phytologist* **168**, 231–240.

Carman, J. G. (1997). Asynchronous expression of duplicate genes in angiosperms may cause apomixis, bispory, tetraspory, and polyembryony. *Biological Journal of the Linnean Society* **61**, 51–94.

Carmona, M. J., Cubas, P. & Martinez-Zapater, J. M. (2002). VFL, the grapevine FLORICAULA/LEAFY ortholog, is expressed in meristematic regions independently of their fate. *Plant Physiology* **130**, 68–77.

Carpenter, R. & Coen, E. S. (1990). Floral homeotic mutations produced by transposon-mutagenesis in *Antirrhinum majus*. *Genes & Development* **4**, 1483–1493.

Carroll, S. B. (2005). Evolution at two levels: on genes and form. *PLOS Biology* **3**, 1159–1166.

Carroll, S. B., Gompel, N., Prud'homme, B., Wittkopp, T. & Kassner, V. (2005). Chance caught on the wing: *cis*-regulatory evolution and the origins of novelty. *Developmental Biology* **283**, 584.

Carter, C., Shafir, S., Yehonatan, L., Palmer, R. G. & Thornburg, R. (2006). A novel role for proline in plant floral nectars. *Naturwissenschaften* **93**, 72–79.

Cartolano, M., Castillo, R., Efremova, N., Kuckenberg, M., Zethof, J., Gerats, T. et al. (2007). A conserved microRNA module exerts homeotic control over *Petunia hybrida* and *Antirrhinum majus* floral organ identity. *Nature Genetics* **39**, 901–905.

Casamitjana-Martinez, E., Hofhuis, H. F., Xu, J., Liu, C.-M., Heidstra, R. & Scheres, B. (2003). Root-specific CLE19 overexpression and the sol1/2. *Current Biology* **13**, 1435–1441.

Causier, B., Castillo, R., Zhou, J. L., Ingram, R., Xue, Y. B., Schwarz-Sommer, Z. et al. (2005). Evolution in action: following function in duplicated floral homeotic genes. *Current Biology* **15**, 1508–1512.

Cecchi, A. F., Palandri, M. R., DiFalco, P. & Tani, G. (1996). Cytological aspects of the hypocotyl correlated to the behavior of the embryo radicle of Tillandsia atmospheric species. *Caryologia* **49**, 113–124.

Chabaud, M., Venard, C., Defaux-Petras, A., Becard, G. & Barker, D. G. (2002). Targeted inoculation of *Medicago truncatula* in vitro root cultures reveals MtENOD11 expression during early stages of infection by arbuscular mycorrhizal fungi. *New Phytologist* **156**, 265–273.

Champagne, C. E. M. & Ashton, N. W. (2001). Ancestry of *KNOX* genes revealed by bryophyte (*Physcomitrella patens*) homologs. *New Phytologist* **150**, 23–36.

Champagne, C. & Sinha, N. (2004). Compound leaves: equal to the sum of their parts? *Development* **131**, 4401–4412.

Charlton, W. A. (1981). Features of the inflorescence of *Triglochin maritima*. *Canadian Journal of Botany-Revue Canadienne De Botanique* **59**, 2108–2115.

Chase, M. W., Fay, M. F. & Savolainen, V. (2000). Higher-level classification in the angiosperms: new insights from the perspective of DNA sequence data. *Taxon* **49**, 685–704.

Chaw, S. M., Parkinson, C. L., Cheng, Y. C., Vincent, T. M. & Palmer, J. D. (2000). Seed plant phylogeny inferred from all three plant genomes: monophyly of extant gymnosperms and origin of Gnetales from conifers. *Proceedings of the National Academy of Sciences of the United States of America* **97**, 4086–4091.

Chen, H., Banerjee, A. K. & Hannapel, D. J. (2004). The tandem complex of BEL and KNOX partners is required for transcriptional repression of ga20ox1. *Plant Journal* **38**, 276–284.

Chen, R., Rosen, E. & Masson, P. H. (1999). Gravitropism in higher plants. *Plant Physiology* **120**, 343–350.

Chen, Y. H., Yang, X. Y., He, K., Liu, M. H., Li, J. G., Gao, Z. F. et al. (2006). The MYB transcription factor superfamily of *Arabidopsis*: expression analysis and phylogenetic comparison with the rice MYB family. *Plant Molecular Biology* **60**, 107–124.

Chilton, M. D., Tepfer, D. A., Petit, A., David, C., Cassedelbart, F. & Tempe, J. (1982). *Agrobacterium rhizogenes* inserts T-DNA into the genomes of the host plant-root cells. *Nature* **295**, 432–434.

Chuang, C. F., Running, M. P., Williams, R. W. & Meyerowitz, E. M. (1999). The *PERIANTHIA* gene encodes a bZIP protein involved in the determination of floral organ number in *Arabidopsis thaliana*. *Genes & Development* **13**, 334–344.

Chumley, T. W., Palmer, J. D., Mower, J. P., Fourcade, H. M., Calie, P. J., Boore, J. L. et al. (2006). The complete chloroplast genome sequence of *Pelargonium* x hortorum: organization and evolution of the largest and most highly rearranged chloroplast genome of land plants. *Molecular Biology and Evolution* **23**, 2175–2190.

Citerne, H. L., Luo, D., Pennington, R. T., Coen, E. & Cronk, Q. C. B. (2003). A phylogenomic investigation of *CYCLOIDEA*-like *TCP* genes in the Leguminosae. *Plant Physiology* **131**, 1042–1053.

Citerne, H. L., Pennington, R. T. & Cronk, Q. C. B. (2006). An apparent reversal in floral symmetry in the legume *Cadia* is a homeotic transformation. *Proceedings of the National Academy of Sciences of the United States of America* **103**, 12017–12020.

Clare, T. S. & Johnston, G. R. (1931). Polyembryony and germination of polyembryonic coniferous seeds. *American Journal of Botany* **18**, 674–683.

Clark, J. I. & Coen, E. S. (2002). The *cycloidea* gene can respond to a common dorsoventral prepattern in *Antirrhinum*. *Plant Journal* **30**, 639–648.

Clark, R. M., Wagler, T. N., Quijada, P. & Doebley, J. (2006). A distant upstream enhancer at the maize domestication gene *tb1* has pleiotropic effects on plant and inflorescent architecture. *Nature Genetics* **38**, 594–597.

Clowes, F. A. L. (1981). The difference between open and closed meristems. *Annals of Botany* **48**, 761–767.

Clowes, F. A. L. (1990). The discrete root epidermis of floating plants. *New Phytologist* **115**, 11–15.

Clowes, F. A. L. (2000). Pattern in root meristem development in angiosperms. *New Phytologist* **146**, 83–94.

Coen, E. (2001). Goethe and the ABC model of flower development. *Comptes Rendus De L Academie Des Sciences Serie Iii-Sciences De La Vie-Life Sciences* **324**, 523–530.

Coen, E. S., Doyle, S., Romero, J. M., Elliott, R., Magrath, R. & Carpenter, R. (1991). Homeotic genes controlling flower development in *Antirrhinum*. *Development* **149**–156.

Coen, E. S. & Meyerowitz, E. M. (1991). The war of the whorls—genetic interactions controlling flower development. *Nature* **353**, 31–37.

Colombo, L., Franken, J., Koetje, E., Vanwent, J., Dons, H. J. M., Angenent, G. C. et al. (1995). The petunia MADS box gene FBP11 determines ovule identity. *Plant Cell* **7**, 1859–1868.

Conti, E., Suring, E., Boyd, D., Jorgensen, J., Grant, J. & Kelso, S. (2000). Phylogenetic relationships and character evolution in *Primula* L.: the usefulness of ITS sequence data. *Plant Biosystems* **134**, 385–392.

Conti, L. & Bradley, D. (2007). TERMINAL FLOWER1 is a mobile signal controlling *Arabidopsis* architecture. *Plant Cell* **19**, 767–778.

Cook, C. D. K. (1969). On the determination of leaf form in *Ranunculus aquatilis*. *New Phytologist* **68**, 469–480.

Cookson, I. C. (1935). On plant-remains from the Silurian of Victoria, Australia, that extend and connect floras hitherto described. *Philosophical Transactions of the Royal Society of London Series B-Biological Sciences* **225**, 127–148.

Corbesier, L., Vincent, C., Jang, S. H., Fornara, F., Fan, Q. Z., Searle, I. et al. (2007). FT protein movement contributes to long-distance signaling in floral induction of *Arabidopsis*. *Science* **316**, 1030–1033.

Corley, S. B., Carpenter, R., Copsey, L. & Coen, E. (2005). Floral asymmetry involves an interplay between TCP and MYB transcription factors in *Antirrhinum*. *Proceedings of the National Academy of Sciences of the United States of America* **102**, 5068–5073.

Corner, E. J. H. (1946). Centrifugal stamens. *Journal of the Arnold Arboretum* **27**, 423–437.

Corner, E. J. H. (1949). The durian theory or the origin of the modern tree. *Annals of Botany* **13**, 367–414.

Corner, E. J. H. (1953). The durian theory extended—I. *Phytomorphology* **3**, 465–476.

Corner, E. J. H. (1954a). The durian theory extended—II. The arillate fruit and the compound leaf. *Phytomorphology* **4**, 152–165.

Corner, E. J. H. (1954b). The durian theory extended—III. Pachycauly and megaspermy—conclusion. *Phytomorphology* **4**, 263–274.

Corner, E. J. H. (1958). Transference of function. *Botanical Journal of the Linnean Society* **56**, 33–40.

Corner, E. J. H. (1959). *The Life of Plants*. Weidenfeld & Nicolson, London.

Corner, E. J. H. (1963). A criticism of the gonophyll theory of the flower. *Phytomorphology* **13**, 290–292.

Corner, E. J. H. (1964). *The Life of Plants*. Weidenfeld & Nicolson, London.

Corner, E. J. H. (1966a). Debunking new morphology. *New Phytologist* **65**, 398–404.

Corner, E. J. H. (1966b). *The Natural History of Palms*. Wiedenfield & Nicolson, London.

Corner, E. J. H. (1976). *The Seeds of Dicotyledons, Volumes 1 and 2*. Cambridge University Press, Cambridge.

Costa, M. M. R., Fox, S., Hanna, A. I., Baxter, C. & Coen, E. (2005). Evolution of regulatory interactions controlling floral asymmetry. *Development* **132**, 5093–5101.

Crane, P. R. (1985). Phylogenetic analysis of seed plants and the origin of angiosperms. *Annals of the Missouri Botanical Garden* **72**, 716–793.

Crawford, B. C. W., Nath, U., Carpenter, R. & Coen, E. S. (2004). CINCINNATA controls both cell differentiation and growth in petal lobes and leaves of *Antirrhinum*. *Plant Physiology* **135**, 244–253.

Crespel, L., Chirollet, M., Durel, C. E., Zhang, D., Meynet, J. & Gudin, S. (2002). Mapping of qualitative and quantitative phenotypic traits in *Rosa* using AFLP markers. *Theoretical and Applied Genetics* **105**, 1207–1214.

Crespi, M., Messens, E., Caplan, A. B., van Montagu, M. & Desomer, J. (1992). Fasciation induction by the phytopathogen *Rhodococcus fascians* depends upon a linear plasmid encoding a cytokinin synthase gene. *EMBO Journal* **11**, 795–804.

Croizat, L. (1961). *Principia Botanica*. Published Privately, Caracas.

Cronk, Q. C. B. & Moller, M. (1997). Genetics of floral symmetry revealed. *Trends in Ecology & Evolution* **12**, 85–86.

Cronk, Q. C. B. (1981). *Senecio redivivus* and its successful conservation in St-Helena. *Environmental Conservation* **8**, 125–126.

Cronk, Q. C. B. (2001). Plant evolution and development in a post-genomic context. *Nature Reviews Genetics* **2**, 607–619.

Cronk, Q. C. B. (2002). Perspectives and paradigms in plant evo-devo. In *Developmental Genetics and Plant Evolution* (eds Q. C. B. Cronk, R. M. Bateman and J. Hawkins), pp. 1–14. Taylor & Francis, London.

Cronk, Q. C. B. (2006). Legume flowers bear fruit. *Proceedings of the National Academy of Sciences of the United States of America* **103**, 4801–4802.

Crook, M. J., Ennos, A. R. & Banks, J. R. (1997). The function of buttress roots: a comparative study of the anchorage systems of buttressed (*Aglaia* and *Nephelium ramboutan* species) and non-buttressed (*Mallotus wrayi*) tropical trees. *Journal of Experimental Botany* **48**, 1703–1716.

Croxdale, J. G. (1978). *Salvinia* leaves. 1. Origin and early differentiation of floating and submerged leaves. *Canadian Journal of Botany-Revue Canadienne De Botanique* **56**, 1982–1991.

Cubas, P. (2004). Floral zygomorphy, the recurring evolution of a successful trait. *Bioessays* **26**, 1175–1184.

CUBAS, P., COEN, E. & ZAPATER, J. M. M. (2001). Ancient asymmetries in the evolution of flowers. *Current Biology* **11**, 1050–1052.

CUBAS, P., LAUTER, N., DOEBLEY, J. & COEN, E. (1999). The TCP domain: a motif found in proteins regulating plant growth and development. *Plant Journal* **18**, 215–222.

CUBAS, P., VINCENT, C. & COEN, E. (1999). An epigenetic mutation responsible for natural variation in floral symmetry. *Nature* **401**, 157–161.

CUI, H., LEVESQUE, M. P., VERNOUX, T., JUNG, J. W., PAQUETTE, A. J., GALLAGHER, K. L. et al. (2007). An evolutionarily conserved mechanism delimiting SHR movement defines a single layer of endodermis in plants. *Science* **316**, 421–425.

CULLEN, J. (1978). A preliminary survey of ptyxis (venation) in the angiosperms. *Notes of the Royal Botanic Garden Edinburgh* **37**, 161–214.

DARWIN, C. R. (1859). *On the Origin of Species*, First edition. Murray, London.

DARWIN, C. R. (1868). *The Variation of Animals and Plants Under Domestication*, First edition. Murray, London.

DARWIN, C. R. (1872). *On the Origin of Species*, Sixth edition. Murray, London.

DARWIN, C. R. & DARWIN, F. (1880). *The Power of Movement in Plants*. John Murray, London.

DAVIES, K. L., STPICZYNSKA, M. & GREGG, A. (2005). Nectar-secreting floral stomata in *Maxillaria anceps* Ames & C. Schweinf. (Orchidaceae). *Annals of Botany* **96**, 217–227.

DAVIS, G. L. (1966). *Systematic Embryology of the Angiosperms*. John Wiley & Sons, New York.

DE VRIES, H. (1872). Über einige Ursachen der Richtung bilateralsymmetrischer pflanzentheile. *Arbeite Botanische Institut Würzburg* **1**, 223–277.

DECKEREISEL, C. & HAGEMANN, W. (1978). Origin of fertile spike of *Ophioglossum pedunculosum*. *Plant Systematics and Evolution* **130**, 143–155.

DELEVORYAS, T. (1968). Some aspects of cycadeoid evolution. *Journal of the Linnean Society of London (Botany)* **61**, 137–146.

DELEVORYAS, T. (1982). Perspectives on the origin of cycads and cycadeoids. *Review of Palaeobotany and Palynology* **37**, 115–132.

DELLAPORTA, S. L. & CALDERON-URREA, A. (1993). Sex determination in flowering plants. *Plant Cell* **5**, 1241–1251.

DELONG, A., CALDERON-URREA, A. & DELLAPORTA, S. L. (1993). Sex determination gene *Tasselseed2* of maize encodes a short-chain alcohol-dehydrogenase required for stage-specific floral organ abortion. *Cell* **74**, 757–768.

DES MARAIS, D. L., SMITH, A. R., BRITTON, D. M. & PRYER, K. M. (2003). Phylogenetic relationships and evolution of extant horsetails, equisetum, based on chloroplast DNA sequence data (rbcL and trnL-F). *International Journal of Plant Sciences* **164**, 737–751.

DEYHOLOS, M. K., CORDNER, G., BEEBE, D. & SIEBURTH, L. E. (2000). The *SCARFACE* gene is required for cotyledon and leaf vein patterning. *Development* **127**, 3205–3213.

DICKINSON, T. A. (1978). Epiphylly in angiosperms. *Botanical Review* **44**, 181–232.

DICKINSON, T. A. & SATTLER, R. (1975). Development of the epiphyllous inflorescence of *Helwingia japonica*. *American Journal of Botany* **62**, 962–973.

DILCHER, D. L. & CRANE, P. R. (1984). *Archaeanthus*—an early angiosperm from the Cenomanian of the western interior of North America. *Annals of the Missouri Botanical Garden* **71**, 351–383.

DINNENY, J. R., YADEGARI, R., FISCHER, R. L., YANOFSKY, M. L. & WEIGEL, D. (2004). The role of JAGGED in shaping lateral organs. *Development* **131**, 1101–1110.

DOEBLEY, J. & LUKENS, L. (1998). Transcriptional regulators and the evolution of plant form. *Plant Cell* **10**, 1075–1082.

DOLAN, L. (2005). Positional information and mobile transcriptional regulators determine cell pattern in the *Arabidopsis* root epidermis. *Journal of Experimental Botany*, **57**, 51–54.

DOLAN, L. & COSTA, S. (2001). Evolution and genetics of root hair stripes in the root epidermis. *Journal of Experimental Botany* **52**, 413–417.

DOMINGUEZ, E., MERCADO, J. A., QUESADA, M. A. & HEREDIA, A. (1999). Pollen sporopollenin: degradation and structural elucidation. *Sexual Plant Reproduction* **12**, 171–178.

DONG, J., KIM, S. T. & LORD, E. M. (2005). Plantacyanin plays a role in reproduction in *Arabidopsis*. *Plant Physiology* **138**, 778–789.

DONNELLY, P. M., BONETTA, D., TSUKAYA, H., DENGLER, R. E. & DENGLER, N. G. (1999). Cell cycling and cell enlargement in developing leaves of *Arabidopsis*. *Developmental Biology* **215**, 407–419.

DONOGHUE, M. J., BELL, C. D. & WINKWORTH, R. C. (2003). The evolution of reproductive characters in Dipsacales. *International Journal of Plant Sciences* **164**, S453–S464.

DONOGHUE, M. J. & DOYLE, J. A. (2000). Seed plant phylogeny: demise of the anthophyte hypothesis? *Current Biology* **10**, R106–R109.

DONOGHUE, M. J., REE, R. H. & BAUM, D. A. (1998). Phylogeny and the evolution of flower symmetry in the *Asteridae*. *Trends in Plant Science* **3**, 311–317.

DOUGLAS, G. (1944). The inferior ovary, 1. *Botanical Review* **10**, 125–186.

DOUGLAS, G. (1957). The inferior ovary, 2. *Botanical Review* **23**, 1–46.

DOYLE, J. A. (2006). Seed ferns and the origin of angiosperms. *Journal of the Torrey Botanical Society* **133**, 169–209.

DOYLE, J. A. & DONOGHUE, M. J. (1986). Seed plant phylogeny and the origin of angiosperms: an experimental cladistic approach. *Botanical Review* **52**, 321–431.

DOYLE, J. A., EKLUND, H. & HERENDEEN, P. S. (2003). Floral evolution in Chloranthaceae: implications of a morphological phylogenetic analysis. *International Journal of Plant Sciences* **164**, S365–S382.

DOYLE, J. A. & ENDRESS, P. K. (2000). Morphological phylogenetic analysis of basal angiosperms: comparison and combination with molecular data. *International Journal of Plant Sciences* **161**, S121–S153.

DRESSELHAUS, T. (2005). Cell-cell communication during double fertilization. *Current Opinion in Plant Biology* **9**, 41–47.

DREWS, G. N., BOWMAN, J. L. & MEYEROWITZ, E. M. (1991). Negative regulation of the *Arabidopsis* homeotic gene *Agamous* by the *Apetala2* product. *Cell* **65**, 991–1002.

DRINNAN, A. N. & CHAMBERS, T. C. (1985). A reassessment of *Taeniopteris daintreei* from the Victorian early Cretaceous—a member of the Pentoxylales and a significant Gondwanan plant. *Australian Journal of Botany* **33**, 89–100.

DUCKETT, J. G. (1970). Spore size in the genus *Equisetum*. *New Phytologist* **69**, 333–346.

DUCKETT, J. G. (1977). Towards an understanding of sex determination in *Equisetum*—analysis of regeneration in gametophytes of subgenus *Equisetum*. *Botanical Journal of the Linnean Society* **74**, 215–242.

DUCKETT, J. G. & PANG, W. C. (1984). The origins of heterospory—a comparative-study of sexual-behavior in the fern *Platyzoma microphyllum* Rbr and the horsetail *Equisetum Giganteum* L. *Botanical Journal of the Linnean Society* **88**, 11–34.

DUHOUX, E., RINAUDO, G., DIEM, H. G., AUGUY, F., FERNANDEZ, D., BOGUSZ, D. et al. (2001). Angiosperm gymnostoma trees produce root nodules colonized by arbuscular mycorrhizal fungi related to glomus. *New Phytologist* **149**, 115–125.

DUPUY, P. & GUÉDÈS, M. (1979). Hypoascidiate bracts in *Pelargonium*. *Botanical Journal of the Linnean Society* **78**, 117–121.

DYMEK, E. E., GODUTI, D., KRAMER, T. & SMITH, E. F. (2006). A kinesin-like calmodulin-binding protein in *Chlamydomonas*: evidence for a role in cell division and flagellar functions. *Journal of Cell Science* **119**, 3107–3116.

EAMES, A. J. (1931). The vascular anatomy of the flower with refutation of the theory of carpel polymorphism. *American Journal of Botany* **18**, 147–188.

EAMES, A. J. (1936). *Morphology of vascular plants. Lower groups (Psylophytales et Filiciales)*. McGraw-Hill, New York.

EAMES, A. J. (1961). *Morphology of Angiosperms*. McGraw-Hill, New York.

EDGAR, B. A. & LEHNER, C. F. (1996). Developmental control of cell cycle regulators: a fly's perspective. *Science* **274**, 1646–1652.

EDWARDS, D. (2003). Xylem in early tracheophytes. *Plant Cell and Environment* **26**, 57–72.

EDWARDS, D., DUCKETT, J. G. & RICHARDSON, J. B. (1995). Hepatic characters in the earliest land plants. *Nature* **374**, 635–636.

EDWARDS, D. S. (1986). *Aglaophyton major*, a non-vascular land-plant from the Devonian Rhynie chert. *Botanical Journal of the Linnean Society* **93**, 173–204.

EICHLER, A. W. (1861). *Zur Entwickelungsgeschichte des Blattes mit Besonderer Berücksichtigung der Nebenblatt-Bildungen*. Inaug. Diss., University of Marburg.

EICHLER, A. W. (1876). *Syllabus der Vorlesungen über spezielle und medizinisch-pharmazeutische Botanik*. Borntraeger, Berlin.

ELLIOTT, R. C., BETZNER, A. S., HUTTNER, E., OAKES, M. P., TUCKER, W. Q. J., GERENTES, D. et al. (1996). *AINTEGUMENTA*, an *APETALA2*-like gene of *Arabidopsis* with pleiotropic roles in ovule development and floral organ growth. *Plant Cell* **8**, 155–168.

ELLMORE, G. S. (1981a). Root dimorphism in *Ludwigia peploides* (Onagraceae)—development of two root types from similar primordia. *Botanical Gazette* **142**, 525–533.

ELLMORE, G. S. (1981b). Root dimorphism in *Ludwigia peploides* (Onagraceae)—structure and gas content of mature roots. *American Journal of Botany* **68**, 557–568.

EMERY, J. F., FLOYD, S. K., ALVAREZ, J., ESHED, Y., HAWKER, N. P., IZHAKI, A. et al. (2003). Radial patterning of *Arabidopsis* shoots by class *IIIHD-ZIP* and *KANADI* genes. *Current Biology* **13**, 1768–1774.

ENDRE, G., KERESZT, A., KEVEI, Z., MIHACEA, S., KALO, P. & KISS, G. B. (2002). A receptor kinase gene regulating symbiotic nodule development. *Nature* **417**, 962–966.

ENDRESS, P. K. (1996). Structure and function of female and bisexual organ complexes in gnetales. *International Journal of Plant Sciences* **157**, S113–S125.

Endress, P. K. (1999). Symmetry in flowers: diversity and evolution. *International Journal of Plant Sciences* **160**, S3–S23.

Endress, P. K. (2001a). Origins of flower morphology. *Journal of Experimental Zoology* **291**, 105–115.

Endress, P. K. (2001b). The flowers in extant basal angiosperms and inferences on ancestral flowers. *International Journal of Plant Sciences* **162**, 1111–1140.

Endress, P. K. (2004). Structure and relationships of basal relictual angiosperms. *Australian Systematic Botany* **17**, 343–366.

Endress, P. K. (2005). Carpels in *Brasenia* (Cabombaceae) are completely ascidiate despite a long stigmatic crest. *Annals of Botany* **96**, 209–215.

Endress, P. K. (2006). Angiosperm floral evolution: morphological developmental framework. *Advances in Botanical Research Incorporating Advances in Plant Pathology*, 44, 1–61.

Endress, P. K. & Doyle, J. A. (2007). Floral phyllotaxis in basal angiosperms—development and evolution. *Current Opinion in Plant Biology* **10**, 52–57.

Endress, P. K. & Igersheim, A. (2000). Reproductive structures of the basal angiosperm *Amborella trichopoda* (Amborellaceae). *International Journal of Plant Sciences* **161**, S237–S248.

Engler, A. (1897). Nachträge. In *Die natürlichen Pflanzenfamilien* (eds A. Engler and K. Prantl). Engelmann, Leipzig.

Erbar, C., Kusma, S. & Leins, P. (1998). Development and interpretation of nectary organs in Ranunculaceae. *Flora* **194**, 317–332.

Escobar-Restrepo, J. M., Huck, N., Kessler, S., Gagliardini, V., Gheyselinck, J., Yang, W. C. et al. (2007). The FERONIA receptor-like kinase mediates male–female interactions during pollen tube reception. *Science* **317**, 656–660.

Eshed, Y., Baum, S. F. & Bowman, J. L. (1999). Distinct mechanisms promote polarity establishment in carpels of *Arabidopsis*. *Cell* **99**, 199–209.

Eshed, Y., Izhaki, A., Baum, S. F., Floyd, S. K. & Bowman, J. L. (2004). Asymmetric leaf development and blade expansion in *Arabidopsis* are mediated by KANADI and YABBY activities. *Development* **131**, 2997–3006.

Esmon, C. A., Pedmale, U. V. & Liscum, E. (2005). Plant tropisms: providing the power of movement to a sessile organism. *International Journal of Developmental Biology* **49**, 665–674.

Esmon, C. A., Tinsley, A. G., Ljung, K., Sandberg, G., Hearne, L. B. & Liscum, E. (2006). A gradient of auxin and auxin-dependent transcription precedes tropic growth responses. *Proceedings of the National Academy of Sciences of the United States of America* **103**, 236–241.

Euripides. (431BCE). *Medea*.

Fairon-Demaret, M. & Li, C. S. (1993). *Lorophyton goense* gen et sp. nov. from the Lower Givetian of Belgium and a discussion of the Middle Devonian Cladoxylopsida. *Review of Palaeobotany and Palynology* **77**, 1–22.

Farnsworth, E. (2000). The ecology and physiology of viviparous and recalcitrant seeds. *Annual Review of Ecology and Systematics* **31**, 107–138.

Favaro, R., Pinyopich, A., Battaglia, R., Kooiker, M., Borghi, L., Ditta, G. et al. (2003). MADS-box protein complexes control carpel and ovule development in *Arabidopsis*. *Plant Cell* **15**, 2603–2611.

Feng, X. Z., Zhao, Z., Tian, Z. X., Xu, S. L., Luo, Y. H., Cai, Z. G. et al. (2006). Control of petal shape and floral zygomorphy in *Lotus japonicus*. *Proceedings of the National Academy of Sciences of the United States of America* **103**, 4970–4975.

Feurtado, J. A., Ambrose, S. J., Cutler, A. J., Ross, A. R. S., Abrams, S. R. & Kermode, A. R. (2004). Dormancy termination of western white pine (*Pinus monticola* Dougl. Ex D. Don) seeds is associated with changes in abscisic acid metabolism. *Planta* **218**, 630–639.

Fey, B. S. & Endress, P. K. (1983). Development and morphological interpretation of the cupule in the Fagaceae. *Flora* **173**, 451–468.

Fisher, J. B. (1974). Axillary and dichotomous branching in the palm Chamaedorea. *American Journal of Botany* **61**, 1046–1056.

Fisher, J. B. (1976). Development of axillary branching and axillary buds in *Strelitzia*. *Canadian Journal of Botany-Revue Canadienne De Botanique* **54**, 578–592.

Fisher, J. B. (2002). Indeterminate leaves of *Chisocheton* (Meliaceae): survey of structure and development. *Botanical Journal of the Linnean Society* **139**, 207–221.

Fitter, A., Williamson, L., Linkohr, B. & Leyser, O. (2002). Root system architecture determines fitness in an *Arabidopsis* mutant in competition for immobile phosphate ions but not for nitrate ions. *Proceedings of the Royal Society of London Series B-Biological Sciences* **269**, 2017–2022.

Fitter, J. T., Thomas, M. R., Niu, C. & Rose, R. J. (2005). Investigation of *Nicotiana tabacum* (plus) *N. suaveolens* cybrids with carpelloid stamens. *Journal of Plant Physiology* **162**, 225–235.

Florin, R. (1948). On the morphology and relationships on the Taxaceae. *Botanical Gazette* **110**, 31–39.

Florin, R. (1951). Evolution in cordaites and conifers. *Acta Horti Bergiani* **15**, 285–388.

FLORIN, R. (1954). The female reproductive organs of conifers and taxads. *Biological Reviews* **29**, 367–389.

FLOYD, S. K. & BOWMAN, J. L. (2004). Gene regulation: ancient microRNA target sequences in plants. *Nature* **428**, 485–486.

FOSTER, A. S. (1939). Structure and growth of the shoot apex of *Cycas revoluta*. *American Journal of Botany* **26**, 372–385.

FOSTER, A. S. (1941). Zonal structure of the shoot apex of *Dioon edule* Lindl. *American Journal of Botany*, 557–564.

FOSTER, T., HAY, A., JOHNSTON, R. & HAKE, S. (2004). The establishment of axial patterning in the maize leaf. *Development* **131**, 3921–3929.

FOURQUIN, C., VINAUGER-DOUARD, M., CHAMBRIER, P., BERNE-DEDIEU, A. & SCUTT, C. P. (2007). Functional conservation between CRABS CLAW orthologues from widely diverged angiosperms. *Annals of Botany* **100**, 651–657.

FOURQUIN, C., VINAUGER-DOUARD, M., FOGLIANI, B., DUMAS, C. & SCUTT, C. P. (2005). Evidence that CRABS CLAW and TOUSLED have conserved their roles in carpel development since the ancestor of the extant angiosperms. *Proceedings of the National Academy of Sciences of the United States of America* **102**, 4649–4654.

FRIEDMAN, W. E. (1986). Growth and development of the male gametophyte of *Ginkgo biloba*. *American Journal of Botany* **73**, 627.

FRIEDMAN, W. E. (1987a). Growth and development of the male gametophyte of *Ginkgo biloba* within the ovule (in vivo). *American Journal of Botany* **74**, 1797–1815.

FRIEDMAN, W. E. (1987b). Morphogenesis and experimental aspects of growth and development of the male gametophyte of *Ginkgo biloba* in vitro. *American Journal of Botany* **74**, 1816–1830.

FRIEDMAN, W. E. (1990). Double fertilization in *Ephedra*, a nonflowering seed plant—its bearing on the origin of angiosperms. *Science* **247**, 951–954.

FRIEDMAN, W. E. (1992). Evidence of a pre-angiosperm origin of endosperm—implications for the evolution of flowering plants. *Science* **255**, 336–339.

FRIEDMAN, W. E. (2006). Embryological evidence for developmental lability during early angiosperm evolution. *Nature* **441**, 337–340.

FRIEDMAN, W. E. & GIFFORD, E. M. (1988). Division of the generative cell and late development in the male gametophyte of *Ginkgo biloba*. *American Journal of Botany* **75**, 1434–1442.

FRIEDMAN, W. E. & WILLIAMS, J. H. (2004). Developmental evolution of the sexual process in ancient flowering plant lineages. *The Plant Cell* **16**, S119–S132.

FRIIS, E. M., DOYLE, J. A., ENDRESS, P. K. & LENG, Q. (2003). *Archaefructus*—angiosperm precursor or specialized early angiosperm? *Trends in Plant Science* **8**, 369–373.

FRIIS, E. M., PEDERSEN, K. R. & CRANE, P. R. (2005). When Earth started blooming: insights from the fossil record. *Current Opinion in Plant Biology* **8**, 5–12.

FRIIS, E. M., PEDERSEN, K. R. & CRANE, P. R. (2006). Cretaceous angiosperm flowers: innovation and evolution in plant reproduction. *Palaeogeography Palaeoclimatology Palaeoecology* **232**, 251–293.

FRIML, J. (2003). Auxin transport—shaping the plant. *Current Opinion in Plant Biology* **6**, 7–12.

FRIML, J., VIETEN, A., SAUER, M., WEIJERS, D., SCHWARZ, H., HAMANN, T. et al. (2003). Efflux-dependent auxin gradients establish the apical–basal axis of *Arabidopsis*. *Nature* **426**, 147–153.

FRIML, J., WIŚNIEWSKA, J., BENKOVÁ, E., MENDGEN, K. & PALME, K. (2002). Lateral relocation of auxin efflux regulator PIN3 mediates tropism in *Arabidopsis*. *Nature* **415**, 806–809.

FRIML, J., YANG, X., MICHNIEWICZ, M., WEIJERS, D., QUINT, A., TIETZ, O. et al. (2004). A PINOID-dependent binary switch in apical–basal PIN polar targeting directs auxin efflux. *Science* **306**, 862–865.

FROHLICH, M. W. (2003). An evolutionary scenario for the origin of flowers. *Nature Reviews Genetics* **4**, 559–566.

FROHLICH, M. W. & PARKER, D. S. (2000). The mostly male theory of flower evolutionary origins: from genes to fossils. *Systematic Botany* **25**, 155–170.

FUKAKI, H., TAMEDA, S., MASUDA, H. & TASAKA, M. (2002). Lateral root formation is blocked by a gain-of-function mutation in the *SOLITARY-ROOT/IAA14* gene of *Arabidopsis*. *Plant Journal* **29**, 153–168.

FUKUDA, H. (2004). Signals that control plant vascular cell differentiation. *Nature Reviews Molecular Cell Biology* **5**, 379–391.

FURMAN, T. E. (1970). The nodular mycorrhizae of *Podocarpus rospigliosii*. *American Journal of Botany* **57**, 910–915.

GAILING, O. & BACHMANN, K. (2003a). QTL mapping reveals a two-step model for the evolutionary reduction of inner microsporangia within the asteracean genus *Microseris*. *Theoretical and Applied Genetics* **107**, 893–901.

GAILING, O. & BACHMANN, K. (2003b). The anthers of *Senecio vulgaris* (Asteraceae): saltatory evolution caught in the act. *Plant Systematics and Evolution* **240**, 1–10.

GALEGO, L. & ALMEIDA, J. (2002). Role of DIVARICATA in the control of dorsoventral asymmetry in *Antirrhinum* flowers. *Genes & Development* **16**, 880–891.

GALINHA, C., HOFHUIS, H., LUIJTEN, M., WILLEMSEN, V., BLILOU, I., HEIDSTRA, R. et al. (2007). PLETHORA proteins as dose-dependent master regulators of *Arabidopsis* root development. *Nature* **449**, 1053–1057.

GALLOIS, J. L., NORA, F. R., MIZUKAMI, Y. & SABLOWSKI, R. (2004). WUSCHEL induces shoot stem cell activity and developmental plasticity in the root meristem. *Genes & Development* **18**, 375–380.

GANONG, W. F. (1901). The cardinal principles of morphology. *Botanical Gazette* **31**, 426–434.

GARCÊS, H. M., CHAMPAGNE, C. E., TOWNSLEY, B. T., PARK, S., MALHÓ, R., PEDROSO, M. C. et al. (2007). Evolution of asexual reproduction in leaves of the genus *Kalanchoë*. *Proceedings of the National Academy of Sciences of the United States of America* **104**, 15578–15583.

GENRE, A., CHABAUD, M., TIMMERS, T., BONFANTE, P. & BARKER, D. G. (2005). Arbuscular mycorrhizal fungi elicit a novel intracellular apparatus in *Medicago truncatula* root epidermal cells before infection. *Plant Cell* **17**, 3489–3499.

GENSEL, P. G. & BERRY, C. M. (2001). Early lycophyte evolution. *American Fern Journal* **91**, 74–98.

GENSEL, P. G., KOTYK, M. & BASINGER, J. F. (2001). Morphology of above- and below-ground structures in early Devonian (Pragian-Emsian) plants. In *Plants Invade the Land: Evolutionary and Environmental Perspectives* (eds P. G. Gensel and D. Edwards), pp. 83–102. Columbia University Press, New York.

GERRATH, J. M. & POSLUSZNY, U. (1989). Morphological and anatomical development in the Vitaceae. 3. Vegetative development in *Parthenocissus inserta*. *Canadian Journal of Botany-Revue Canadienne De Botanique* **67**, 803–816.

GERRIENNE, P., MEYER-BERTHAUD, B., FAIRON-DEMARET, M., STREEL, M. & STEEMANS, P. (2004). Runcaria, a middle devonian seed plant precursor. *Science* **306**, 856–858.

GEURTS, R., FEDOROVA, E. & BISSELING, T. (2005). Nod factor signaling genes and their function in the early stages of *Rhizobium* infection. *Current Opinion in Plant Biology* **8**, 346–352.

GEUTEN, K., BECKER, A., KAUFMANN, K., CARIS, P., JANSSENS, S., VIAENE, T. et al. (2006). Petaloidy and petal identity MADS-box genes in the balsaminoid genera *Impatiens* and *Marcgravia*. *Plant Journal* **47**, 501–518.

GIOVANNETTI, M., SBRANA, C., AVIO, L., CITERNESI, A. S. & LOGI, C. (1993). Differential hyphal morphogenesis in arbuscular mycorrhizal fungi during preinfection stages. *New Phytologist* **125**, 587–593.

GIULINI, A., WANG, J. & JACKSON, D. (2004). Control of phyllotaxy by the cytokinin-inducible response regulator homologue ABPHYL1. *Nature* **430**, 1031–1034.

GLEISSBERG, S., GROOT, E. P., SCHMALZ, M., EICHERT, M., KOLSCH, A. & HUTTER, S. (2005). Developmental events leading to peltate leaf structure in *Tropaeolum majus* (Tropaeolaceae) are associated with expression domain changes of a *YABBY* gene. *Development Genes and Evolution* **215**, 313–319.

GLOVER, B. J., BUNNEWELL, S. & MARTIN, C. (2004). Convergent evolution within the genus *Solanum*: the specialised anther cone develops through alternative pathways. *Gene* **331**, 1–7.

GLOVER, B. J. & MARTIN, C. (1998). The role of petal cell shape and pigmentation in pollination success in *Antirrhinum majus*. *Heredity* **80**, 778–784.

GOETHE, J. W. v. (1790). *Versuch die Metamorphose der Pflanzen zu erklären*. Ettinger, Gotha.

GOLZ, J. F., KECK, E. J. & HUDSON, A. (2002). Spontaneous mutations in *KNOX* genes give rise to a novel floral structure in *Antirrhinum*. *Current Biology* **12**, 515–522.

GOLZ, J. F., ROCCARO, M., KUZOFF, R. & HUDSON, A. (2004). GRAMINIFOLIA promotes growth and polarity of *Antirrhinum* leaves. *Development* **131**, 3661–3670.

GOMEZ, E., ROYO, J., GUO, Y., THOMPSON, R. & HUEROS, G. (2002). Establishment of cereal endosperm expression domains: identification and properties of a maize transfer cell-specific transcription factor, ZmMRP-1. *Plant Cell* **14**, 599–610.

GOMPEL, N., PRUD'HOMME, B., WITTKOPP, P. J., KASSNER, V. A. & CARROLL, S. B. (2005). Chance caught on the wing: *cis*-regulatory evolution and the origin of pigment patterns in *Drosophila*. *Nature* **433**, 481–487.

GOULD, K. S. (1993). Leaf heteroblasty in *Pseudopanax crassifolius*—functional significance of leaf morphology and anatomy. *Annals of Botany* **71**, 61–70.

GOULD, R. E. & DELEVORYAS, T. (1977). The biology of *Glossopteris*: evidence from petrified seed-bearing and pollen-bearing organs. *Alcheringa: An Australasian Journal of Palaeontology* **1**, 387–399.

GOULD, S. J. (1979). Importance of heterochrony for evolutionary biology. *Systematic Zoology* **28**, 224–226.

GOULD, S. J. (2000). Of coiled oysters and big brains: how to rescue the terminology of heterochrony, now gone astray. *Evolution & Development* **2**, 241–248.

GRAHAM, L. E. & KANEKO, Y. (1991). Subcellular structures of relevance to the origin of land plants (embryophytes) from green-algae. *Critical Reviews in Plant Sciences* **10**, 323–342.

Graham, L. E. & Wilcox, L. W. (1983). The occurrence and phylogenetic significance of putative placental-transfer cells in the green-alga Coleochaete. *American Journal of Botany* **70**, 113–120.

Graham, L. K. E. & Wilcox, L. W. (2000). The origin of alternation of generations in land plants: a focus on matrotrophy and hexose transport. *Philosophical Transactions of the Royal Society of London Series B-Biological Sciences* **355**, 757–766.

Grant, V. (1950a). The pollination of *Calycanthus occidentalis. American Journal of Botany* **37**, 294–297.

Grant, V. (1950b). The protection of the ovules in flowering plants. *Evolution* **4**, 179–201.

Grbić, V. & Bleecker, A. B. (2000) Axillary meristem development in *Arabidopsis thaliana. The Plant Journal* **21**, 215–223.

Green, P. B., Steele, C. S. & Rennich, S. C. (1996). Phyllotactic patterns: a biophysical mechanism for their origin. *Annals of Botany* **77**, 515–527.

Greenberg, J. T. (1996). Programmed cell death: a way of life for plants. *Proceedings of the National Academy of Sciences of the United States of America* **93**, 12094–12097.

Greig, N. & Mauseth, J. D. (1991). Structure and function of dimorphic prop roots in *Piper auritum* L. *Bulletin of the Torrey Botanical Club* **118**, 176–183.

Grigg, S. P., Canales, C., Hay, A. & Tsiantis, M. (2005). SERRATE coordinates shoot meristem function and leaf axial patterning in *Arabidopsis. Nature* **437**, 1022–1026.

Grimanelli, D., Garcia, M., Kaszas, E., Perotti, E. & Leblanc, O. (2003). Heterochronic expression of sexual reproductive programs during apomictic development in *Tripsacum. Genetics* **165**, 1521–1531.

Groff, P. A. & Kaplan, D. R. (1988). The relation of root systems to shoot systems in vascular plants. *Botanical Review* **54**, 387–422.

Groot, E. P., Doyle, J. A., Nichol, S. A.& Rost, T. L. (2004). Phylogenetic distribution and evolution of root apical meristem organization in dicotyledonous angiosperms. *International Journal of Plant Sciences* **165**, 97–105.

Groot, E. P., Sweeney, E. J. & Rost, T. L. (2003). Development of the adhesive pad on climbing fig (*Ficus pumila*) stems from clusters of adventitious roots. *Plant and Soil* **248**, 85–96.

Grubb, P. J. (1970). Observations on the structure and biology of *Haplomitrium* and *Takakia*, hepatics with roots. *New Phytologist* **69**, 303–326.

Guédès, M. (1966). Stamen tepal and corona in Narcissus. *Advancing Frontiers of Plant Science* **14**, 43–108.

Guédès, M. (1971). Carpel peltation and syncarpy in *Coriaria ruscifolia* L. *New Phytologist* **70**, 213–227.

Guédès, M. (1972). Stamen-carpel homologies. *Flora* **161**, 184–208.

Guédès, M. (1979). *Morphology of Seed-Plants*. J. Cramer, Vaduz.

Guédès, M. & Dupuy, P. (1970). Further remarks on the leaflet theory of the ovule. *New Phytologist* **69**, 1081–1092.

Guédès, M. & Schmid, R. (1978). Peltate (ascidiate) carpel theory and carpel peltation in *Actinidia chinensis* (Actinidiaceae). *Flora* **167**, 525–543.

Guitton, A. E. & Berger, F. (2005). Loss of function of MULTICOPY SUPPRESSOR OF IRA 1 produces nonviable parthenogenetic embryos in *Arabidopsis. Current Biology* **15**, 750–754.

Guitton, A. E., Page, D. R., Chambrier, P., Lionnet, C., Faure, J. E., Grossniklaus, U. et al. (2004). Identification of new members of Fertilisation Independent Seed Polycomb Group pathway involved in the control of seed development in *Arabidopsis thaliana. Development* **131**, 2971–2981.

Gunawardena, A., Sault, K., Donnelly, P., Greenwood, J. S. & Dengler, N. G. (2005). Programmed cell death and leaf morphogenesis in *Monstera obliqua* (Araceae). *Planta* **221**, 607–618.

Gustafson-brown, C., Savidge, B. & Yanofsky, M. F. (1994). Regulation of the *Arabidopsis* floral homeotic gene *Apetala1. Cell* **76**, 131–143.

Gustafsson, M. H. G. & Albert, V. A. (1999). Inferior ovaries and angiosperm diversification. In *Molecular Systematics and Plant Evolution* (eds P. M. Hollingsworth, R. M. Bateman and R. J. Gornall), pp. 403–431. Taylor & Francis, London.

Haecker, A., Gross-Hardt, R., Geiges, B., Sarkar, A., Breuninger, H., Herrmann, M. et al. (2004). Expression dynamics of *WOX* genes mark cell fate decisions during early embryonic patterning in *Arabidopsis thaliana. Development* **131**, 657–668.

Hagemann, W. & Gleissberg, S. (1996). Organogenetic capacity of leaves: the significance of marginal blastozones in angiosperms. *Plant Systematics and Evolution* **199**, 121–152.

Haig, D. & Wilczek, A. (2006). Sexual conflict and the alternation of haploid and diploid generations. *Philosophical Transactions of the Royal Society B-Biological Sciences* **361**, 335–343.

Hajibabaei, M., Xia, J. N. & Drouin, G. (2006). Seed plant phylogeny: gnetophytes are derived conifers and a sister group to Pinaceae. *Molecular Phylogenetics and Evolution* **40**, 208–217.

Hallé, F., Oldeman, R. A. A. & Tomlinson, P. B. (1978). *Tropical Trees and Forests: An Architectural Analysis.* Springer-Verlag, Berlin.

Hallier, H. (1905). Provisional scheme of the natural (phylogenetic) system of flowering plants. *New Phytologist* **4**, 151–162.

Hao, S., Beck, C. B. & Wang, D. (2003). Structure of the earliest leaves: adaptations to high concentrations of atmospheric CO_2. *International Journal of Plant Sciences* **164**, 71–75.

Hao, S. G., Wang, D. M. & Wang, Q. (2004). A new species of *Estinnophyton* from the Lower Devonian Posongchong formation, Yunnan, China; its phylogenetic and palaeophytogeograpical signficance. *Botanical Journal of the Linnean Society* **146**, 201–216.

Hareven, D., Gutfinger, T., Parnis, A., Eshed, Y. & Lifschitz, E. (1996). The making of a compound leaf: genetic manipulation of leaf architecture in tomato. *Cell* **84**, 735–744.

Harper, L. & Freeling, M. (1996). Interactions of liguleless1 and liguleless2 function during ligule induction in maize. *Genetics* **144**, 1871–1882.

Harrison, C. J., Corley, S. B., Moylan, E. C., Alexander, D. L., Scotland, R. W. & Langdale, J. A. (2005a). Independent recruitment of a conserved developmental mechanism during leaf evolution. *Nature* **434**, 509–514.

Harrison, C. J., Rezvani, M. & Langdale, J. A. (2007). Growth from two transient apical initials in the meristem of *Selaginella kraussiana. Development* **134**, 881–889.

Harrison, J., Moller, M., Langdale, J., Cronk, Q. & Hudson, A. (2005b). The role of *KNOX* genes in the evolution of morphological novelty in *Streptocarpus. Plant Cell* **17**, 430–443.

Hawes, M. C., Gunawardena, U., Miyasaka, S. & Zhao, X. W. (2000). The role of root border cells in plant defense. *Trends in Plant Science* **5**, 128–133.

Hay, A., Kaur, H., Phillips, A., Hedden, P., Hake, S. & Tsiantis, M. (2002). The gibberellin pathway mediates KNOTTED1-type homeobox function in plants with different body plans. *Current Biology* **12**, 1557–1565.

Hay, A. & Tsiantis, M. (2006). The genetic basis for differences in leaf form between *Arabidopsis thaliana* and its wild relative *Cardamine hirsuta. Nature Genetics* **38**, 942–947.

Hayashi, S. (1996). A Cdc2 dependent checkpoint maintains diploidy in *Drosophila. Development* **122**, 1051–1058.

He, C. Y. & Saedler, H. (2005). Heterotopic expression of MPF2 is the key to the evolution of the Chinese lantern of *Physalis*, a morphological novelty in Solanaceae. *Proceedings of the National Academy of Sciences of the United States of America* **102**, 5779–5784.

He, C. Y. & Saedler, H. (2007). Hormonal control of the inflated calyx syndrome, a morphological novelty, in *Physalis. Plant Journal* **49**, 935–946.

He, C. Y., Sommer, H., Grosardt, B., Huijser, P. & Saedler, H. (2007). PFMAGO, a MAGO NASHI-like factor, interacts with the MADS-domain protein MPF2 from *Physalis floridana. Molecular Biology and Evolution* **24**, 1229–1241.

Heide-jorgensen, H. S. & Kuijt, J. (1995). The haustorium of the root parasite *Triphysaria* (Scrophulariaceae), with special reference to xylem bridge ultrastructure. *American Journal of Botany* **82**, 782–797.

Heil, M., Rattke, J. & Boland, W. (2005). Postsecretory hydrolysis of nectar sucrose and specialization in ant/plant mutualism. *Science* **308**, 560–563.

Henslow, G. (1890). On the vascular systems of floral organs, and their importance in the interpretation of the morphology of flowers. *Botanical Journal of Linnean Society* **28**, 151–197.

Hepworth, S. R., Klenz, J. E. & Haughn, G. W. (2006). UFO in the *Arabidopsis* inflorescence apex is required for floral-meristem identity and bract suppression. *Planta* **223**, 769–778.

Hepworth, S. R., Zhang, Y. L., McKim, S., Li, X. & Haughn, G. (2005). BLADE-ON-PETIOLE-dependent signaling controls leaf and floral patterning in *Arabidopsis. Plant Cell* **17**, 1434–1448.

Hermann, P. M. & Palser, B. F. (2000). Stamen development in the Ericaceae. I. Anther wall, microsporogenesis, inversion, and appendages. *American Journal of Botany* **87**, 934–957.

Hernandez, L. F. & Green, P. B. (1993). Transductions for the expression of structural pattern—analysis in sunflower. *Plant Cell* **5**, 1725–1738.

Herr, J. M. (1995). The origin of the ovule. *American Journal of Botany* **82**, 547–564.

Heslop-Harrison, Y. & Shivanna, K. R. (1977). The receptive surface of the angiosperm stigma. *Annals of Botany* **41**, 1233–1258.

Hickey, L. J. (1973). Classification of the architecture of dicotyledonous leaves. *American Journal of Botany* **60**, 17–33

Hickey, L. J. (1974). A revised classification of the architecture of dicotyledonous leaves. In *Anatomy of the Dicotyledons* (eds C. R. Metcalfe and L. Chalk),

Volume I, Second edition, pp. 25–39. Clarendon Press, Oxford.

Hickey, L. J. & Wolfe, J. A. (1975). The bases of angiosperm phylogeny: vegetative morphology. *Annals of the Missouri Botanical Garden* **62**, 538–589.

Hicks, G. S. (1973). Studies on tobacco petal primordia in culture—petal duplication induced by surgery. *Botanical Gazette* **134**, 154–160.

Higashiyama, T., Kuroiwa, H., Kawano, S. & Kuroiwa, T. (1998). Guidance in vitro of the pollen tube to the naked embryo sac of *Torenia fournieri*. *Plant Cell* **10**, 2019–2031.

Higashiyama, T., Kuroiwa, H. & Kuroiwa, T. (2003). Pollen-tube guidance: beacons from the female gametophyte. *Current Opinion in Plant Biology* **6**, 36–41.

Higashiyama, T., Yabe, S., Sasaki, N., Nishimura, Y., Miyagishima, S., Kuroiwa, H. et al. (2001). Pollen tube attraction by the synergid cell. *Science* **293**, 1480–1483.

Hileman, L. C. & Baum, D. A. (2003). Why do paralogs persist? Molecular evolution of CYCLOIDEA and related floral symmetry genes in Antirrhineae (Veronicaceae). *Molecular Biology and Evolution* **20**, 591–600.

Hileman, L. C., Kramer, E. M. & Baum, D. A. (2003). Differential regulation of symmetry genes and the evolution of floral morphologies. *Proceedings of the National Academy of Sciences of the United States of America* **100**, 12814–12819.

Hintz, M., Bartholmes, C., Nutt, P., Ziermann, J., Hameister, S., Neuffer, B. et al. (2006). Catching a 'hopeful monster': shepherd's purse (*Capsella bursa-pastoris*) as a model system to study the evolution of flower development. *Journal of Experimental Botany* **57**, 3531–3542.

Hishi, T. & Takeda, H. (2005). Dynamics of heterorhizic root systems: protoxylem groups within the fine-root system of *Chamaecyparis obtusa*. *New Phytologist* **167**, 509–521.

Hobe, M., Müller, R., Grünewald, M., Brand, U. & Simon, R. (2003). Loss of CLE40, a protein functionally equivalent to the stem cell restricting signal CLV3, enhances root waving in *Arabidopsis*. *Development Genes and Evolution* **213**, 371–381.

Hofmeister, W. (1851). *Vergleichende Untersuchungen der Keimung, Entfaltung und Fruchtbuildung hoherer Kryptogamen (Moose, Farne, Equisetaceen, Rhizokarpeen und Lycopodiaceen) und der Samenbildung der Coniferen*. F. Hofmeister, Leipzig.

Hofmeister, W. (1857). Beitrage zur Kenntniss der Gefasskryptogamen. II. *Abhandlungen der Mathematisch-Physischen Classe der Königlich Sächsischen Gesellschaft der Wissenschaften* **3**, 601–682.

Honys, D. & Twell, D. (2004). Transcriptome analysis of haploid male gametophyte development in *Arabidopsis*. *Genome Biology* **5** [epub.].

Hord, C. L., Chen, C., Deyoung, B. J., Clark, S. E. & Ma, H. (2006). The BAM1/BAM2 receptor-like kinases are important regulators of *Arabidopsis* early anther development. *Plant Cell* **18**, 1667–1680.

Horiguchi, G., Kim, G. T. & Tsukaya, H. (2005). The transcription factor AtGRF5 and the transcription coactivator AN3 regulate cell proliferation in leaf primordia of *Arabidopsis thaliana*. *Plant Journal* **43**, 68–78.

Horner, H. T. (1966). Developmental aspects of heterospory in genus *Selaginella*. A light and electron microscope study. *American Journal of Botany* **53**, 610 [abstract].

Horner, H. T. & Arnott, H. A. (1963). Sporangial arrangement in North American species of *Selaginella*. *Botanical Gazette* **124**, 371–383.

Hou, G. C. & Hill, J. P. (2002). Heteroblastic root development in *Ceratopteris richardii* (Parkeriaceae). *International Journal of Plant Sciences* **163**, 341–351.

Hou, G. C. & Hill, J. P. (2004). Developmental anatomy of the fifth shoot-borne root in young sporophytes of *Ceratopteris richardii*. *Planta* **219**, 212–220.

Hou, G. C., Hill, J. P. & Blancaflor, E. B. (2004). Developmental anatomy and auxin response of lateral root formation in *Ceratopteris richardii*. *Journal of Experimental Botany* **55**, 685–693.

Houliston, G. J. & Chapman, H. M. (2004). Reproductive strategy and population variability in the facultative apomict *Hieracium pilosella* (Asteraceae). *American Journal of Botany* **91**, 37–44.

Hsu, T. C., Liu, H. C., Wang, J. S., Chen, R. W., Wang, Y. C. & Lin, B. L. (2001). Early genes responsive to abscisic acid during heterophyllous induction in *Marsilea quadrifolia*. *Plant Molecular Biology* **47**, 703–715.

Hu, Y., Xie, Q. & Chua, N. (2003) The *Arabidopsis* auxin-inducible gene *ARGOS* controls lateral organ size. *The Plant Cell* **15**, 1951–1961.

Huang, T., Bohlenius, H., Eriksson, S., Parcy, F. & Nilsson, O. (2005). The mRNA of the *Arabidopsis* gene *FT* moves from leaf to shoot apex and induces flowering. *Science* **309**, 1694–1696.

Hufford, L. (1996). The morphology and evolution of male reproductive structures of gnetales. *International Journal of Plant Sciences* **157**, S95–S112.

Hughes, J. S. & Otto, S. P. (1999). Ecology and the evolution of biphasic life cycles. *American Naturalist* **154**, 306–320.

Huijser, P., Klein, J., Lonnig, W. E., Meijer, H., Saedler, H. & Sommer, H. (1992). Bracteomania, an inflorescence anomaly, is caused by the loss of function of the MADS-box gene *squamosa* in *Antirrhinum majus*. *EMBO Journal* **11**, 1239–1249.

Hussey, P. J. (2002). Cytoskeleton—microtubules do the twist. *Nature* **417**, 128–129.

Imaizumi-Anraku, H., Takeda, N., Charpentier, M., Perry, J., Miwa, H., Umehara, Y. et al. (2005). Plastid proteins crucial for symbiotic fungal and bacterial entry into plant roots. *Nature* **433**, 527–531.

Inada, S., Ohgishi, M., Mayama, T., Okada, K. & Sakai, T. (2004). RPT2 is a signal transducer involved in phototropic response and stomatal opening by association with phototropin 1 in *Arabidopsis thaliana*. *Plant Cell* **16**, 887–896.

Ingram, G. C., Boisnard-Lorig, C., Dumas, C. & Rogowsky, P. M. (2000). Expression patterns of genes encoding HD-ZipIV homeo domain proteins define specific domains in maize embryos and meristems. *Plant Journal* **22**, 401–414.

Ishiguro, S., Kawai-Oda, A., Ueda, J., Nishida, I. & Okada, K. (2001). The *DEFECTIVE IN ANTHER DEHISCENCE1* gene encodes a novel phospholipase A1 catalyzing the initial step of jasmonic acid biosynthesis, which synchronizes pollen maturation, anther dehiscence, and flower opening in *Arabidopsis*. *Plant Cell* **13**, 2191–2209.

Ito, T., Wellmer, F., Yu, H., Das, P., Ito, N., Alves-Ferreira, M. et al. (2004). The homeotic protein AGAMOUS controls microsporogenesis by regulation of SPOROCYTELESS. *Nature* **430**, 356–360.

Itoh, J. I., Kitano, H., Matsuoka, M. & Nagato, Y. (2000). *SHOOT ORGANIZATION* genes regulate shoot apical meristem organization and the pattern of leaf primordium initiation in rice. *Plant Cell* **12**, 2161–2174.

Iwakawa, H., Shinmyo, A. & Sekine, M. (2006). *Arabidopsis* CDKA;1, a cdc2 homologue, controls proliferation of generative cells in male gametogenesis. *Plant Journal* **45**, 819–831.

Jack, T. (2004). Molecular and genetic mechanisms of floral control. *Plant Cell* **16**, S1–S17.

Jackson, D. & Hake, S. (1999). Control of phyllotaxy in maize by the *abphyl1* gene. *Development* **126**, 315–323.

Jackson, R. B., Mooney, H. A. & Schulze, E. D. (1997). A global budget for fine root biomass, surface area, and nutrient contents. *Proceedings of the National Academy of Sciences of the United States of America* **94**, 7362–7366.

Jackson, R. B., Moore, L. A., Hoffmann, W. A., Pockman, W. T. & Linder, C. R. (1999). Ecosystem rooting depth determined with caves and DNA. *Proceedings of the National Academy of Sciences of the United States of America* **96**, 11387–11392.

Jacobs, W. P. (1951). Auxin relationship in an intercalary meristem. Further studies on the gynophore of *Arachis hypogaea* L. *American Journal of Botany* **38**, 307–310.

Jaeger, I. (1963). Die hypopeltaten Sepalen von *Viola arvensis* und *V. mirabilis*. *Österreichischen botanischen Zeitschrift* **110**, 417.

Jaffe, F. W., Tattersall, A. & Glover, B. J. (2007). A truncated MYB transcription factor from *Antirrhinum majus* regulates epidermal cell outgrowth. *Journal of Experimental Botany* **58**, 1515–1524.

Jaramillo, M. A. & Kramer, E. M. (2004). APETALA3 and PISTILLATA homologs exhibit novel expression patterns in the unique perianth of *Aristolochia* (Aristolochiaceae). *Evolution & Development* **6**, 449–458.

Jaramillo, M. A. & Kramer, E. M. (2007). The role of developmental genetics in understanding homology and morphological evolution in plants. *International Journal of Plant Sciences* **168**, 61–72.

Jasinski, S., Piazza, P., Craft, J., Hay, A., Woolley, L., Rieu, L. et al. (2005). KNOX action in *Arabidopsis* is mediated by coordinate regulation of cytokinin and gibberellin activities. *Current Biology* **15**, 1560–1565.

Jenik, P. D. & Barton, M. K. (2005). Surge and destroy: the role of auxin in plant embryogenesis. *Development* **132**, 3577–3585.

Jiang, K. & Feldman, L. J. (2005). Regulation of root apical meristem development. *Annual Review of Cell and Developmental Biology* **21**, 485–509.

Johnson, M. A., von Besser, K., Zhou, Q., Smith, E., Aux, G., Patton, D. et al. (2004). *Arabidopsis* hapless mutations define essential gametophytic functions. *Genetics* **168**, 971–982.

Jonsson, H., Heisler, M., Reddy, G. V., Agrawal, V., Gor, V., Shapiro, B. E. et al. (2005). Modeling the organization of the WUSCHEL expression domain in the shoot apical meristem. *Bioinformatics* **21**, I232–I240.

Jullien, P. E., Katz, A., Oliva, M., Ohad, N. & Berger, F. (2006a). Polycomb group complexes self-regulate imprinting of the polycomb group gene *MEDEA* in *Arabidopsis*. *Current Biology* **16**, 486–492.

Jullien, P. E., Kinoshita, T., Ohad, N. & Berger, F. (2006b). Maintenance of DNA methylation during the *Arabidopsis* life cycle is essential for parental imprinting. *Plant Cell* **18**, 1360–1372.

Kamisugi, Y. & Cuming, A. C. (2005). The evolution of the abscisic acid-response in land plants: comparative analysis of group 1 *LEA* gene expression in moss and cereals. *Plant Molecular Biology* **59**, 723–737.

Kamiya, N., Nagasaki, H., Morikami, A., Sato, Y. & Matsuoka, M. (2003). Isolation and characterization of a rice *WUSCHEL*-type homeobox gene that is specifically expressed in the central cells of a quiescent center in the root apical meristem. *Plant Journal* **35**, 429–441.

Kang, H. G., Jeon, J. S., Lee, S. & An, G. H. (1998). Identification of class B and class C floral organ identity genes from rice plants. *Plant Molecular Biology* **38**, 1021–1029.

Kaplan, D. R. (1973). The problem of leaf morphology and evolution in the monocotyledons. *Quarterly Review of Biology* **48**, 437–457.

Kaplan, D. R. (2001). The science of plant morphology: definition, history, and role in modern biology. *American Journal of Botany* **88**, 1711–1741.

Kaplan, D. R. & Cooke, T. J. (1996). The genius of Wilhelm Hofmeister: the origin of causal-analytical research in plant development. *American Journal of Botany* **83**, 1647–1660.

Kasahara, R. D., Portereiko, M. F., Sandaklie-Nikolova, L., Rabiger, D. S. & Drews, G. N. (2005). MYB98 is required for pollen tube guidance and synergid cell differentiation in *Arabidopsis*. *Plant Cell* **17**, 2981–2992.

Kater, M. M., Franken, J., Carney, K. J., Colombo, L. & Angenent, G. C. (2001). Sex determination in the monoecious species cucumber is confined to specific floral whorls. *Plant Cell* **13**, 481–493.

Kato, M. & Akiyama, H. (2005). Interpolation hypothesis for origin of the vegetative sporophyte of land plants. *Taxon* **54**, 443–450.

Kaufmann, K., Anfang, N., Saedler, H. & Theissen, G. (2005). Mutant analysis, protein–protein interactions and subcellular localization of the *Arabidopsis* B-sister (ABS) protein. *Molecular Genetics and Genomics* **274**, 103–118.

Kaufmann, K., Melzer, R. & Theissen, G. (2005). MIKC-type MADS-domain proteins: structural modularity, protein interactions and network evolution in land plants. *Gene* **347**, 183–198.

Kaya, H., Shibahara, K. I., Taoka, K. I., Iwabuchi, M., Stillman, B. & Araki, T. (2001). *FASCIATA* genes for chromatin assembly factor-1 in *Arabidopsis* maintain the cellular organization of apical meristems. *Cell* **104**, 131–142.

Keck, E., McSteen, P., Carpenter, R. & Coen, E. (2003). Separation of genetic functions controlling organ identity in flowers. *EMBO Journal* **22**, 1058–1066.

Keeley, J. E., Osmond, C. B. & Raven, J. A. (1984). Stylites, a vascular land plant without stomata absorbs CO_2 via its roots. *Nature* **310**, 694–695.

Kellogg, E. A. (2002). Are macroevolution and microevolution qualitatively different? Evidence from Poaceae and other families. In *Developmental Genetics and Plant Evolution* (eds Q. C. B. Cronk, R. M. Bateman and J. Hawkins), pp. 70–84. Taylor & Francis, London.

Kenrick, P. (1994). Alternation of generations in land plants—new phylogenetic and paleobotanical evidence. *Biological Reviews of the Cambridge Philosophical Society* **69**, 293–330.

Kenrick, P. & Crane, P. R. (1997c). *The Origin and Early Diversification of Land Plants: A Cladistic Study.* Smithsonian Institution Press, Washington, DC.

Kenrick, P. & Crane, P. R. (1997a). *The Origin and Early Evolution of Land Plants: A Cladistic Study.* Smithsonian Institution Press, Washington, DC.

Kenrick, P. & Crane, P. R. (1997b). The origin and early evolution of plants on land. *Nature* **389**, 33–39.

Kidner, C. A. & Martienssen, R. A. (2004). Spatially restricted microRNA directs leaf polarity through ARGONAUTE1. *Nature* **428**, 81–84.

Kim, G. T., Shoda, K., Tsuge, T., Cho, K. H., Uchimiya, H., Yokoyama, R. et al. (2002). The *ANGUSTIFOLIA* gene of *Arabidopsis*, a plant *CtBP* gene, regulates leaf-cell expansion, the arrangement of cortical microtubules in leaf cells and expression of a gene involved in cell-wall formation. *EMBO Journal* **21**, 1267–1279.

Kim, M., McCormick, S., Timmermans, M. & Sinha, N. (2003a). The expression domain of PHANTASTICA determines leaflet placement in compound leaves. *Nature* **424**, 438–443.

Kim, M., Pham, T., Hamidi, A., McCormick, S., Kuzoff, R. K. & Sinha, N. (2003b). Reduced leaf complexity in tomato wiry mutants suggests a role for *PHAN* and *KNOX* genes in generating compound leaves. *Development* **130**, 4405–4415.

Kim, S., Koh, J., Yoo, M. J., Kong, H., Hu, Y., Ma, H. et al. (2005). Expression of floral MADS-box genes in basal angiosperms: implications for the evolution of floral regulators. *Plant Journal* **43**, 724–744.

Kim, S., Mollet, J. C., Dong, J., Zhang, K. L., Park, S. Y. & Lord, E. M. (2003c). Chemocyanin, a small basic protein from the lily stigma, induces pollen tube chemotropism. *Proceedings of the National Academy of Sciences of the United States of America* **100**, 16125–16130.

King, M. C. & Wilson, A. C. (1975). Evolution at 2 levels in humans and chimpanzees. *Science* **188**, 107–116.

Kinoshita, T., Miura, A., Choi, Y. H., Kinoshita, Y., Cao, X. F., Jacobsen, S. E. et al. (2004). One-way control of FWA imprinting in *Arabidopsis* endosperm by DNA methylation. *Science* **303**, 521–523.

Kirchoff, B. K. (1983). Floral organogenesis in 5 genera of the Marantaceae and in *Canna* (Cannaceae). *American Journal of Botany* **70**, 508–523.

Kirchoff, B. K. (1991). Homeosis in the flowers of the Zingiberales. *American Journal of Botany* **78**, 833–837.

Kitazawa, D., Hatakeda, Y., Kamada, M., Fujii, N., Miyazawa, Y., Hoshino, A. et al. (2005). Shoot circumnutation and winding movements require gravisensing cells. *Proceedings of the National Academy of Sciences of the United States of America* **102**, 18742–18747.

Kiyosue, T., Ohad, N., Yadegari, R., Hannon, M., Dinneny, J., Wells, D. et al. (1999). Control of fertilization-independent endosperm development by the *MEDEA* polycomb gene *Arabidopsis*. *Proceedings of the National Academy of Sciences of the United States of America* **96**, 4186–4191.

Knopf, R. R. & Trebitsh, T. (2006). The female-specific Cs-*ACS1G* gene of cucumber. A case of gene duplication and recombination between the non-sex-specific 1-aminocyclopropane-1-carboxylate synthase gene and a branched-chain amino acid transaminase gene. *Plant and Cell Physiology* **47**, 1217–1228.

Koi, S. & Kato, M. (2003). Comparative developmental anatomy of the root in three species of *Cladopus* (Podostemaceae). *Annals of Botany* **91**, 927–937.

Koornneef, M., Bentsink, L. & Hilhorst, H. (2002). Seed dormancy and germination. *Current Opinion in Plant Biology* **5**, 33–36.

Korall, P. & Kenrick, P. (2002). Phylogenetic relationships in Selaginellaceae based on RBCL sequences. *American Journal of Botany* **89**, 506–517.

Kosuta, S., Chabaud, M., Lougnon, G., Gough, C., Denarie, J., Barker, D. G. et al. (2003). A diffusible factor from arbuscular mycorrhizal fungi induces symbiosis-specific MtENOD11 expression in roots of *Medicago truncatula*. *Plant Physiol* **131**, 952–962.

Krings, M. (2003). A pteridosperm stem with clusters of shoot-borne roots from the Namurian B (Upper Carboniferous) of Hagen-Vorhalle (Germany). *Review of Palaeobotany and Palynology* **123**, 289–301.

Krings, M., Kerp, H., Taylor, T. N. & Taylor, E. L. (2003). How Paleozoic vines and lianas got off the ground: on scrambling and climbing Carboniferous-early Permian pteridosperms. *Botanical Review* **69**, 204–224.

Krizek, B. A. & Fletcher, J. C. (2005). Molecular mechanisms of flower development: an armchair guide. *Nature Reviews Genetics* **6**, 688–698.

Krizek, B. A., Lewis, M. W. & Fletcher, J. C. (2006). RABBIT EARS is a second-whorl repressor of AGAMOUS that maintains spatial boundaries in *Arabidopsis* flowers. *Plant Journal* **45**, 369–383.

Kroken, S. B., Graham, L. E. & Cook, M. E. (1996). Occurrence and evolutionary significance of resistant cell walls in charophytes and bryophytes. *American Journal of Botany* **83**, 1241–1254.

Ku, S. J., Yoon, H., Suh, H. S. & Chung, Y. Y. (2003). Male-sterility of thermosensitive genic male-sterile rice is associated with premature programmed cell death of the tapetum. *Planta* **217**, 559–565.

Kubo, M., Udagawa, M., Nishikubo, N., Horiguchi, G., Yamaguchi, M., Ito, J. et al. (2005). Transcription switches for protoxylem and metaxylem vessel formation. *Genes & Development* **19**, 1855–1860.

Kuijt, J. (1977). Haustoria of phanerogamic parasites. *Annual Review of Phytopathology* **17**, 91–118.

Kurata, T., Ishida, T., Kawabata-Awai, C., Noguchi, M., Hattori, S., Sano, R. et al. (2005). Cell-to-cell movement of the CAPRICE protein in *Arabidopsis* root epidermal cell differentiation. *Development* **132**, 5387–5398.

Kuwabara, A., Ikegami, K., Koshiba, T. & Nagata, T. (2003a). Effects of ethylene and abscisic acid upon heterophylly in *Ludwigia arcuata* (Onagraceae). *Planta* **217**, 880–887.

Kuwabara, A., Ikegami, K., Koshiba, T. & Nagata, T. (2003b). Interactions between ethylene and abscisic acid upon heterophylly in *Ludwigia arcuata* (Onagraceae). *Plant and Cell Physiology* **44**, S10.

Kuwabara, A. & Nagata, T. (2005). Orientation of cell division was affected by ethylene upon heterophyllous changes in *Ludwigia arcuata* (Onagraceae). *Plant and Cell Physiology* **46**, S30.

Kwak, S. H., Shen, R. & Schiefelbein, J. (2005). Positional signaling mediated by a receptor-like kinase in *Arabidopsis*. *Science* **307**, 1111–1113.

Kwon, C. S., Chen, C. B. & Wagner, D. (2005). WUSCHEL is a primary target for transcriptional regulation by SPLAYED in dynamic control of stem cell fate in *Arabidopsis*. *Genes & Development* **19**, 992–1003.

Kwong, R. W., Bui, A. Q., Lee, H., Kwong, L. W., Fischer, R. L., Goldberg, R. B. et al. (2003). LEAFY COTYLEDON1-LIKE defines a class of regulators essential for embryo development. *Plant Cell* **15**, 5–18.

Lam, H. J. (1950). Stachyospory and phyllospory as factors in the natural system of the Cormophyta. *Svensk Botanisk Tidskrift* 44, 517–534.

Lamb, R. S., Hill, T. A., Tan, Q. K. G. & Irish, V. F. (2002). Regulation of *APETALA3* floral homeotic gene expression by meristem identity genes. *Development* **129**, 2079–2086.

LaMotte, C. E. & Pickard, B. G. (2004a). Control of gravitropic orientation. I. Non-vertical orientation by primary roots of maize results from decay of competence for orthogravitropic induction. *Functional Plant Biology* **31**, 93–107.

LaMotte, C. E. & Pickard, B. G. (2004b). Control of gravitropic orientation. II. Dual receptor model for gravitropism. *Functional Plant Biology* **31**, 109–120.

Larkin, J. C., Oppenheimer, D. G., Pollock, S. & Marks, M. D. (1993). *Arabidopsis Glabrous1* gene requires downstream sequences for function. *Plant Cell* **5**, 1739–1748.

Layzell, D. B., Gaito, S. T. & Hunt, S. (1988). Model of gas-exchange and diffusion in legume nodules.1. Calculation of gas-exchange rates and the energy-cost of N_2 fixation. *Planta* **173**, 117–127.

Lee, I., Wolfe, D. S., Nilsson, O. & Weigel, D. (1997). A LEAFY co-regulator encoded by UNUSUAL FLORAL ORGANS. *Current Biology* **7**, 95–104.

Lee, J. Y., Baum, S. F., Alvarez, J., Patel, A., Chitwood, D. H. & Bowman, J. L. (2005a). Activation of CRABS CLAW in the nectaries and carpels of *Arabidopsis*. *Plant Cell* **17**, 25–36.

Lee, J. Y., Baum, S. F., Oh, S. H., Jiang, C. Z., Chen, J. C. & Bowman, J. L. (2005b). Recruitment of CRABS CLAW to promote nectary development within the eudicot clade. *Development* **132**, 5021–5032.

Lee, J. E., Oliveira, R. S., Dawson, T. E. & Fung, I. (2005c). Root functioning modifies seasonal climate. *Proceedings of the National Academy of Sciences of the United States of America* **102**, 17576–17581.

Lee, J. Y., Mummenhoff, K. & Bowman, J. L. (2002). Allopolyploidization and evolution of species with reduced floral structures in *Lepidium* L. (Brassicaceae). *Proceedings of the National Academy of Sciences of the United States of America* **99**, 16835–16840.

Lee, M. M. & Schiefelbein, J. (2001). Developmentally distinct *MYB* genes encode functionally equivalent proteins in *Arabidopsis*. *Development* **128**, 1539–1546.

Leebens-Mack, J. H., Wall, K., Duarte, J., Zheng, Z. G., Oppenheimer, D. & Depamphilis, C. (2006). A genomics approach to the study of ancient polyploidy and floral developmental genetics. *Advances in Botanical Research Incorporating Advances in Plant Pathology* **44**, 527–549.

Lemon, G. D. & Posluszny, U. (1997). Shoot morphology and organogenesis of the aquatic floating fern *Salvinia molesta* D.S. Mitchell, examined with the aid of laser scanning confocal microscopy. *International Journal of Plant Sciences* **158**, 693–703.

Levin, J. Z., Fletcher, J. C., Chen, X. M. & Meyerowitz, E. M. (1998). A genetic screen for modifiers of UFO meristem activity identifies three novel FUSED *FLORAL ORGANS* genes required for early flower development in *Arabidopsis*. *Genetics* **149**, 579–595.

Levin, J. Z. & Meyerowitz, E. M. (1995). *Ufo*—an *Arabidopsis* gene involved in both floral meristem and floral organ development. *Plant Cell* **7**, 529–548.

Levy, J., Bres, C., Geurts, R., Chalhoub, B., Kulikova, O., Duc, G. et al. (2004). A putative Ca^{2+} and calmodulin-dependent protein kinase required for bacterial and fungal symbioses. *Science* **303**, 1361–1364.

Leyser, H. M. O. & Furner, I. J. (1992). Characterisation of three shoot apical meristem mutants of *Arabidopsis thaliana*. *Development* **116**, 397–403.

Li, C. S. & Edwards, D. (1995). A reinvestigation of Halle *Drepanophycus spinaeformis* Gopp from the Lower Devonian of Yunnan Province, Southern China. *Botanical Journal of the Linnean Society* **118**, 163–192.

Li, Y. H., Beisson, F., Koo, A. J. K., Molina, I., Pollard, M. & Ohlrogge, J. (2007). Identification of acyltransferases required for cutin biosynthesis and production of cutin with suberin-like monomers. *Proceedings of the National Academy of Sciences of the United States of America* **104**, 18339–18344.

Lieu, S. M. (1979). Organogenesis in *Triglochin striata*. *Canadian Journal of Botany-Revue Canadienne De Botanique* **57**, 1418–1438.

Ligrone, R., Duckett, J. G. & Renzaglia, K. S. (1993). The gametophyte–sporophyte junction in land plants. *Advances in Botanical Research Incorporating Advances in Plant Pathology* **19**, 231–317.

Liljegren, S. J., Gustafson-Brown, C., Pinyopich, A., Ditta, G. S. & Yanofsky, M. F. (1999). Interactions among APETALA1, LEAFY, and TERMINAL FLOWER1 specify meristem fate. *Plant Cell* **11**, 1007–1018.

Lincoln, C., Long, J., Yamaguchi, J., Serikawa, K. & Hake, S. (1994). A *Knotted1*-Like homeobox gene in *Arabidopsis* is expressed in the vegetative meristem and dramatically alters leaf morphology when overexpressed in transgenic plants. *Plant Cell* **6**, 1859–1876.

Linnaeus, C. & Rudberg, D. (1749). De Peloria [reprint of Rudberg's 1744 dissertation]. In *Amoenitates academicae* [*Botanical delights*], Vol. 1, pp. 280–298.

Litt, A. & Stevenson, D. W. (2003). Floral development and morphology of Vochysiaceae. II. The position of the single fertile stamen. *American Journal of Botany* **90**, 1548–1559.

Liu, C. M., McElver, J., Tzafrir, I., Joosen, R., Wittich, P., Patton, D. et al. (2002). Condensin and cohesin knockouts in *Arabidopsis* exhibit a titan seed phenotype. *Plant Journal* **29**, 405–415.

Liu, Z. Y., Moore, P. H., Ma, H., Ackerman, C. M., Ragiba, M., Yu, Q. Y. et al. (2004). A primitive Y chromosome in papaya marks incipient sex chromosome evolution. *Nature* **427**, 348–352.

Lohmann, J. U., Hong, R. L., Hobe, M., Busch, M. A., Parcy, F., Simon, R. et al. (2001). A molecular link between stem cell regulation and floral patterning in *Arabidopsis*. *Cell* **105**, 793–803.

Lu, P. Z. & Jernstedt, J. A. (1996). Rhizophore and root development in *Selaginella martensii*: meristem transitions and identity. *International Journal of Plant Sciences* **157**, 180–194.

Luo, D., Carpenter, R., Vincent, C., Copsey, L. & Coen, E. (1996). Origin of floral asymmetry in *Antirrhinum*. *Nature* **383**, 794–799.

Luo, D., Coen, E. S., Doyle, S. & Carpenter, R. (1991). Pigmentation mutants produced by transposon mutagenesis in *Antirrhinum majus*. *Plant Journal* **1**, 59–69.

Luo, M., Bilodeau, P., Koltunow, A., Dennis, E. S., Peacock, W. J. & Chaudhury, A. M. (1999). Genes controlling fertilization-independent seed development in *Arabidopsis thaliana*. *Proceedings of the National Academy of Sciences of the United States of America* **96**, 296–301.

Lupia, R., Schneider, H., Moeser, G. M., Pryer, K. M. & Crane, P. R. (2000). Marsileaceae sporocarps and spores from the Late Cretaceous of Georgia, USA. *International Journal of Plant Sciences* **161**, 975–988.

Lynch, M. & Force, A. (2000). The probability of duplicate gene preservation by subfunctionalization. *Genetics* **154**, 459–473.

Maas, P. J. M. & Rubsamen, T. (1986). Triuridaceae. *Flora Neotropica* **40**, 1–55.

Mabee, P. M., Ashburner, M., Cronk, Q. C. B., Gkoutos, G. V., Haendel, M., Segerdell, E. et al. (2007). Phenotype ontologies: the bridge between genomics and evolution. *Trends in Ecology and Evolution* **22**, 345–350.

Mable, B. K. & Otto, S. P. (1998). The evolution of life cycles with haploid and diploid phases. *Bioessays* **20**, 453–462.

Mable, B. K. & Otto, S. P. (2001). Masking and purging mutations following EMS treatment in haploid, diploid and tetraploid yeast (*Saccharomyces cerevisiae*). *Genetical Research* **77**, 9–26.

Madsen, E. B., Madsen, L. H., Radutoiu, S., Olbryt, M., Rakwalska, M., Szczyglowski, K. et al. (2003). A receptor kinase gene of the LysM type is involved in legume perception of rhizobial signals. *Nature* **425**, 637–640.

Maheshwari, P. (1950). *An Introduction to the Embryology of Angiosperms*. McGraw-Hill, New York.

Maksymowych, R. & Wochok, Z. S. (1969). Activity of marginal and plate meristems during leaf development of *Xanthium pennsylvanicum*. *American Journal of Botany* **56**, 26–30.

Malcomber, S. T. & Kellogg, E. A. (2006). Evolution of unisexual flowers in grasses (Poaceae) and the putative sex-determination gene, *TASSELSEED2* (*TS2*). *New Phytologist* **170**, 885–899.

Marshall, C. R., Raff, E. C. & Raff, R. A. (1994). Dollo's law and the death and resurrection of genes. *Proceedings of the National Academy of Sciences of the United States of America* **91**, 12283–12287.

Marton, M. L., Cordts, S., Broadhvest, J. & Dresselhaus, T. (2005). Micropylar pollen tube guidance by egg apparatus 1 of maize. *Science* **307**, 573–576.

Masiero, S., Li, M. A., Will, I., Hartmann, U., Saedler, H., Huijser, P. et al. (2004). *INCOMPOSITA*: a MADS-box gene controlling prophyll development and floral meristem identity in *Antirrhinum*. *Development* **131**, 5981–5990.

Masters, M. T. (1869). *Vegetable Teratology*. Ray Society, London.

Masucci, J. D., Rerie, W. G., Foreman, D. R., Zhang, M., Galway, M. E., Marks, M. D. et al. (1996). The homeobox gene *GLABRA 2* is required for position-dependent cell differentiation in the root epidermis of *Arabidopsis thaliana*. *Development* **122**, 1253–1260.

Mattsson, J., Ckurshumova, W. & Berleth, T. (2003). Auxin signaling in *Arabidopsis* leaf vascular development. *Plant Physiology* **131**, 1327–1339.

Mayer, K. F. X., Schoof, H., Haecker, A., Lenhard, M., Jurgens, G. & Laux, T. (1998). Role of WUSCHEL in regulating stem cell fate in the *Arabidopsis* shoot meristem. *Cell* **95**, 805–815.

McAbee, J. M., Hill, T. A., Skinner, D. J., Izhaki, A., Hauser, B. A., Meister, R. J. et al. (2006). ABERRANT TESTA SHAPE encodes a KANADI family member, linking polarity determination to separation and growth of *Arabidopsis* ovule integuments. *Plant Journal* **46**, 522–531.

McAbee, J. M., Kuzoff, R. K. & Gasser, C. S. (2005). Mechanisms of derived unitegmy among *Impatiens* species. *Plant Cell* **17**, 1674–1684.

McConnell, J. R., Emery, J., Eshed, Y., Bao, N., Bowman, J. & Barton, M. K. (2001). Role of PHABULOSA and PHAVOLUTA in determining radial patterning in shoots. *Nature* **411**, 709–713.

McGee, P. A. (1988). Growth-response to and morphology of mycorrhizas of *Thysanotus* (Anthericaceae, Monocotyledonae). *New Phytologist* **109**, 459–463.

Meeuse, A. D. J. (1973). Neomorphology and angiosperm phylogeny. *Acta Botanica Neerlandica* **22**, 246.

Meeuse, A. D. J. (1975). Changing floral concepts—anthocorms, flowers, and anthoids. *Acta Botanica Neerlandica* **24**, 23–36.

Mehltreter, K. & Palacios-Rios, M. (2003). Phenological studies of *Acrostichum danaeifolium* (Pteridaceae, Pteridophyta) at a mangrove site on the Gulf of Mexico. *Journal of Tropical Ecology* **19**, 155–162.

Meinke, D. W. (1992). A homeotic mutant of *Arabidopsis thaliana* with leafy cotyledons. *Science* **258**, 1647–1650.

Meinke, D. W., Franzmann, L. H., Nickle, T. C. & Yeung, E. C. (1994). Leafy cotyledon mutants of *Arabidopsis*. *Plant Cell* **6**, 1049–1064.

Meloche, C. G. & Diggle, P. K. (2001). Preformation, architectural complexity, and developmental flexibility in *Acomastylis rossii* (Rosaceae). *American Journal of Botany* **88**, 980–991.

Melville, R. (1962). A new theory of the angiosperm flower. I. The gynoecium. *Kew Bulletin* **16**, 1–50.

Melville, R. (1963). A new theory of the angiosperm flower: II. The androecium. *Kew Bulletin* **17**, 1–63.

Melzer, R., Kaufmann, K. & Theissen, G. (2006). Missing links: DNA-binding and target gene specificity of floral homeotic proteins. *Advances in Botanical Research Incorporating Advances in Plant Pathology*, **44**, 209–236.

Menand, B., Yi, K., Jouannic, S., Hoffmann, L., Ryan, E., Linstead, P. et al. (2007). An ancient mechanism controls the development of cells with a rooting function in land plants. *Science* **316**, 1477–1480.

Meyer, V. G. (1966). Flower abnormalities. *Botanical Review* **32**, 165–218.

Meyerowitz, E. M., Smyth, D. R. & Bowman, J. L. (1989). Abnormal flowers and pattern-formation in floral development. *Development* **106**, 209–217.

Mibus, H. & Tatlioglu, T. (2004). Molecular characterization and isolation of the *F/f* gene for femaleness in cucumber (*Cucumis sativus* L.). *Theoretical and Applied Genetics* **109**, 1669–1676.

Ming, R. & Moore, P. H. (2007). Genomics of sex chromosomes. *Current Opinion in Plant Biology* **10**, 123–130.

Ming, R., Wang, J. P., Moore, P. H. & Paterson, A. H. (2007). Sex chromosomes in flowering plants. *American Journal of Botany* **94**, 141–150.

Mitra, R. M., Gleason, C. A., Edwards, A., Hadfield, J., Downie, J. A., Oldroyd, G. E. et al. (2004). A Ca^{2+}/calmodulin-dependent protein kinase required for symbiotic nodule development: gene identification by transcript-based cloning. *Proceedings of the National Academy of Sciences of the United States of America* **101**, 4701–4705.

Mitsuda, N., Seki, M., Shinozaki, K. & Ohme-Takagi, M. (2005). The NAC transcription factors NST1 and NST2 of *Arabidopsis* regulate secondary wall thickenings and are required for anther dehiscence. *Plant Cell* **17**, 2993–3006.

Mizukami, Y. & Ma, H. (1992). Ectopic expression of the floral homeotic gene agamous in transgenic *Arabidopsis* plants alters floral organ identity. *Cell* **71**, 119–131.

Molinero-Rosales, N., Jamilena, M., Zurita, S., Gomez, P., Capel, J. & Lozano, R. (1999). FALSIFLORA, the tomato orthologue of FLORICAULA and LEAFY, controls flowering time and floral meristem identity. *Plant Journal* **20**, 685–693.

Möller, M. & Cronk, Q. C. B. (2001). Evolution of morphological novelty: a phylogenetic analysis of growth patterns in *Streptocarpus* (Gesneriaceae). *Evolution* **55**, 918–929.

Mori, T., Kuroiwa, H., Higashiyama, T. & Kuroiwa, T. (2006). GENERATIVE CELL SPECIFIC 1 is essential for angiosperm fertilization. *Nature Cell Biology* **8**, 64–71.

Motchoulski, A. & Liscum, E. (1999). *Arabidopsis* NPH3: a NPH1 photoreceptor-interacting protein essential for phototropism. *Science* **286**, 961–964.

Motose, H., Sugiyama, M. & Fukuda, H. (2004). A proteoglycan mediates inductive interaction during plant vascular development. *Nature* **429**, 873–878.

Muehlbauer, G. J., Fowler, J. E. & Freeling, M. (1997). Sectors expressing the homeobox gene *liguleless3* implicate a time-dependent mechanism for cell fate acquisition along the proximal–distal axis of the maize leaf. *Development* **124**, 5097–5106.

Muehlbauer, G. J., Fowler, J. E., Girard, L., Tyers, R., Harper, L. & Freeling, M. (1999). Ectopic expression of the maize homeobox gene *Liguleless3* alters cell fates in the leaf. *Plant Physiology* **119**, 651–662.

Nagasawa, N., Miyoshi, M., Sano, Y., Satoh, H., Hirano, H., Sakai, H. et al. (2003). *SUPERWOMAN1* and *DROOPING LEAF* genes control floral organ identity in rice. *Development* **130**, 705–718.

Nakahigashi, K., Jasencakova, Z., Schubert, I. & Goto, K. (2005). The *Arabidopsis* HETEROCHROMATIN PROTEIN1 homolog (TERMINAL FLOWER2) silences genes within the euchromatic region but not genes positioned in heterochromatin. *Plant and Cell Physiology* **46**, 1747–1756.

Nakajima, K., Furutani, I., Tachimoto, H., Matsubara, H. & Hashimoto, T. (2004). SPIRAL1 encodes a plant-specific microtubule-localized protein required for directional control of rapidly expanding *Arabidopsis* cells. *Plant Cell* **16**, 1178–1190.

Nakajima, K., Sena, G., Nawy, T. & Benfey, P. N. (2001). Intercellular movement of the putative transcription factor SHR in root patterning. *Nature* **413**, 307–311.

Nakamura, T., Fukuda, T., Nakano, M., Hasebe, M., Kameya, T. & Kanno, A. (2005). The modified ABC model explains the development of the petaloid perianth of *Agapanthus praecox* ssp *orientalis* (Agapanthaceae) flowers. *Plant Molecular Biology* **58**, 435–445.

Nakazato, T., Jung, M. K., Housworth, E. A., Rieseberg, L. H. & Gastony, G. J. (2006). Genetic map-based analysis of genome structure in the homosporous fern *Ceratopteris richardii*. *Genetics* **173**, 1585–1597.

Narita, N. N., Moore, S., Horiguchi, G., Kubo, M., Demura, T., Fukuda, H. et al. (2004). Overexpression of a novel small peptide ROTUNDIFOLIA4 decreases cell proliferation and alters leaf shape in *Arabidopsis thaliana*. *Plant Journal* **38**, 699–713.

Nath, U., Crawford, B. C. W., Carpenter, R. & Coen, E. (2003). Genetic control of surface curvature. *Science* **299**, 1404–1407.

Newman, E. I. & Reddell, P. (1987). The distribution of mycorrhizas among families of vascular plants. *New Phytologist* **106**, 745–751.

Nickerson, J. & Drouin, G. (2004). The sequence of the largest subunit of RNA polymerase II is a useful marker for inferring seed plant phylogeny. *Molecular Phylogenetics and Evolution* **31**, 403–415.

Niklas, K. J. (1998). The mechanical roles of clasping leaf sheaths: evidence from *Arundinaria tecta* (Poaceae) shoots subjected to bending and twisting forces. *Annals of Botany* **81**, 23–34.

Niklas, K. J., Cobb, E. D. & Marler, T. (2006). A comparison between the record height-to-stem diameter allometries of pachycaulis and leptocaulis species. *Annals of Botany* **97**, 79–83.

Nikovics, K., Blein, T., Peaucelle, A., Ishida, T., Morin, H., Aida, M. et al. (2006). The balance between the *miR164a* and *CUC2* genes controls leaf margin serration in *Arabidopsis*. *Plant Cell* **18**, 2929–2945.

Nishimura, T., Wada, T., Yamamoto, K. T. & Okada, K. (2005). The *Arabidopsis* STV1 protein, responsible for translation reinitiation, is required for auxin-mediated gynoecium patterning. *Plant Cell* **17**, 2940–2953.

Noda, K., Glover, B. J., Linstead, P. & Martin, C. (1994). Flower color intensity depends on specialized cell-shape controlled by a Myb-related transcription factor. *Nature* **369**, 661–664.

Nogler, G. A. (1984). Genetics of apospory in apomictic *Ranunculus auricomus*. 5. Conclusion. *Botanica Helvetica* **94**, 411–422.

Nonomura, K. I., Miyoshi, K., Eiguchi, M., Suzuki, T., Miyao, A., Hirochika, H. et al. (2003). The *MSP1* gene is necessary to restrict the number of cells entering into male and female sporogenesis and to initiate anther wall formation in rice. *Plant Cell* **15**, 1728–1739.

Norberg, M., Holmlund, M. & Nilsson, O. (2005). The *BLADE ON PETIOLE* genes act redundantly to control the growth and development of lateral organs. *Development* **132**, 2203–2213.

Nowak, J., Dengler, N. G. & Posluszny, U. (2007). The role of abscission during leaflet separation in *Chamaedorea elegans* (Arecaceae). *International Journal of Plant Sciences* **168**, 533–545.

Noyes, R. D. (2000). Diplospory and parthenogenesis in sexual × agamospermous (apomictic) *Erigeron* (Asteraceae) hybrids. *International Journal of Plant Sciences* **161**, 1–12.

Noyes, R. D. & Rieseberg, L. H. (2000). Two independent loci control agamospermy (apomixis) in the triploid flowering plant *Erigeron annuus*. *Genetics* **155**, 379–390.

Nuismer, S. L. & Otto, S. P. (2004). Host–parasite interactions and the evolution of ploidy. *Proceedings of the National Academy of Sciences of the United States of America* **101**, 11036–11039.

Ohad, N., Yadegari, R., Margossian, L., Hannon, M., Michaeli, D., Harada, J. J. et al. (1999). Mutations in *FIE*, a WD polycomb group gene, allow endosperm development without fertilization. *Plant Cell* **11**, 407–415.

Ohashi, Y., Oka, A., Rodrigues-Pousada, R., Possenti, M., Ruberti, I., Morelli, G. et al. (2003). Modulation of phospholipid signaling by

GLABRA2 in root-hair pattern formation. *Science* **300**, 1427–1430.

Ohashi, Y., Oka, A., Ruberti, I., Morelli, G. & Aoyama, T. (2002). Entopically additive expression of GLABRA2 alters the frequency and spacing of trichome initiation. *Plant Journal* **29**, 359–369.

Ohno, S. (1970). *Evolution by Gene Duplication*. Springer-Verlag, New York.

Oliver, M. J., Velten, J. & Mishler, B. D. (2005). Desiccation tolerance in bryophytes: a reflection of the primitive strategy for plant survival in dehydrating habitats? *Integrative and Comparative Biology* **45**, 788–799.

Olsen, O. A. (2004). Nuclear endosperm development in cereals and *Arabidopsis thaliana*. *Plant Cell* **16**, S214–S227.

Ono, T., Kaya, H., Takeda, S., Abe, M., Ogawa, Y., Kato, M. et al. (2006). Chromatin assembly factor 1 ensures the stable maintenance of silent chromatin states in *Arabidopsis*. *Genes Cells* **11**, 153–162.

Ordidge, M., Chiurugwi, T., Tooke, F. & Battey, N. H. (2005). LEAFY, TERMINAL FLOWER1 and AGAMOUS are functionally conserved but do not regulate terminal flowering and floral determinacy in *Impatiens balsamina*. *Plant Journal* **44**, 985–1000.

Osmont, K. S., Jesaitis, L. A. & Freeling, M. (2003). The extended *auricle1* (*eta1*) gene is essential for the genetic network controlling postinitiation maize leaf development. *Genetics* **165**, 1507–1519.

Otegui, M. S., Verbrugghe, K. J. & Skop, A. R. (2005). Midbodies and phragmoplasts: analogous structures involved in cytokinesis. *Trends in Cell Biology* **15**, 404–413.

Otto, S. P. & Goldstein, D. B. (1992). Recombination and the evolution of diploidy. *Genetics* **131**, 745–751.

Palanivelu, R., Brass, L., Edlund, A. F. & Preuss, D. (2003). Pollen tube growth and guidance is regulated by *POP2*, an *Arabidopsis* gene that controls GABA levels. *Cell* **114**, 47–59.

Palanivelu, R. & Preuss, D. (2000). Pollen tube targeting and axon guidance: parallels in tip growth mechanisms. *Trends in Cell Biology* **10**, 517–524.

Palanivelu, R. & Preuss, D. (2006). Distinct short-range ovule signals attract or repel *Arabidopsis thaliana* pollen tubes in vitro. *BioMed Central Plant Biology* **6**, 7 [epub.].

Parcy, F., Bomblies, K. & Weigel, D. (2002). Interaction of LEAFY, AGAMOUS and TERMINAL FLOWER1 in maintaining floral meristem identity in *Arabidopsis*. *Development* **129**, 2519–2527.

Parcy, F., Nilsson, O., Busch, M. A., Lee, I. & Weigel, D. (1998). A genetic framework for floral patterning. *Nature* **395**, 561–566.

Park, J. H., Ishikawa, Y., Yoshida, R., Kanno, A. & Kameya, T. (2003). Expression of *AODEF*, a B-functional MADS-box gene, in stamens and inner tepals of the dioecious species *Asparagus officinalis* L. *Plant Molecular Biology* **51**, 867–875.

Park, J. H., Ishikawa, Y., Ochiai, T., Kanno, A. & Kameya, T. (2004). Two *GLOBOSA*-like genes are expressed in second and third whorls of homochlamydeous flowers in *Asparagus officinalis* L. *Plant and Cell Physiology* **45**, 325–332.

Park, J. H., Kanno, A. & Kameya, T. (2003). Isolation and characterization of class B floral organ identity gene from garden *Asparagus*. *Plant and Cell Physiology* **44**, S183.

Park, S. O., Hwang, S. & Hauser, B. A. (2004). The phenotype of *Arabidopsis* ovule mutants mimics the morphology of primitive seed plants. *Proceedings of the Royal Society of London Series B-Biological Sciences* **271**, 311–316.

Parker, G., Schofield, R., Sundberg, B. & Turner, S. (2003). Isolation of *COV1*, a gene involved in the regulation of vascular patterning in the stem of *Arabidopsis*. *Development* **130**, 2139–2148.

Parkin, J. (1957). The unisexual flower again—a criticism. *Phytomorphology* **7**, 7–9.

Parniske, M. (2004). Molecular genetics of the arbuscular mycorrhizal symbiosis. *Current Opinion in Plant Biology* **7**, 414–421.

Pate, J. S., Kuo, J., Dixon, K. W. & Crisp, M. D. (1989). Anomalous secondary thickening in roots of *Daviesia* (Fabaceae) and its taxonomic significance. *Botanical Journal of the Linnean Society* **99**, 175–193.

Patterson, C. (1988). Homology in classical and molecular biology. *Molecular Biology and Evolution* **5**, 603–625.

Payne, T., Clement, J., Arnold, D. & Lloyd, A. (1999). Heterologous *myb* genes distinct from GL1 enhance trichome production when overexpressed in *Nicotiana tabacum*. *Development* **126**, 671–682.

Pelaz, S., Ditta, G. S., Baumann, E., Wisman, E. & Yanofsky, M. F. (2000). B and C floral organ identity functions require *SEPALLATA* MADS-box genes. *Nature* **405**, 200–203.

Pelaz, S., Gustafson-Brown, C., Kohalmi, S. E., Crosby, W. L. & Yanofsky, M. F. (2001). APETALA1 and SEPALLATA3 interact to promote flower development. *Plant Journal* **26**, 385–394.

Peng, J., Richards, D. E., Hartley, N. M., Murphy, G. P., Devos, K. M., Flintham, J. E. et al. (1999). 'Green revolution' genes encode mutant gibberellin response modulators. *Nature* **400**, 256–261.

Perez-Rodriguez, M., Jaffe, F. W., Butelli, E., Glover, B. J. & Martin, C. (2005). Development of three different cell types is associated with the activity of a specific MYB transcription factor in the ventral petal of *Antirrhinum majus* flowers. *Development* **132**, 359–370.

Pettitt, J. M. (1977). Megaspore wall in gymnosperms—ultrastructure in some zooidogamous forms. *Proceedings of the Royal Society of London Series B-Biological Sciences* **195**, 497–515.

Phelps-Durr, T. L., Thomas, J., Vahab, P. & Timmermans, M. C. (2005). Maize rough sheath2 and its *Arabidopsis* orthologue ASYMMETRIC LEAVES1 interact with HIRA, a predicted histone chaperone, to maintain *knox* gene silencing and determinacy during organogenesis. *Plant Cell* **17**, 2886–2898.

Philipson, W. R. & Philipson, M. N. (1979). Leaf vernation in *Nothofagus*. *New Zealand Journal of Botany* **17**, 417–421.

Phillips, P. K. & Heath, J. E. (1992). Heat-exchange by the pinna of the African elephant (*Loxodonta africana*). *Comparative Biochemistry and Physiology a-Physiology* **101**, 693–699.

Phillips, W. S. (1963). Depth of roots in soil. *Ecology* **44**, 424–424.

Phipps, C. J. & Taylor, T. N. (1996). Mixed arbuscular mycorrhizae from the Triassic of Antarctica. *Mycologia* **88**, 707–714.

Pineau, C., Freydier, A., Ranocha, P., Jauneau, A., Turner, S., Lemonnier, G. et al. (2005). *hca*: an *Arabidopsis* mutant exhibiting unusual cambial activity and altered vascular patterning. *Plant Journal* **44**, 271–289.

Pinyopich, A., Ditta, G. S., Savidge, B., Liljegren, S. J., Baumann, E., Wisman, E. et al. (2003). Assessing the redundancy of MADS-box genes during carpel and ovule development. *Nature* **424**, 85–88.

Pnueli, L., Carmel-Goren, L., Hareven, D., Gutfinger, T., Alvarez, J., Ganal, M. et al. (1998). The *SELF-PRUNING* gene of tomato regulates vegetative to reproductive switching of sympodial meristems and is the ortholog of CEN and TFL1. *Development* **125**, 1979–1989.

Portereiko, M. F., Lloyd, A., Steffen, J. G., Punwani, J. A., Otsuga, D. & Drews, G. N. (2006). AGL80 is required for central cell and endosperm development in *Arabidopsis*. *Plant Cell* **18**, 1862–1872.

Pouteau, S., Nicholls, D., Tooke, F., Coen, E. & Battey, N. (1997). The induction and maintenance of flowering in *Impatiens*. *Development* **124**, 3343–3351.

Pray, T. R. (1955). Foliar venation of angiosperms. II. Histogenesis of the venation of *Liriodendron*. *American Journal of Botany* **42**, 18–27.

Preston, J. C. & Kellogg, E. A. (2007). Conservation and divergence of *APETALA1/FRUITFULL*-like gene function in grasses: evidence from gene expression analyses. *Plant Journal* **52**, 69–81.

Prusinkiewicz, P., Erasmus, Y., Lane, B., Harder, L. D. & Coen, E. (2007). Evolution and development of inflorescence architectures. *Science* **316**, 1452–1456.

Pryer, K. M., Schneider, H., Smith, A. R., Cranfill, R., Wolf, P. G., Hunt, J. S. et al. (2001). Horsetails and ferns are a monophyletic group and the closest living relatives to seed plants. *Nature* **409**, 618–622.

Pryer, K. M., Smith, A. R. & Skog, J. E. (1995). Phylogenetic relationships of extant ferns based on evidence from morphology and rbcL sequences. *American Fern Journal* **85**, 205–282.

Puri, V. (1952). Floral anatomy and inferior ovary. *Phytomorphology* **2**, 122–129.

Putz, N., Huning, G. & Froebe, H. A. (1995). Cost and advantage of soil channel formation by contractile roots in successful plant movement. *Annals of Botany* **75**, 633–639.

Qiu, Y. L., Dombrovska, O., Lee, J., Li, L. B., Whitlock, B. A., Bernasconi-Quadroni, F. et al. (2005). Phylogenetic analyses of basal angiosperms based on nine plastid, mitochondrial, and nuclear genes. *International Journal of Plant Sciences* **166**, 815–842.

Qiu, Y. L., Li, L. B., Wang, B., Chen, Z. D., Dombrovska, O., Lee, J. et al. (2007). A nonflowering land plant phylogeny inferred from nucleotide sequences of seven chloroplast, mitochondrial, and nuclear genes. *International Journal of Plant Sciences* **168**, 691–708.

Qiu, Y. L., Li, L. B., Wang, B., Chen, Z. D., Knoop, V., Groth-Malonek, M. et al. (2006). The deepest divergences in land plants inferred from phylogenomic evidence. *Proceedings of the National Academy of Sciences of the United States of America* **103**, 15511–15516.

Quarin, C. L., Espinoza, F., Martinez, E. J., Pessino, S. C. & Bovo, O. A. (2001). A rise of ploidy level induces the expression of apomixis in *Paspalum notatum*. *Sexual Plant Reproduction* **13**, 243–249.

Radutoiu, S., Madsen, L. H., Madsen, E. B., Felle, H. H., Umehara, Y., Gronlund, M. et al. (2003). Plant recognition of symbiotic bacteria requires two LysM receptor-like kinases. *Nature* **425**, 585–592.

Ratcliffe, O. J., Bradley, D. J. & Coen, E. S. (1999). Separation of shoot and floral identity in *Arabidopsis*. *Development* **126**, 1109–1120.

Raunkiær, C. (1934). *The Life Forms of Plants and Statistical Plant Geography*. Oxford University Press, Oxford.

Raven, J. A. & Edwards, D. (2001). Roots: evolutionary origins and biogeochemical significance. *Journal of Experimental Botany* **52**, 381–401.

Redecker, D., Kodner, R. & Graham, L. E. (2000). Glomalean fungi from the Ordovician. *Science* **289**, 1920–1921.

Reinhardt, B., Hänggi, E., Müller, S., Bauch, M., Wyrzykowska, J., Kerstetter, R. et al. (2007). Restoration of DWF4 expression to the leaf margin of a *dwf4* mutant is sufficient to restore leaf shape but not size: the role of the margin in leaf development. *Plant Journal* **52**, 1094–1104.

Reinhardt, D., Frenz, M., Mandel, T. & Kuhlemeier, C. (2005). Microsurgical and laser ablation analysis of leaf positioning and dorsoventral patterning in tomato. *Development* **132**, 15–26.

Reinhardt, D., Mandel, T. & Kuhlemeier, C. (2000). Auxin regulates the initiation and radial position of plant lateral organs. *Plant Cell* **12**, 507–518.

Reinhardt, D., Pesce, E. R., Stieger, P., Mandel, T., Baltensperger, K., Bennett, M. et al. (2003). Regulation of phyllotaxis by polar auxin transport. *Nature* **426**, 255–260.

Remy, W., Gensel, P. G. & Hass, H. (1993). The gametophyte generation of some early Devonian land plants. *International Journal of Plant Sciences* **154**, 35–58.

Rensing, S. A., Lang, D., Zimmer, A. D., Terry, A., Salamov, A., Shapiro, H. et al. (2008). The physcomitrella genome reveals evolutionary insights into the conquest of land by plants. *Science* **319**, 64–69.

Renzaglia, K. S., Duff, R. J., Nickrent, D. L. & Garbary, D. J. (2000). Vegetative and reproductive innovations of early land plants: implications for a unified phylogeny. *Philosophical Transactions of the Royal Society of London Series B-Biological Sciences* **355**, 769–793.

Renzaglia, K. S. & Garbary, D. J. (2001). Motile gametes of land plants: diversity, development, and evolution. *Critical Reviews in Plant Sciences* **20**, 107–213.

Renzaglia, K. S. & Maden, A. R. (2000). Microtubule organizing centers and the origin of centrioles during spermatogenesis in the pteridophyte *Phylloglossum*. *Microscopy Research and Technique* **49**, 496–505.

Richerd, S., Couvet, D. & Valero, M. (1993). Evolution of the alternation of haploid and diploid phases in life-cycles. 2. Maintenance of the haplo-diplontic cycle. *Journal of Evolutionary Biology* **6**, 263–280.

Rickett, H. W. (1955). Materials for a dictionary of botanical terms, III. Inflorescences. *Bulletin of the Torrey Botanical Club* **82**, 419–445.

Riechmann, J. L., Krizek, B. A. & Meyerowitz, E. M. (1996). Dimerization specificity of *Arabidopsis* MADS domain homeotic proteins APETALA1, APETALA3, PISTILLATA, and AGAMOUS. *Proceedings of the National Academy of Sciences of the United States of America* **93**, 4793–4798.

Riechmann, J. L., Wang, M. Q. & Meyerowitz, E. M. (1996). DNA-binding properties of *Arabidopsis* MADS domain homeotic proteins APETALA1, APETALA3, PISTILLATA and AGAMOUS. *Nucleic Acids Research* **24**, 3134–3141.

Robles, P. & Pelaz, S. (2005). Flower and fruit development in *Arabidopsis thaliana*. *International Journal of Developmental Biology* **49**, 633–643.

Roe, J. L., Nemhauser, J. L. & Zambryski, P. C. (1997). TOUSLED participates in apical tissue formation during gynoecium development in *Arabidopsis*. *Plant Cell* **9**, 335–353.

Roe, J. L., Rivin, C. J., Sessions, R. A., Feldmann, K. A. & Zambryski, P. C. (1993). The *TOUSLED* gene in *A. thaliana* encodes a protein-kinase homolog that is required for leaf and flower development. *Cell* **75**, 939–950.

Roels, P. & Smets, E. (1994). A comparative floral ontogenic study between *Adoxa moschatellina* and *Sambucus ebulus*. *Belgian Journal of Botany* **127**, 157–170.

Romano, C. P., Cooper, M. L. & Klee, H. J. (1993). Uncoupling auxin and ethylene effects in transgenic tobacco and *Arabidopsis* plants. *Plant Cell* **5**, 181–189.

Ronse De Craene, L. P. (2003). The evolutionary significance of homeosis in flowers: a morphological perspective. *International Journal of Plant Sciences* **164**, S225–S235.

Ronse De Craene, L. P. (2004). Floral development of *Berberidopsis corallina*: a crucial link in the evolution of flowers in the core eudicots. *Annals of Botany* **94**, 741–751.

Ronse De Craene, L. P. (2007). Are petals sterile stamens or bracts? The origin and evolution of petals in the core eudicots. *Annals of Botany* **100**, 621–630.

Ronse De Craene, L. P. & Smets, E. F. (2001). Staminodes: their morphological and evolutionary significance. *Botanical Review* **67**, 351–402.

Ronse De Craene, L. P., Soltis, P. S. & Soltis, D. E. (2003). Evolution of floral structures in basal angiosperms. *International Journal of Plant Sciences* **164**, S329–S363.

Rosenberg, O. (1906). Über die Embryobildung in der Gattung Hieracium. *Berichte der Deutschen Botanischen Gesellschaft* **24**, 157–161.

Roth, A. & Mosbrugger, V. (1996). Numerical studies of water conduction in land plants: evolution of early stele types. *Paleobiology* **22**, 411–421.

Rothwell, G. W. (1984). The apex of Stigmaria (Lycopsida), rooting organ of Lepidodendrales. *American Journal of Botany* **71**, 1031–1034.

Rothwell, G. W. & Erwin, D. M. (1985). The rhizomorph apex of Paurodendron—implications for homologies among the rooting organs of Lycopsida. *American Journal of Botany* **72**, 86–98.

Rothwell, G. W. & Lev-Yadun, S. (2005). Evidence of polar auxin flow in 375 million-year-old fossil wood. *American Journal of Botany* **92**, 903–906.

Rothwell, G. W., Scheckler, S. E. & Gillespie, W. H. (1989). Elkinsia gen. nov., a late Devonian gymnosperm with cupulate ovules. *Botanical Gazette* **150**, 170–189.

Rudall, P. J. (2002). Homologies of inferior ovaries and septal nectaries in monocotyledons. *International Journal of Plant Sciences* **163**, 261–276.

Rudall, P. J. (2003). Monocot Pseudanthia revisited: floral structure of the mycoheterotrophic family Triuridaceae. *International Journal of Plant Sciences* **164**, S307–S320.

Rudall, P. J. & Bateman, R. M. (2003). Evolutionary change in flowers and inflorescences: evidence from naturally occurring terata. *Trends in Plant Science* **8**, 76–82.

Rudall, P. J. & Bateman, R. M. (2004). Evolution of zygomorphy in monocot flowers: iterative patterns and developmental constraints. *New Phytologist* **162**, 25–44.

Rudall, P. J. & Bateman, R. M. (2006). Morphological phylogenetic analysis of Pandanales: testing contrasting hypotheses of floral evolution. *Systematic Botany* **31**, 223–238.

Rudall, P. J., Manning, J. C. & Goldblatt, P. (2003). Evolution of floral nectaries in Iridaceae. *Annals of the Missouri Botanical Garden* **90**, 613–631.

Rumsey, F. J., Vogel, J. C., Russell, S. J., Barrett, J. A. & Gibby, M. (1998). Climate, colonisation and celibacy: population structure in central European *Trichomanes speciosum* (Pteridophyta). *Botanica Acta* **111**, 481–489.

Rumsey, F. J., Vogel, J. C., Russell, S. J., Barrett, J. A. & Gibby, M. (1999). Population structure and conservation biology of the endangered fern *Trichomanes speciosum* Willd. (Hymenophyllaceae) at its northern distributional limit. *Biological Journal of the Linnean Society* **66**, 333–344.

Rutishauser, R. (1995). Developmental patterns of leaves in Podostemaceae compared with more typical flowering plants—saltational evolution and fuzzy morphology. *Canadian Journal of Botany-Revue Canadienne De Botanique* **73**, 1305–1317.

Rutishauser, R. (1997). Structural and developmental diversity in Podostemaceae (river-weeds). *Aquatic Botany* **57**, 29–70.

Rutishauser, R. & Isler, B. (2001). Developmental genetics and morphological evolution of flowering plants, especially bladderworts (*Utricularia*): fuzzy arberian morphology complements classical morphology. *Annals of Botany* **88**, 1173–1202.

Ruzin, S. E. (1979). Root contraction in *Freesia* (Iridaceae). *American Journal of Botany* **66**, 522–531.

Rydin, C., Kallersjo, M. & Friist, E. M. (2002). Seed plant relationships and the systematic position of Gnetales based on nuclear and chloroplast DNA: conflicting data, rooting problems, and the monophyly of conifers. *International Journal of Plant Sciences* **163**, 197–214.

Ryu, K. H., Kang, Y. H., Park, Y. H., Hwang, D., Schiefelbein, J. & Lee, M. M. (2005). The WEREWOLF MYB protein directly regulates CAPRICE transcription during cell fate specification in the *Arabidopsis* root epidermis. *Development* **132**, 4765–4775.

Saarela, J. M., Rai, H. S., Doyle, J. A., Endress, P. K., Mathews, S., Marchant, A. D. et al. (2007). Hydatellaceae identified as a new branch near the base of the angiosperm phylogenetic tree. *Nature* **446**, 312–315.

Sabatini, S., Heidstra, R., Wildwater, M. & Scheres, B. (2003). SCARECROW is involved in positioning the stem cell niche in the *Arabidopsis* root meristem. *Genes Dev* **17**, 354–358.

Sablowski, R. (2004). Root development: the embryo within? *Current Biology* **14**, R1054–R1055.

Sablowski, R. (2007). Flowering and determinacy in *Arabidopsis*. *Journal of Experimental Botany* **58**, 899–907.

Sachs, T. (1981). The control of the patterned differentiation of vascular tissues. *Advances in Botanical Research Incorporating Advances in Plant Pathology* **9**, 151–262.

Sachs, T. (2000). Integrating cellular and organismic aspects of vascular differentiation. *Plant and Cell Physiology* **41**, 649–656.

Sachs, T. & Cohen, D. (1982). Circular vessels and the control of vascular differentiation in plants. *Differentiation* **21**, 22–26.

Sahni, B. (1923). On the theoretical significance of certain so-called abnormalities in the sporangiophores of *Psilotum* and *Tmespteris*. *Journal of the Indian Botanical Society* **3**, 185–191.

SAHNI, B. (1932). A petrified *Williamsonia* (*W. sewardiana* sp. nov.) from the Rajmahal Hills, India. *Memoir Geological Survey India, Palaeontologia indica,* New Series 20, 1–19.

SAHNI, B. (1948). The Pentoxyleae: a group of Jurassic gymnosperms from the Rajmahal Hills of India. *Botanical Gazette* **110**, 47–80.

SAJO, M. G. & RUDALL, P. J. (1999). Development of ensiform leaves and other vegetative structures in *Xyris*. *Botanical Journal of the Linnean Society* **130**, 171–182.

SAKAI, H., KRIZEK, B. A., JACOBSEN, S. E. & MEYEROWITZ, E. N. (2000). Regulation of SUP expression identifies multiple regulators involved in *Arabidopsis* floral meristem development. *Plant Cell* **12**, 1607–1618.

SAKAMOTO, T., KAMIYA, N., UEGUCHI-TANAKA, M., IWAHORI, S. & MATSUOKA, M. (2001). KNOX homeodomain protein directly suppresses the expression of a gibberellin biosynthetic gene in the tobacco shoot apical meristem. *Genes & Development* **15**, 581–590.

SANDVIK, S. M. & TOTLAND, O. (2003). Quantitative importance of staminodes for female reproductive success in *Parnassia palustris* under contrasting environmental conditions. *Canadian Journal of Botany-Revue Canadienne De Botanique* **81**, 49–56.

SATINA, S., BLAKESLEE, A. & AVERY, A. (1940). Demonstration of the three germ layers in the shoot apex of *Datura* by means of induced polyploidy in periclinal chimeras. *American Journal of Botany* **27**, 895–905.

SATTERTHWAIT, D. F. & SCHOPF, J. W. (1972). Structurally preserved phloem zone tissue in *Rhynia*. *American Journal of Botany* **59**, 373–376.

SATTLER, R. (1996). Classical morphology and continuum morphology: opposition and continuum. *Annals of Botany* **78**, 577–581.

SATTLER, R. & LACROIX, C. (1988). Development and evolution of basal cauline placentation—*Basella rubra*. *American Journal of Botany* **75**, 918–927.

SATTLER, R. & MAIER, U. (1977). Development of epiphyllous appendages of *Begonia hispida* var. *cucullifera*: implications for comparative morphology. *Canadian Journal of Botany* **55**, 411–425.

SAUNDERS, E. R. (1923). A reversionary character in the stock (*Matthiola incana*) and its significance in regard to the structure and evolution of the gynoecium in the Rhoedales, the Orchidaceae, and other families. *Annals of Botany* **37**, 451–482.

SAVOLAINEN, V. & CHASE, M. W. (2003). A decade of progress in plant molecular phylogenetics. *Trends in Genetics* **19**, 717–724.

SAVOLAINEN, V., CHASE, M. W., HOOT, S. B., MORTON, C. M., SOLTIS, D. E., BAYER, C. et al. (2000). Phylogenetics of flowering plants based on combined analysis of plastid *atpB* and *rbcL* gene sequences. *Systematic Biology* **49**, 306–362.

SCARPELLA, E., FRANCIS, P. & BERLETH, T. (2004). Stage-specific markers define early steps of procambium development in *Arabidopsis* leaves and correlate termination of vein formation with mesophyll differentiation. *Development* **131**, 3445–3455.

SCARPELLA, E., MARCOS, D., FRIML, J. & BERLETH, T. (2006). Control of leaf vascular patterning by polar auxin transport. *Genes & Development* **20**, 1015–1027.

SCHATZ, G. E., WILLIAMSON, G. B., COGSWELL, C. M. & STERN, A. C. (1985). Stilt roots and growth of arboreal palms. *Biotropica* **17**, 206–220.

SCHENK, H. J. & JACKSON, R. B. (2002). The global biogeography of roots. *Ecological Monographs* **72**, 311–328.

SCHERES, B., WOLKENFELT, H., WILLEMSEN, V., TERLOVW, M., LAWSON, E., DEAN, C. & WEISBEEK, P. (1994). Embryonic origin of the *Arabidopsis* primary root and root meristem initials. *Development* **120**, 2475–2487.

SCHERES, B. & XU, L. (2006). Polar auxin transport and patterning: grow with the flow. *Genes & Development* **20**, 922–926.

SCHICHNES, D. E. & FREELING, M. (1993). A mutant affecting the developmental timing of the leaves and shoots of maize. *Journal of Cellular Biochemistry Supplement 17B*, 35–35.

SCHICHNES, D., SCHNEEBERGER, R. & FREELING, M. (1997). Induction of leaves directly from leaves in the maize mutant *Lax midrib1-O*. *Developmental Biology* **186**, 36–45.

SCHICHNES, D. E. & FREELING, M. (1998). *Lax midrib1-O* heterochronic mutant of maize. *American Journal of Botany* **85**, 481–491.

SCHIEFELBEIN, J. (2003). Cell-fate specification in the epidermis: a common patterning mechanism in the root and shoot. *Current Opinion in Plant Biology* **6**, 74–78.

SCHIEFTHALER, U., BALASUBRAMANIAN, S., SIEBER, P., CHEVALIER, D., WISMAN, E. & SCHNEITZ, K. (1999). Molecular analysis of *NOZZLE*, a gene involved in pattern formation and early sporogenesis during sex organ development in *Arabidopsis thaliana*. *Proceedings of the National Academy of Sciences of the United States of America* **96**, 11664–11669.

SCHLEIDEN, M. J. (1842). *Grundzüge der wissenschaftliche Botanik*. Wilhelm Engelmann, Leipzig.

SCHMID, R. (1972). Floral bundle fusion and vascular conservatism. *Taxon* **21**, 429–446.

SCHMIDT, A. (1924). Histologische Studien an phanerogamen Vegetationspunkten. *Botanische Archiv* **8**, 345–404.

SCHNEIDER, H. (2000). Morphology and anatomy of roots in the filmy fern tribe Trichomaneae *H. Schneider* (Hymenophyllaceae, Filicatae) and the evolution of rootless taxa. *Botanical Journal of the Linnean Society* **132**, 29–46.

SCHNEIDER, H. & PRYER, K. M. (2002). Structure and function of spores in the aquatic heterosporous fern family Marsileaceae. *International Journal of Plant Sciences* **163**, 485–505.

SCHNEIDER, H., SCHUETTPELZ, E., PRYER, K. M., CRANFILL, R., MAGALLON, S. & LUPIA, R. (2004). Ferns diversified in the shadow of angiosperms. *Nature* **428**, 553–557.

SCHOOF, H., LENHARD, M., HAECKER, A., MAYER, K. F. X., JURGENS, G. & LAUX, T. (2000). The stem cell population of *Arabidopsis* shoot meristems is maintained by a regulatory loop between the *CLAVATA* and *WUSCHEL* genes. *Cell* **100**, 635–644.

SCHREIBER, D. N., BANTIN, J. & DRESSELHAUS, T. (2004). The MADS box transcription factor ZmMADS2 is required for anther and pollen maturation in maize and accumulates in apoptotic bodies during anther dehiscence. *Plant Physiology* **134**, 1069–1079.

SCHUETTPELZ, E. & HOOT, S. B. (2004). Phylogeny and biogeography of *Caltha* (Ranunculaceae) based on chloroplast and nuclear DNA sequences. *American Journal of Botany* **91**, 247–253.

SCHUMAKER, K. S. & DIETRICH, M. A. (1998). Hormone-induced signaling during moss development. *Annual Review of Plant Physiology and Plant Molecular Biology* **49**, 501–523.

SCHWARZ-SOMMER, Z., HUIJSER, P., NACKEN, W., SAEDLER, H. & SOMMER, H. (1990). Genetic control of flower development by homeotic genes in *Antirrhinum majus*. *Science* **250**, 931–936.

SCOTT, A. C. & HEMSLEY, A. R. (1993). The spores of the Dinantian lycopsid cone *Flemingites scottii* from Pettycur, Fife, Scotland. In *Studies in Palaeobotany and Palynology in Honour of Professor W. G. Chaloner, vol. 49. Special Papers in Palaeontology*, pp. 31–41. Palaeontological Association, London.

SCOTT, R. J., SPIELMAN, M. & DICKINSON, H. G. (2004). Stamen structure and function. *Plant Cell* **16**, S46–S60.

SCUTT, C. P., VINAUGER-DOUARD, M., FOURQUIN, C., FINET, C. & DUMAS, C. (2006). An evolutionary perspective on the regulation of carpel development. *Journal of Experimental Botany* **57**, 2143–2152.

SEKINE, M., IWAKAWA, H. & HARASHIMA, H. (2006). *Arabidopsis* CDKA;1, a Cdc2 homolog, controls proliferation of generative cells in male gametogenesis. *Plant and Cell Physiology* **47**, S12.

SHAH, J. J. & DAVE, Y. S. (1971). Morpho-histogenic studies on tendrils of *Passiflora*. *Annals of Botany* **35**, 627–635.

SHEFFIELD, E. (1994). Alternation of generations in ferns—mechanisms and significance. *Biological Reviews of the Cambridge Philosophical Society* **69**, 331–343.

SHERWOOD, R. T., BERG, C. C. & YOUNG, B. A. (1994). Inheritance of apospory in Buffelgrass. *Crop Science* **34**, 1490–1494.

SHIFRISS, O. (1961). Sex control in cucumbers. *Journal of Heredity* **52**, 5–12.

SIEBER, P., GHEYSELINCK, J., GROSS-HARDT, R., LAUX, T., GROSSNIKLAUS, U. & SCHNEITZ, K. (2004). Pattern formation during early ovule development in *Arabidopsis thaliana*. *Developmental Biology* **273**, 321–334.

SIEBURTH, L. E. & DEYHOLOS, M. K. (2005). Vascular development: the long and winding road. *Current Opinion in Plant Biology* **9**, 48–54.

SIEBURTH, L. E., MUDAY, G. K., KING, E. J., BENTON, G., KIM, S., METCALF, K. E. et al. (2006). SCARFACE encodes an ARF-GAP that is required for normal auxin efflux and vein patterning in *Arabidopsis*. *Plant Cell* **18**, 1396–1411.

SINGER, S. D. & ASHTON, N. W. (2007). Revelation of ancestral roles of *KNOX* genes by a functional analysis of *Physcomitrella* homologues. *Plant Cell Reports* **26**, 2039–2054.

SINNOTT, E. W. & BAILEY, I. W. (1914). Investigations on the phylogeny of the angiosperms. 3. Nodal anatomy and the morphology of stipules. *American Journal of Botany* **1**, 441–453.

SKENE, K. R. (2000). Pattern formation in cluster roots: some developmental and evolutionary considerations. *Annals of Botany* **85**, 901–908.

SKENE, K. R. (2001). Cluster roots: model experimental tools for key biological problems. *Journal of Experimental Botany* **52**, 479–485.

SKINNER, D. J., HILL, T. A. & GASSER, C. S. (2004). Regulation of ovule development. *Plant Cell* **16**, S32–S45.

SKOG, J. E. & BANKS, H. P. (1973). Ibyka amphikoma, gen. et sp. n.—new protoarticulate precursor from late Middle Devonian of New York State. *American Journal of Botany* **60**, 366–380.

SMIT, P., RAEDTS, J., PORTYANKO, V., DEBELLE, F., GOUGH, C., BISSELING, T. et al. (2005). NSP1 of the GRAS protein family is essential for rhizobial Nod factor-induced transcription. *Science* **308**, 1789–1791.

Smith, F. H. (1930). Corm and contractile roots of *Brodiaea lactea*. *American Journal of Botany* **17**, 916–927.

Smith, F. H. & Smith, E. C. (1942). Anatomy of the inferior ovary of *Darbya*. *American Journal of Botany* **29**, 464–471.

Smith, G. L. (1963). Studies in *Potentilla* L. II. Cytological aspects of apomixis in *P. crantzii* (Cr.) Beck ex Fritsch. *New Phytologist* **62**, 283–300.

Smith, J. (1841). Notice of a plant which produces perfect seeds without any apparent action of pollen. *Transactions of the Linnean Society of London* **18**, 509–512.

Smoot, E. L. & Taylor, T. N. (1986). Evidence of simple polyembryony in Permian seeds from Antarctica. *American Journal of Botany* **73**, 1079–1081.

Snow, M. & Snow, R. (1952). Minimum areas and leaf determination. *Proceedings of the Royal Society of London B* **139**, 545–566.

Sokoloff, D., Rudall, P. J. & Remizowa, M. (2006). Flower-like terminal structures in racemose inflorescences: a tool in morphogenetic and evolutionary research. *Journal of Experimental Botany* **57**, 3517–3530.

Soltis, D. E., Chanderbali, A. S., Kim, S., Buzgo, M. & Soltis, P. S. (2007). The ABC model and its applicability to basal angiosperms. *Annals of Botany* **100**, 155–163.

Soltis, D. E., Fishbein, M. & Kuzoff, R. K. (2003). Reevaluating the evolution of epigyny: data from phylogenetics and floral ontogeny. *International Journal of Plant Sciences* **164**, S251–S264.

Soltis, D. E., Gitzendanner, M. A. & Soltis, P. S. (2007). A 567-taxon data set for angiosperms: the challenges posed by bayesian analyses of large data sets. *International Journal of Plant Sciences* **168**, 137–157.

Soltis, D. E. & Hufford, L. (2002). Ovary position diversity in Saxifragaceae: clarifying the homology of epigyny. *International Journal of Plant Sciences* **163**, 277–293.

Soltis, P. S. & Soltis, D. E. (2004). The origin and diversification of angiosperms. *American Journal of Botany* **91**, 1614–1626.

Sorensen, M. B., Chaudhury, A. M., Robert, H., Bancharel, E. & Berger, F. (2001). Polycomb group genes control pattern formation in plant seed. *Current Biology* **11**, 277–281.

Sorensen, M. B., Mayer, U., Lukowitz, W., Robert, H., Chambrier, P., Jurgens, G. et al. (2002). Cellularisation in the endosperm of *Arabidopsis thaliana* is coupled to mitosis and shares multiple components with cytokinesis. *Development* **129**, 5567–5576.

Speck, T. & Vogellehner, D. (1988). Biophysical examinations of the bending stability of various stele types and the upright axes of early vascular land plants. *Botanica Acta* **101**, 262–268.

Spielman, M., Preuss, D., Li, F. L., Browne, W. E., Scott, R. J. & Dickinson, H. G. (1997). TETRASPORE is required for male meiotic cytokinesis in *Arabidopsis thaliana*. *Development* **124**, 2645–2657.

Spielman, M., Vinkenoog, R., Dickinson, H. G. & Scott, R. J. (2001). The epigenetic basis of gender in flowering plants and mammals. *Trends in Genetics* **17**, 705–711.

Sporne, K. R. (1948). A note on a rapid clearing technique of wide application. *New Phytologist* **47**, 290–291.

Sporne, K. R. (1949). A new approach to the problem of the primitive flower. *New Phytologist* **48**, 259–276.

Sporne, K. R. (1962). *The Morphology of Pteridophytes: The Structure of Ferns and Allied Plants.* Hutchinson, London.

Sporne, K. R. (1965). *The Morphology of Gymnosperms: The Structure and Evolution of Primitive Seed-plants.* Hutchinson, London.

Sporne, K. R. (1974). *The Morphology of Angiosperms: The Structure and Evolution of Flowering Plants.* Hutchinson, London.

Sporne, K. R. (1980). A reinvestigation of character correlations among dicotyledons. *New Phytologist* **85**, 419–449.

Sporne, K. R. (1982). The advancement index vindicated. *New Phytologist* **91**, 137–145.

Sprent, J. I. (2001). *Nodulation in Legumes.* Royal Botanic Gardens, Kew.

St. John, E. P. (1952). The evolution of the Ophioglossaceae of the eastern United States. III. The evolution of the leaf. *Quarterly Journal of the Florida Academy of Sciences* **15**, 1–19.

Stacey, G., Libault, M., Brechenmacher, L., Wan, J. & May, G. D. (2006). Genetics and functional genomics of legume nodulation. *Current Opinion in Plant Biology* **9**, 110–121.

Staedler, Y. M., Weston, P. H & Endress, P. K. (2007). Floral phyllotaxis and floral architecture in Calycanthaceae (Laurales). *International Journal of Plant Sciences* **168**, 285–306.

Stebbins, G. L. (1970). Transference of function as a factor in evolution of seeds and their accessory structures. *Israel Journal of Botany* **19**, 59–70.

STEIN, W. (1993). Modeling the evolution of stelar architecture in vascular plants. *International Journal of Plant Sciences* **154**, 229–263.

STEINBORN, K., MAULBETSCH, C., PRIESTER, B., TRAUTMANN, S., PACHER, T., GEIGES, B. et al. (2002). The *Arabidopsis PILZ* group genes encode tubulin-folding cofactor orthologs required for cell division but not cell growth. *Genes & Development* **16**, 959–971.

STEINGRAEBER, D. & FISHER, J. B. (1986). Indeterminate growth of leaves in *Guarea* (Meliaceae): a twig analogue. *American Journal of Botany* **73**, 852–862.

STEVENSON, D. W. (1981). Observations on ptyxis, phenology, and trichomes in the Cycadales and their systematic implications. *American Journal of Botany* **68**, 1104–1114.

STEWART, K. D. & GIFFORD, E. M. (1967). Ultrastructure of developing megaspore mother cell of *Ginkgo biloba*. *American Journal of Botany* **54**, 375–383.

STEWART, W. N. & ROTHWELL, G. W. (1993). *Palaeobotany and the Evolution of Plants*, Second edition. Cambridge University Press, Cambridge.

STIEGER, P. A., MEYER, A. D., KATHMANN, P., FRUNDT, C., NIEDERHAUSER, I., BARONE, M. et al. (2004). The *orf13 T-DNA* gene of *Agrobacterium rhizogenes* confers meristematic competence to differentiated cells. *Plant Physiology* **135**, 1798–1808.

STRACKE, S., KISTNER, C., YOSHIDA, S., MULDER, L., SATO, S., KANEKO, T. et al. (2002). A plant receptor-like kinase required for both bacterial and fungal symbiosis. *Nature* **417**, 959–962.

SUN, G., DILCHER, D. L., ZHENG, S. L. & ZHOU, Z. K. (1998). In search of the first flower: a Jurassic angiosperm, *Archaefructus*, from northeast China. *Science* **282**, 1692–1695.

SUSSEX, I. M. (1954). Experiments on the cause of dorsiventrality in leaves. *Nature* **174**, 351–352.

SUSSEX, I. M. (1955). Morphogenesis in *Solanum tuberosum* L.: experimental investigation of leaf dorsiventrality and orientation in the juvenile shoot. *Phytomorphology* **5**, 286–300.

SUZUKI, M., KETTERLING, M. G., LI, Q. B. & MCCARTY, D. R. (2003). Viviparous1 alters global gene expression patterns through regulation of abscisic acid signaling. *Plant Physiology* **132**, 1664–1677.

SUZUKI, M., TAKAHASHI, T. & KOMEDA, Y. (2002). Formation of corymb-like inflorescences due to delay in bolting and flower development in the *corymbosa2* mutant of *Arabidopsis*. *Plant and Cell Physiology* **43**, 298–306.

SWENSEN, S. M. (1996). The evolution of actinorhizal symbioses: evidence for multiple origins of the symbiotic association. *American Journal of Botany* **83**, 1503–1512.

TAKIGUCHI, Y., IMAICHI, R. & KATO, M. (1997). Cell division patterns in the apices of subterranean axis and aerial shoot of *Psilotum nudum* (Psilotaceae): morphological and phylogenetic implications for the subterranean axis. *American Journal of Botany* **84**, 588–596.

TANURDZIC, M. & BANKS, J. A. (2004). Sex-determining mechanisms in land plants. *Plant Cell* **16**, S61–S71.

TATTERSALL, A. D., TURNER, L., KNOX, M. R., AMBROSE, M. J., ELLIS, T. H. N. & HOFER, J. M. I. (2005). The mutant *crispa* reveals multiple roles for PHANTASTICA in pea compound leaf development. *Plant Cell* **17**, 1046–1060.

TAYLOR, T. N., KERP, H. & HASS, H. (2005). Life history biology of early land plants: deciphering the gametophyte phase. *Proceedings of the National Academy of Sciences of the United States of America* **102**, 5892–5897.

TAYLOR, T. N., REMY, W., HASS, H. & KERP, H. (1995). Fossil arbuscular mycorrhizae from the Early Devonian. *Mycologia* **87**, 560–573.

THEISSEN, G. (2001). Development of floral organ identity: stories from the MADS house. *Current Opinion in Plant Biology* **4**, 75–85.

THEISSEN, G. & BECKER, A. (2004). Gymnosperm orthologues of class B floral homeotic genes and their impact on understanding flower origin. *Critical Reviews in Plant Sciences* **23**, 129–148.

THEISSEN, G., BECKER, A., DI ROSA, A., KANNO, A., KIM, J. T., MUNSTER, T. et al. (2000). A short history of MADS-box genes in plants. *Plant Molecular Biology* **42**, 115–149.

THEISSEN, G., BECKER, A., KIRCHNER, C., MÜNSTER, T., WINTER, K-U & SAEDLER, H. (2002) How land plants learned their floral ABCS: the role of MADS-box genes in the evolutionary origin of flowers. In *Developmental Genetics and Plant Evolution* (eds Q. C. B. Cronk, R. M. Bateman and J. Hawkins), pp. 173–205. Taylor & Francis, London.

THEISSEN, G. & SAEDLER, H. (2001). Floral quartets. *Nature* **409**, 469–471.

THITAMADEE, S., TSUCHIHARA, K. & HASHIMOTO, T. (2002). Microtubule basis for left-handed helical growth in *Arabidopsis*. *Nature* **417**, 193–196.

THOMPSON, K. (1986). Are unisexual flowers primitive? *New Phytologist* **103**, 597–601.

TICCONI, C. A. & ABEL, S. (2004). Short on phosphate: plant surveillance and countermeasures. *Trends in Plant Science* **9**, 548–555.

TIMMERMANS, M. C. P., SCHULTES, N. P., JANKOVSKY, J. P. & NELSON, T. (1998). Leafbladeless1 is required for dorsoventrality of lateral organs in maize. *Development* **125**, 2813–2823.

Tomilov, A. A., Tomilova, N. B., Abdallah, I. & Yoder, J. I. (2005). Localized hormone fluxes and early haustorium development in the hemiparasitic plant *Triphysaria versicolor. Plant Physiology* **138**, 1469–1480.

Tomlinson, P. B. (1984). Homology in modular organisms—concepts and consequences. Introduction. *Systematic Botany* **9**, 373.

Tomlinson, P. B. & Takaso, T. (2002). Seed cone structure in relation to development and pollination: a biological approach. *Canadian Journal of Botany* **80**, 1250–1273.

Tooke, F. & Battey, N. H. (2000). A leaf-derived signal is a quantitative determinant of floral form in *Impatiens. Plant Cell* **12**, 1837–1847.

Tooke, F., Ordidge, M., Chiurugwi, T. & Battey, N. (2005). Mechanisms and function of flower and inflorescence reversion. *Journal of Experimental Botany* **56**, 2587–2599.

Torrey, J. G. (1950). The induction of lateral roots by indoleacetic acid and root decapitation. *American Journal of Botany* **37**, 257–264.

Trebitsh, T., Staub, J. E. & O'Neill, S. D. (1997). Identification of a 1-aminocyclopropane-1-carboxylic acid synthase gene linked to the female (F) locus that enhances female sex expression in cucumber. *Plant Physiology* **113**, 987–995.

Trinick, M. J. (1973). Symbiosis between *Rhizobium* and the non-legume, *Trema aspera*. [Note: "*Trema aspera*" is now considered a *Parasponia*]. *Nature* **244**, 459–460.

Truscott, F. H. (1966). Some aspects of morphogenesis in *Cuscuta gronovii. American Journal of Botany* **53**, 739–750.

Tryon, A. F. (1964). *Platyzoma*—a Queensland fern with incipient heterospory. *American Journal of Botany* **51**, 939–942.

Tsuge, T., Tsukaya, H. & Uchimiya, H. (1996). Two independent and polarized processes of cell elongation regulate leaf blade expansion in *Arabidopsis thaliana* (L) Heynh. *Development* **122**, 1589–1600.

Tsukaya, H. (2002). Leaf development. *The Arabidopsis Book* (online resource).

Tsukaya, H. (2005). Leaf shape: genetic controls and environmental factors. *International Journal of Developmental Biology* **49**, 547–555.

Tsukaya, H., Shoda, K., Kim, G. T. & Uchimiya, H. (2000). Heteroblasty in *Arabidopsis thaliana* (L.) Heynh. *Planta* **210**, 536–542.

Tucker, M. R., Araujo, A. C. G., Paech, N. A., Hecht, V., Schmidt, E. D. L., Rossell, J. B. et al. (2003). Sexual and apomictic reproduction in *Hieracium* subgenus *Pilosella* are closely interrelated developmental pathways. *Plant Cell* **15**, 1524–1537.

Tucker, M. R., Paech, N. A., Willemse, M. T. M. & Koltunow, A. M. G. (2001). Dynamics of callose deposition and beta-1,3-glucanase expression during reproductive events in sexual and apomictic *Hieracium. Planta* **212**, 487–498.

Tucker, S. C. (2001). The ontogenetic basis for missing petals in *Crudia* (Leguminosae: Caesalpinioideae: Detarieae). *International Journal of Plant Sciences* **162**, 83–89.

Tucker, S. C. & Bourland, J. A. (1994). Ontogeny of staminate and carpellate flowers of *Schisandra glabra* (*Schisandra*). *Plant Systematics and Evolution* **Suppl. 8**, 137–158.

Tzafrir, I., McElver, J. A., Liu, C. M., Yang, L. J., Wu, J. Q., Martinez, A. et al. (2002). Diversity of TITAN functions in *Arabidopsis* seed development. *Plant Physiology* **128**, 38–51.

Uchida, N., Townsley, B., Chung, K. H. & Sinha, N. (2007). Regulation of *SHOOT MERISTEMLESS* genes via an upstream-conserved noncoding sequence coordinates leaf development. *Proceedings of the National Academy of Sciences of the United States of America* **104**, 15953–15958.

Uhde-Stone, C., Liu, J. Q., Zinn, K. E., Allan, D. L. & Vance, C. P. (2005). Transgenic proteoid roots of white lupin: a vehicle for characterizing and silencing root genes involved in adaptation to P stress. *Plant Journal* **44**, 840–853.

Uhde-Stone, C., Zinn, K. E., Ramirez-Yanez, M., Li, A. G., Vance, C. P. & Allan, D. L. (2003). Nylon filter arrays reveal differential gene expression in proteoid roots of white lupin in response to phosphorus deficiency. *Plant Physiology* **131**, 1064–1079.

Usami, T., Mochizuki, N., Kondo, M., Nishimura, M. & Nagatani, A. (2004). Cryptochromes and phytochromes synergistically regulate *Arabidopsis* root greening under blue light. *Plant and Cell Physiology* **45**, 1798–1808.

Van der Graaff, E., Den Dulk-Ras, A., Hooykaas, P. J. J. & Keller, B. (2000). Activation tagging of the *LEAFY PETIOLE* gene affects leaf petiole development in *Arabidopsis thaliana. Development* **127**, 4971–4980.

Van Dijk, P. J. & Bakx-Schotman, J. M. T. (2004). Formation of unreduced megaspores (diplospory) in apomictic dandelions (*Taraxacum officinale*, s.l) is controlled by a sex-specific dominant locus. *Genetics* **166**, 483–492.

Van Dijk, P. J., Tas, I. C. Q., Falque, M. & Bakx-Schotman, T. (1999). Crosses between sexual and apomictic dandelions (*Taraxacum*). II. The breakdown of apomixis. *Heredity* **83**, 715–721.

Van Tunen, A. J., Eikelboom, W. & Angenent, G. C. (1993). Floral organogenesis in *Tulipa*. *Flowering Newsletter* **16**, 33–37.

Van Valen, L. (1982). Homology and causes. *Journal of Morphology* **173**, 305–312.

VandenBerg, C., Willemsen, V., Hendriks, G., Weisbeek, P. & Scheres, B. (1997). Short-range control of cell differentiation in the *Arabidopsis* root meristem. *Nature* **390**, 287–289.

Vanneste, S., De Rybel, B., Beemster, G. T. S., Ljung, K., De Smet, I., Van Isterdael, G. et al. (2005). Cell cycle progression in the pericycle is not sufficient for SOLITARY ROOT/IAA14-mediated lateral root initiation in *Arabidopsis thaliana*. *Plant Cell* **17**, 3035–3050.

Veit, B., Schmidt, R. J., Hake, S. & Yanofsky, M. F. (1993). Maize floral development—new genes and old mutants. *Plant Cell* **5**, 1205–1215.

Verduijn, M. H., Van Dijk, P. J. & Van Damme, J. M. M. (2004). The role of tetraploids in the sexual–asexual cycle in dandelions (*Taraxacum*). *Heredity* **93**, 390–398.

Vereecke, D., Cornelis, K., Temmerman, W., Jaziri, M., van Montagu, M., Holsters, M. et al. (2002). Chromosomal locus that affects pathogenicity of *Rhodococcus fascians*. *Journal of Bacteriology* **184**, 1112–1120.

Vergara-Silva, F., Espinosa-Matias, S., Ambrose, B. A., Vazquez-Santana, S., Martinez-Mena, A., Marquez-Guzman, J. et al. (2003). Inside-out flowers characteristic of *Lacandonia schismatica* evolved at least before its divergence from a closely related taxon, *Triuris brevistylis*. *International Journal of Plant Sciences* **164**, 345–357.

Vernon, D. M., Hannon, M. J., Le, M. & Forsthoefel, N. R. (2001). An expanded role for the *TWN1* gene in embryogenesis: defects in cotyledon pattern and morphology in the *TWN1* mutant of *Arabidopsis* (Brassicaceae). *American Journal of Botany* **88**, 570–582.

Vernon, D. M. & Meinke, D. W. (1994). Embryogenic transformation of the suspensor in twin, a polyembryonic mutant of *Arabidopsis*. *Developmental Biology* **165**, 566–573.

Vernoux, T. & Benfey, P. (2005). Signals that regulate stem cell activity during plant development. *Current Opinion in Genetics and Development* **15**, 388–394.

Vessey, J. K., Pawlowski, K. & Bergman, B. (2004). Root-based N-2-fixing symbioses: legumes, actinorhizal plants, *Parasponia* sp and cycads. *Plant and Soil* **266**, 205–230.

Vinkenoog, R. & Scott, R. J. (2001). Autonomous endosperm development in flowering plants: how to overcome the imprinting problem? *Sexual Plant Reproduction* **14**, 189–194.

Vinkenoog, R., Spielman, M., Adams, S., Fischer, R. L., Dickinson, H. G. & Scott, R. J. (2000). Hypomethylation promotes autonomous endosperm development and rescues postfertilization lethality in *fie* mutants. *Plant Cell* **12**, 2271–2282.

Vishnu-Mittre. (1953). A male flower of Pentoxyleae with remarks on structure of the female cones of the group. *Palaeobotanist* **2**, 75–84.

Voechting, H. (1878–1884). *Ueber organbildung im pflanzenreich. Physiologische untersuchungen ueber wachsthumsursachen und lebenseinheiten*. Cohen, Bonn, Germany.

Vyskot, B. & Hobza, R. (2004). Gender in plants: sex chromosomes are emerging from the fog. *Trends in Genetics* **20**, 432–438.

Wada, T., Kurata, T., Tominaga, R., Koshino-Kimura, Y., Tachibana, T., Goto, K. et al. (2002). Role of a positive regulator of root hair development, CAPRICE, in *Arabidopsis* root epidermal cell differentiation. *Development* **129**, 5409–5419.

Waites, R. & Hudson, A. (1995). *Phantastica*—a gene required for dorsoventrality of leaves in *Antirrhinum majus*. *Development* **121**, 2143–2154.

Walker-Larsen, J. & Harder, L. D. (2000). The evolution of staminodes in angiosperms: patterns of stamen reduction, loss, and functional re-invention. *American Journal of Botany* **87**, 1367–1384.

Walker-Larsen, J. & Harder, L. D. (2001). Vestigial organs as opportunities for functional innovation: the example of the Penstemon staminode. *Evolution* **55**, 477–487.

Wang, C. N. & Cronk, Q. C. B. (2003). Meristem fate and bulbil formation in *Titanotrichum* (Gesneriaceae). *American Journal of Botany* **90**, 1696–1707.

Wang, C. N., Moller, M. & Cronk, Q. C. B. (2004a). Altered expression of GFLO, the Gesneriaceae homologue of FLORICAULA/LEAFY, is associated with the transition to bulbil formation in *Titanotrichum oldhamii*. *Development Genes and Evolution* **214**, 122–127.

Wang, C. N., Moller, M. & Cronk, Q. C. B. (2004b). Aspects of sexual failure in the reproductive processes of a rare bulbiliferous plant, *Titanotrichum oldhamii* (Gesneriaceae), in subtropical Asia. *Sexual Plant Reproduction* **17**, 23–31.

Wang, R. L., Stec, A., Hey, J., Lukens, L. & Doebley, J. (1999). The limits of selection during maize domestication. *Nature* **398**, 236–239.

Wang, W., Tanurdzic, M., Luo, M., Sisneros, N., Kim, H. R., Weng, J. K. et al. (2005). Construction of a bacterial artificial chromosome library from the spikemoss *Selaginella moellendorffii*: a new resource for plant comparative genomics. *BMC Plant Biology* **5**, Issue 10.

Wang, Z. L., Mambelli, S. & Setter, T. L. (2002). Abscisic acid catabolism in maize kernels in response to water deficit at early endosperm development. *Annals of Botany* **90**, 623–630.

Wang, Z. Q. (2004). A new Permian gnetalean cone as fossil evidence for supporting current molecular phylogeny. *Annals of Botany* **94**, 281–288.

Wardlaw, C. W. (1950). The comparative investigation of apices of vascular plants by experimental methods. *Philosophical Transactions of the Royal Society, Series B* **234**, 583–604.

Wardlaw, C. W. (1953). A commentary on Turing's diffusion–reaction theory of morphogenesis. *New Phytologist* **52**, 40–47.

Weberling, F. (1989). *Morphology of Flowers and Inflorescences*. Cambridge University Press, Cambridge.

Webster, M. & Zelditch, M. L. (2005). Evolutionary modifications of ontogeny: heterochrony and beyond. *Paleobiology* **31**, 354–372.

Weigel, D. & Nilsson, O. (1995). A developmental switch sufficient for flower initiation in diverse plants. *Nature* **377**, 495–500.

Weigmann, K., Cohen, S. M. & Lehner, C. F. (1997). Cell cycle progression, growth and patterning in imaginal discs despite inhibition of cell division after inactivation of *Drosophila* Cdc2 kinase. *Development* **124**, 3555–3563.

Wenkel, S., Emery, J., Hou, B. H., Evans, M. M. & Barton, M. K. (2007). A feedback regulatory module formed by LITTLE ZIPPER and HD-ZIPIII genes. *Plant Cell* **19**, 3379–3390.

Went, F. W. & Thimann, K. V. (1937). *Phytohormones*. Macmillan, New York.

Wesley, O. C. (1930). Spermatogenesis in *Coleochaete scutata*. *Botanical Gazette* **89**, 180–192.

Weston, P. H. (1999). Process morphology from a cladistic perspective. In *Homology and Systematics* (eds R. Scotland and R. T. Pennington), pp. 124–144. Taylor & Francis, London.

Wettstein, R. (1907). *Handbuch der systematischen Botanik*. Deuticke, Leipzig.

Whipple, C. J., Zanis, M. J., Kellogg, E. A. & Schmidt, R. J. (2007). Conservation of B class gene expression in the second whorl of a basal grass and outgroups links the origin of lodicules and petals. *Proceedings of the National Academy of Sciences of the United States of America* **104**, 1081–1086.

White, P. S. (1983a). Corners rules in Eastern deciduous trees—allometry and its implications for the adaptive architecture of trees. *Bulletin of the Torrey Botanical Club* **110**, 203–212.

White, P. S. (1983b). Evidence that temperate east North-American evergreen woody-plants follow Corners rules. *New Phytologist* **95**, 139–145.

White, R. A. & Turner, M. D. (1995). Anatomy and development of the fern sporophyte. *Botanical Review* **61**, 281–305.

Whitmore, T. C. (1962). Studies in systematic bark morphology. I. Bark morphology in Dipterocarpaceae. *New Phytologist* **61**, 191–207.

Whittier, D. P. (2004). Induced apogamy in *Tmesipteris* (Psilotaceae). *Canadian Journal of Botany-Revue Canadienne De Botanique* **82**, 721–725.

Wigge, P. A., Kim, M. C., Jaeger, K. E., Busch, W., Schmid, M., Lohmann, J. U. et al. (2005). Integration of spatial and temporal information during floral induction in *Arabidopsis*. *Science* **309**, 1056–1059.

Williams, E. G., Sage, T. L. & Thien, L. B. (1993). Functional syncarpy by intercarpellary growth of pollen tubes in a primitive apocarpous angiosperm, *Illicium floridanum* (Illiciaceae). *American Journal of Botany* **80**, 137–142.

Williams, J. H. & Friedman, W. E. (2002). Identification of diploid endosperm in an early angiosperm lineage. *Nature* **415**, 522–526.

Williams, J. H. & Friedman, W. E. (2004). The four-celled female gametophyte of *Illicium* (Illiciaceae; Austrobaileyales): implications for understanding the origin and early evolution of monocots, eumagnoliids, and eudicots. *American Journal of Botany* **91**, 332–351.

Winter, K. U., Becker, A., Münster, T., Kim, J. T., Saedler, H., Theissen, G. (1999) MADS-box genes reveal that gnetophytes are more closely related to conifers than to flowering plants. *Proceedings of the National Academy of Sciences of the United States of America* **96**, 7342–7347.

Winter, K. U., Weiser, C., Kaufmann, K., Bohne, A., Kirchner, C., Kanno, A. et al. (2002). Evolution of class B floral homeotic proteins: obligate heterodimerization originated from homodimerization. *Molecular Biology and Evolution* **19**, 587–596.

Woodward, F. I. (1979). Differential temperature responses of the growth of certain plant species from different altitudes. 2. Analyses of the control and morphology of leaf extension and specific leaf area of *Phleum bertolonii* DC. and *Phleum alpinum* L. *New Phytologist* **82**, 397–405.

WORSDELL, W. C. (1905). The principles of morphology. I. *New Phytologist* **4**, 124–133.

WORSDELL, W. C. (1916). *The Principles of Plant Teratology*. Ray Society, London.

WU, X. L., DABI, T. & WEIGEL, D. (2005). Requirement of homeobox gene *STIMPY/WOX9* for *Arabidopsis* meristem growth and maintenance. *Current Biology* **15**, 436–440.

WUNDERLIN, R. P. (1983). Revision of the *Arborescent Bauhinias* (Fabaceae, Caesalpinioideae, Cercideae) native to middle America. *Annals of the Missouri Botanical Garden* **70**, 95–127.

WURSCHUM, T., GROSS-HARDT, R. & LAUX, T. (2006). APETALA2 regulates the stem cell niche in the *Arabidopsis* shoot meristem. *Plant Cell* **18**, 295–307.

YAMAGUCHI, T., NAGASAWA, N., KAWASAKI, S., MATSUOKA, M., NAGATO, Y. & HIRANO, H. Y. (2004). The *YABBY* gene *DROOPING LEAF* regulates carpel specification and midrib development in *Oryza sativa*. *Plant Cell* **16**, 500–509.

YANAI, O., SHANI, E., DOLEZAL, K., TARKOWSKI, P., SABLOWSKI, R., SANDBERG, G. et al. (2005). *Arabidopsis* KNOXI proteins activate cytokinin biosynthesis. *Current Biology* **15**, 1566–1571.

YANG, C. Y., SPIELMAN, M., COLES, J. P., LI, Y., GHELANI, S., BOURDON, V. et al. (2003a). TETRASPORE encodes a kinesin required for male meiotic cytokinesis in *Arabidopsis*. *Plant Journal* **34**, 229–240.

YANG, S. L., JIANG, L. X., PUAH, C. S., XIE, L. F., ZHANG, X. Q., CHEN, L. Q. et al. (2005). Overexpression of TAPETUM DETERMINANT1 alters the cell fates in the *Arabidopsis* carpel and tapetum via genetic interaction with EXCESS MICROSPOROCYTES1/ EXTRA SPOROGENOUS CELLS. *Plant Physiology* **139**, 186–191.

YANG, S. L., XIEA, L. F., MAO, H. Z., PUAH, C. S., YANG, W. C., JIANG, L. X. et al. (2003b). TAPETUM DETERMINANT1 is required for cell specialization in the *Arabidopsis* anther. *Plant Cell* **15**, 2792–2804.

YANG, W. C., YE, D., XU, J. & SUNDARESAN, V. (1999). The *SPOROCYTELESS* gene of *Arabidopsis* is required for initiation of sporogenesis and encodes a novel nuclear protein. *Genes & Development* **13**, 2108–2117.

YANG, Y. Z., FANNING, L. & JACK, T. (2003). The K domain mediates heterodimerization of the *Arabidopsis* floral organ identity proteins, APETALA3 and PISTILLATA. *Plant Journal* **33**, 47–59.

YANG, Y. Z. & JACK, T. (2004). Defining subdomains of the K domain important for protein–protein interactions of plant MADS proteins. *Plant Molecular Biology* **55**, 45–59.

YU, D. Y., KOTILAINEN, M., POLLANEN, E., MEHTO, M., ELOMAA, P., HELARIUTTA, Y. et al. (1999). Organ identity genes and modified patterns of flower development in *Gerbera hybrida* (Asteraceae). *Plant Journal* **17**, 51–62.

YU, H. J., HOGAN, P. & SUNDARESAN, V. (2005). Analysis of the female gametophyte transcriptome of *Arabidopsis* by comparative expression profiling. *Plant Physiology* **139**, 1853–1869.

YU, L. X. & SETTER, T. L. (2003). Comparative transcriptional profiling of placenta and endosperm in developing maize kernels in response to water deficit. *Plant Physiology* **131**, 568–582.

ZANIS, M. J., SOLTIS, P. S., QIU, Y. L., ZIMMER, E. & SOLTIS, D. E. (2003). Phylogenetic analyses and perianth evolution in basal angiosperms. *Annals of the Missouri Botanical Garden* **90**, 129–150.

ZELDITCH, M. L., SHEETS, H. D. & FINK, W. L. (2000). Spatiotemporal reorganization of growth rates in the evolution of ontogeny. *Evolution* **54**, 1363–1371.

ZHANG, J. Z. & SOMERVILLE, C. R. (1997). Suspensor-derived polyembryony caused by altered expression of valyl-tRNA synthetase in the *twn2* mutant of *Arabidopsis*. *Proceedings of the National Academy of Sciences of the United States of America* **94**, 7349–7355.

ZHANG, P. F., CHOPRA, S. & PETERSON, T. (2000). A segmental gene duplication generated differentially expressed *myb*-homologous genes in maize. *Plant Cell* **12**, 2311–2322.

ZHAO, D. Z., WANG, G. F., SPEAL, B. & MA, H. (2002). The *EXCESS MICROSPOROCYTES1* gene encodes a putative leucine-rich repeat receptor protein kinase that controls somatic and reproductive cell fates in the *Arabidopsis* anther. *Genes & Development* **16**, 2021–2031.

ZHAO, L., KIM, Y. J., DINH, T. T. & CHEN, X. M. (2007). miR172 regulates stem cell fate and defines the inner boundary of APETALA3 and PISTILLATA expression domain in *Arabidopsis* floral meristems. *The Plant Journal* **51**, 840–849.

ZHAO, Y., MEDRANO, L., OHASHI, K., FLETCHER, J. C., YU, H., SAKAI, H. et al. (2004). HANABA TARANU is a GATA transcription factor that regulates shoot apical meristem and flower development in *Arabidopsis*. *Plant Cell* **16**, 2586–2600.

Index

Note on the Index: terms are indexed to the sections in which they occur; as each section is approximately one page long this is equivalent to a page reference but more faithfully locates the term and related material. Figures are indexed both by the figure number and (in brackets) the section in which it occurs. Terms are categorized by font style as follows: [**bold**] - technical or descriptive term relating to plant organ; [*italic lower case*] - latin name of organism or taxon; [*ITALIC UPPER CASE*] - gene name; [lower case plain text] - more general term, non-botanical term or English name of organism; [UPPER CASE PLAIN TEXT]: protein.